Wave Energy Convertors in Tropical Seas

Tay provides a comprehensive description of the state of the art for recent conceptual designs on the integration of wave energy convertor (WEC) systems with floating structures in tropical seas, discusses the unique environmental challenges, and provides a guide for readers to develop WECs that are optimised for both wave patterns and the structures to which they are attached.

Through this book, readers will gain a deep understanding of the unique environmental characteristics of tropical climates, discover how WECs can be integrated into various floating structures, and learn how WECs can be adapted and optimised for use in tropical climates. This book also focuses on the hydroelasticity of large floating structures and how the structural deformation under wave action affects the energy generation of the WEC. Tay provides the theoretical formulation and numerical framework for the modelling of WEC systems when integrated with large floating structures. Different wave models and mathematical optimisation schemes are presented to enable readers to determine the best design and application parameters depending on requirements. In the book, Tay includes case studies to illustrate how power generation of the integrated WEC arrays can be enhanced. This includes power enhancement via sloshing resonance, wave run-up, structural deformation, and various optimisation schemes. These case studies offer reader practical guidance on applying the information in this book to real-world scenarios.

The combination of presented case studies, discussion of the challenges of tropical seas, and design optimisation parameters make this book invaluable to researchers and industry players working on the concepts of power enhancement and integrated WECs on floating structures.

Zhi Yung Tay is an Associate Professor at Singapore Institute of Technology. Previously, he was a Postdoctoral Research Associate at the Institute for Energy Systems, University of Edinburgh and a Research Fellow at National University of Singapore (NUS). Dr Tay also held a Senior Research Engineer position at the Keppel Offshore and Marine Technology Centre. He received his Ph.D. from NUS in 2009.

Wave Energy Convertors in Tropical Seas

Large Floating Structures and Power Enhancement

Zhi Yung Tay

CRC Press
Taylor & Francis Group
Boca Raton London New York

CRC Press is an imprint of the
Taylor & Francis Group, an **informa** business

Designed cover image: Architect Michael Baluarte

First edition published 2026
by CRC Press
2385 NW Executive Center Drive, Suite 320, Boca Raton FL 33431

and by CRC Press
4 Park Square, Milton Park, Abingdon, Oxon, OX14 4RN

CRC Press is an imprint of Taylor & Francis Group, LLC

ISBN: 978-1-032-48200-2 (hbk)
ISBN: 978-1-032-48201-9 (pbk)
ISBN: 978-1-003-38785-5 (ebk)

DOI: 10.1201/9781003387855

Typeset in Palatino
by SPi Technologies India Pvt Ltd (Straive)

To my mum, Chua Kim Sang (10.11.1955–10.08.2024).

Remembering you, always and forever.

Contents

About the Author

Zhi Yung Tay is an Associate Professor at Singapore Institute of Technology. Previously, he was a Postdoctoral Research Associate at the Institute for Energy Systems, University of Edinburgh, and a Research Fellow at National University of Singapore (NUS). Dr Tay also held a Senior Research Engineer position at the Keppel Offshore and Marine Technology Centre. He received his Ph.D. from NUS in 2009.

1

How Can We Make WECS Feasible in Tropical Climates?

Renewable energy such as wave energy has gained popularity as a means of reducing greenhouse gases (GHGs). However, the high cost and lack of available sea space in some countries have hindered the deployment of wave energy converters (WECs) as an alternative means of sustainable energy production. By combining WECs with infrastructures such as floating breakwaters or piers, the idea of electricity generated from WECs will be more appealing. This chapter considers some of the concepts of the integration of WECs with floating structures as a means to generate electricity and attenuate wave force in tropical seas. The state of the art of such a concept proposed by various academician and industry players will be presented to demonstrate the potential of tapping wave energy as alternative sources for the achieving of a sustainable solution. This includes the proposed concept of integrating WECs with oil and gas platform and seafront land infrastructure such as coastal defence structures. The chapter will present the various applications of floating solution ideas with the aim of achieving the United Nations (UN) Sustainable Development Goals (SDGs) and will explain how renewable energy, especially various types of WEC (i.e., point absorber, attenuator, terminator, and oscillating water column WECs) could be used in extracting energy. Novel methodologies to enhance the WEC arrays when integrated with floating infrastructure will be presented.

1.1 Tropical Zone

1.1.1 Tropical Climate

The tropical zone is located at the equator and is defined as the region of Earth located between the Tropic of Cancer in the north (at 23.5° north latitude) and the Tropic of Capricorn in the south (at 23.5° south latitude). It receives a fairly constant amount of sunlight throughout the year, with a mean temperature of between 25° C to 28° C. The tropical zone encompasses parts of North America, South America, Africa, Asia, and Australia, covering about 36% of Earth's landmass and housing about a third of the world's

DOI: 10.1201/9781003387855-1

population.[1] About 63% of the countries in the worlds are located or partially located at the tropics, amount to 122 countries, with 95 countries located at the tropics and 27 partially in the tropics as shown in Figure 1.1.[2]

According to the Köppen Climate Classification,[3] the tropical climate can be further classified into the tropical rainforest climate (Af), tropical monsoon climate (Am) and tropical savannas climate (Aw) as shown in Figure 1.2. The partially tropical countries are close to the temperate zone, thus having a tropical savannas climate with a lower mean temperature greater than 18 °C and annual precipitation between 762 mm (30 inch) to 1,016 mm (40 inch).[4] Fully tropical countries, on the other hand, are closer to the equator and have either rainforest or monsoon climate. The tropical rainforest climate has an average annual precipitation exceeds about 1,800–2,500 mm with no distinct wet or dry season due to the high rainfall throughout the months,[5] whereas the tropical monsoon climate is characterised with an abundant annual precipitation during the rainy/monsoon season and may have seasonal dryness during the dry season. As compared to temperate countries such as the United Kingdom, the annual precipitation is in the range of 1,000mm to 1,400mm according to the statistics compiled from 2014 to 2022.[6]

The tropical climate has significant economic, ecological and cultural importance, and plays a crucial role in the functioning of the planet's ecosystems and weather patterns. The tropical climate contributes to the biodiversity of the ecosystem due to its warm and humid conditions, which support a high diversity of plant and animal species, many of which are found nowhere else on Earth. It also has importance for the global food and commodity market by providing ideal conditions for crops such as coffee, cocoa, banana, rubber, and palm oil. As the tropical zone is home to rainforest in the world, the rainforest serves as major carbon sinks, helping to regulate Earth's climate, and also contains valuable natural resources such as timbers and minerals. The ideal warm and humid weather for mangroves is able to store five times more carbon than rainforest, whereas seagrasses can capture carbon up to 35 times faster than rainforest.[7] These are important ecosystems found in the tropical climate as means to fight against climate change. Also, the tropical climate plays a crucial role in global weather patterns, as it is home to the Intertropical Convergence Zone (ITCZ) – a low-pressure zone generally encircles Earth near the thermal equator where its specific position varies seasonally.[8] The ITCZ appears as a band of clouds consisting of showers, with occasional thunderstorms. The ITCZ creates a band of low pressure where trade winds from the northern and southern hemispheres converge and produce heavy rainfall; it is thereby responsible for the dry and wet seasons in the tropics.[9]

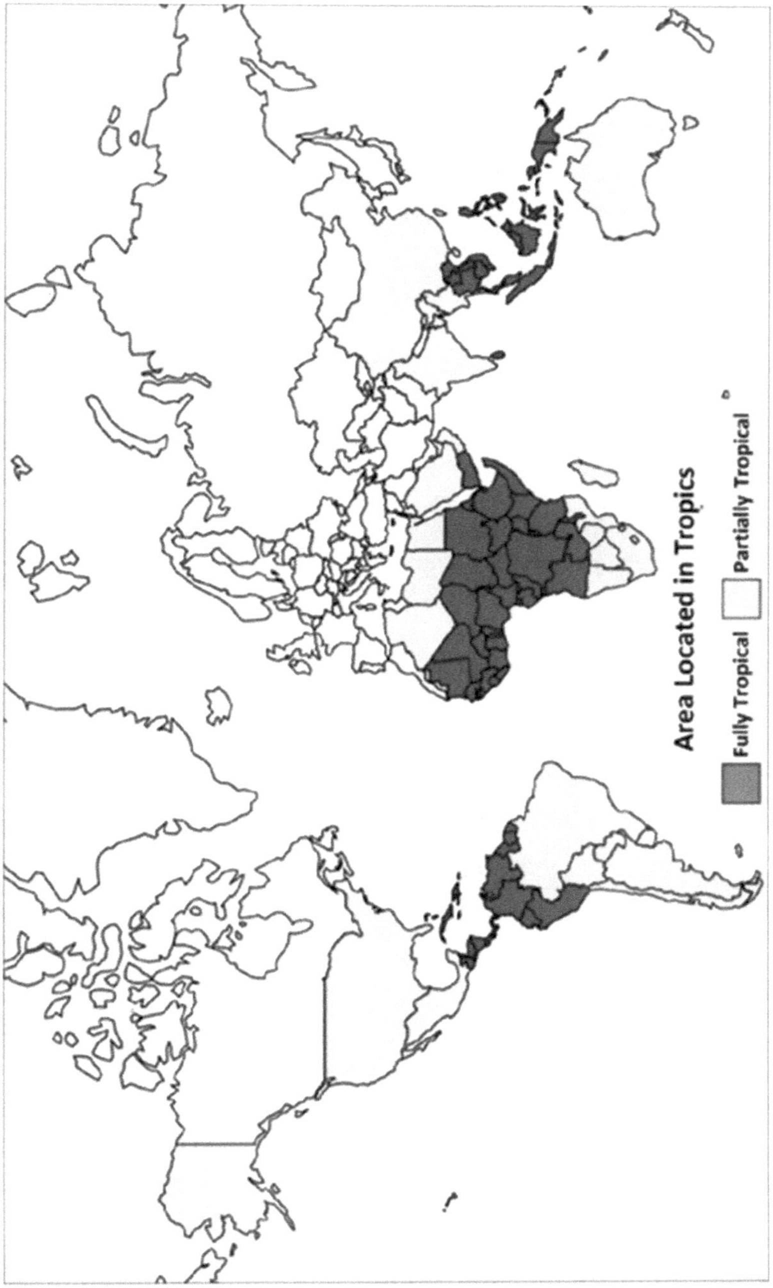

FIGURE 1.1
Countries in the tropical zone.[2]

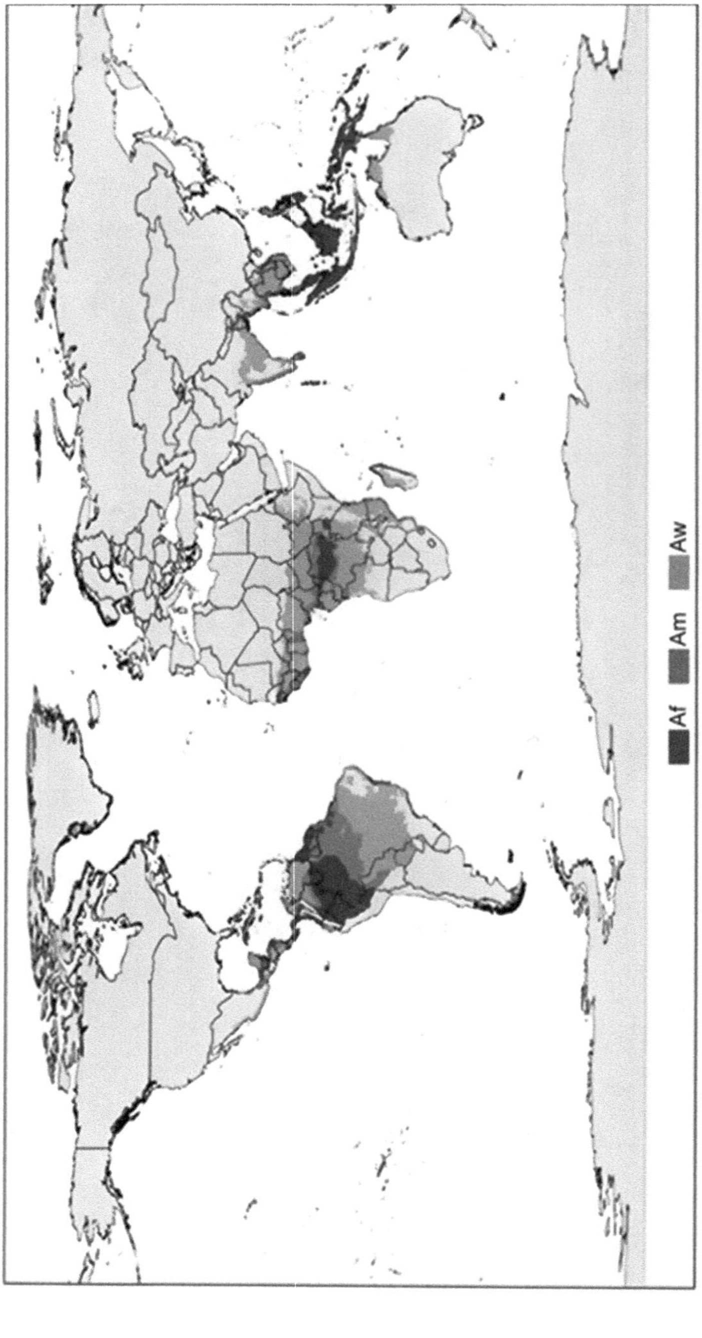

FIGURE 1.2
Tropical climate from 1991 to 2020 according to the Köppen Climate Classification System.[3]

1.1.2 Human Migration to Coastal Cities

According to historical records, coastal cities are situated on the boundary or interface between land and sea, which also encompasses extensive interior lakes. The migration of the human population towards coastal cities has a rich history, and it has taken place over several thousands of years in different parts of the world. People with different backgrounds, races, and religions have been drawn to coastal areas for various reasons, such as greater economic prospects, improved quality of life, and cultural diversity. During ancient times, cities like Palembang of the Srivijaya Empire and Malacca under the Malacca Sultanate emerged as coastal city-states and became significant trade, cultural, and political centres. Additionally, the sea has been instrumental in supporting the cities, offering water, food, employment, recreational opportunities, and access to the world through maritime transportation. The onset of industrialisation in the 18th and 19th centuries marked a significant change in human migration to coastal cities. People flocked to these cities in search of employment with the expansion of industrialisation and the growth of factory jobs, leading to the swift urbanisation of many coastal cities, such as Batavia, Ho Chi Minh City, Calcutta, Mumbai, Rio de Janeiro, and Buenos Aires.

The 21st century has seen the migration of people to coastal cities due to economic opportunities, improved social and educational prospects, and better infrastructure. Coastal cities are centres of trade and industry, offering jobs in tourism, shipping, and other sectors. They also provide a diverse cultural scene, educational institutions, and improved public services, leading to an increase in population. According to a study by the UN, by 2030, the urban areas will house around 60% of the world's population, and that number is expected to increase to 68% by 2050. It is thus not surprising to see that six out of the ten largest cities in the world are coastal cities (see Figure 1.3).

Countries located in tropical climates are particularly susceptible to the risk of rising sea levels and overpopulation issues. Despite these risks, tropical coastal cities continue to attract a significant portion of the population due to their warm climate, easy access to the sea and its resources, and opportunities for trade and commerce. The increasing vulnerability of coastal cities to flooding due to sea-level rise and overcrowding issues has gained attention from various parties such as the UN in recent decades. Figure 1.4 shows the world population growth from 1950 and the projection to 2035, in accordance with different climate zones. Most of the cities are located along the coastal area with their coastal cities projected to grow larger by 2035 with more than half of the cities located in the tropical zone, as shaded in Table 1.1. The tropical coastal cities such as Delhi, Kinshasa, and Dhaka have a high absolute growth exceeding ten million from 2015 to 2035.

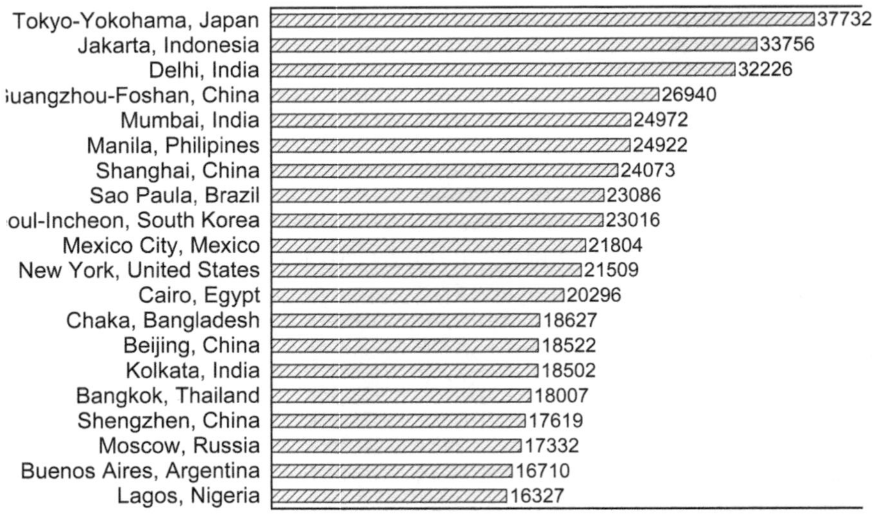

FIGURE 1.3
Largest cities worldwide in 2022, by population.

Source: Wendell Cox Consultancy@Statista 2022.

1.2 Floating Solutions/Issues Faced by Tropical Coastal Cities

In April 2019, the success of the first UN Roundtable on Sustainable Coastal Cities led to the signing of a Memorandum of Understanding (MoU) between the Metropolitan City of Busan of the Republic of Korea, UN-Habitat, and OCEANIX to build the world's first sustainable floating city (see Figure 1.5). This was followed by the second UN Roundtable in April 2022 to share cutting-edge ideas, collaboration models, and research in this frontier space.[11] The design of the floating city prototype incorporates the UN SDGs where they envisage a sustainable floating city powered by renewable as an extension of existing coastal cities. This initiative is a bold step to transition to more resilient and sustainable cities to counter the threat of seawater level rise impacting on cities especially the coastal cities and small island development states (SIDS).

According to the statistics provided by the World Population Review, most countries located at the tropical climate zone are developing countries, thereby making them relatively slow in seeking solutions adaptive to climate change. Figure 1.6 shows the list of developed versus developing countries categorised according to the Human Development Index (HDI) with HDI greater than 0.8 classified as developed countries. The figure shows

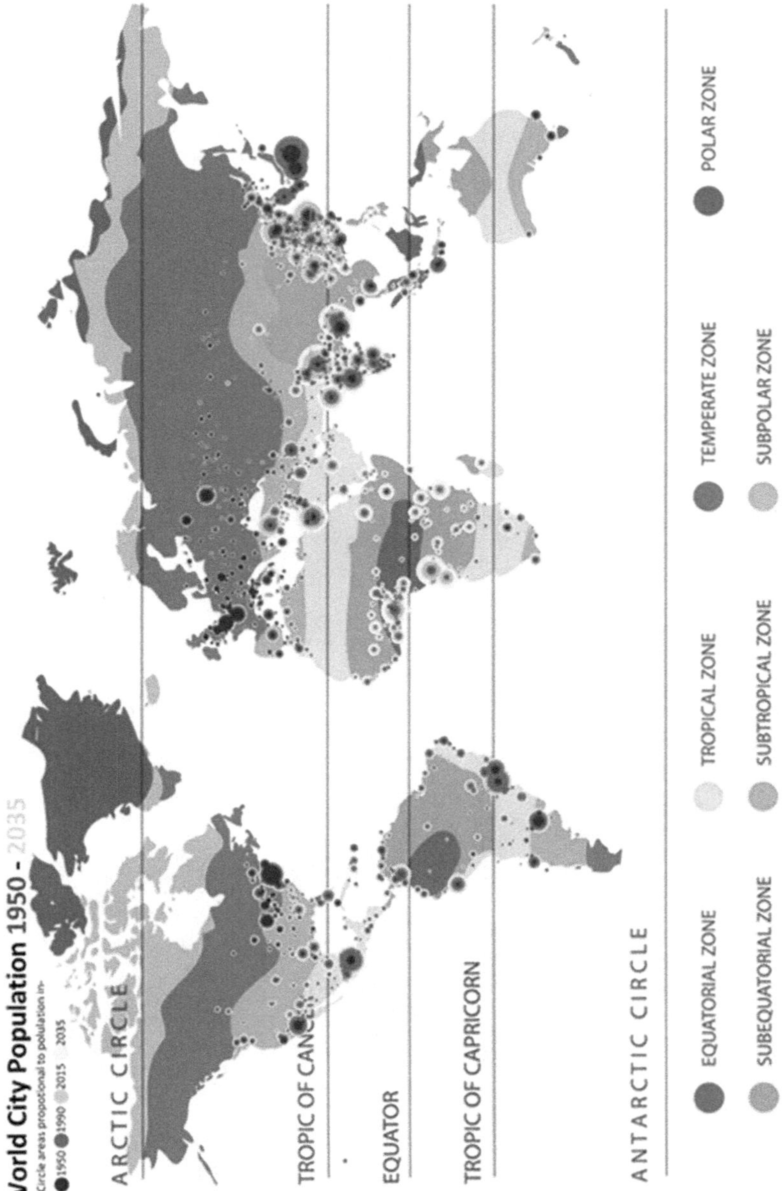

FIGURE 1.4
World population growth by climate zone, data: UN (2018).[10]

TABLE 1.1
Top Largest Cities in 2035 by Population, UN (2018)[10]

Ranking	City	Population	Absolute Growth from 2015 to 2035
1	Delhi	43.3 million	17.5 million
2	Tokyo	36.0 million	–1.2 million
3	Shanghai	34.3 million	10.9 million
4	Dhaka	31.2 million	13.6 million
5	Cairo	28.5 million	9.7 million
6	Mumbai	27.3 million	8.3 million
7	Kinshasa	26.7 million	15.1 million
8	Mexico City	25.4 million	4.1 million
9	Beijing	25.3 million	7.0 million
10	Sao Paulo	24.5 million	3.6 million

FIGURE 1.5
OCEANIX Busan, world's first prototype floating city.[11]

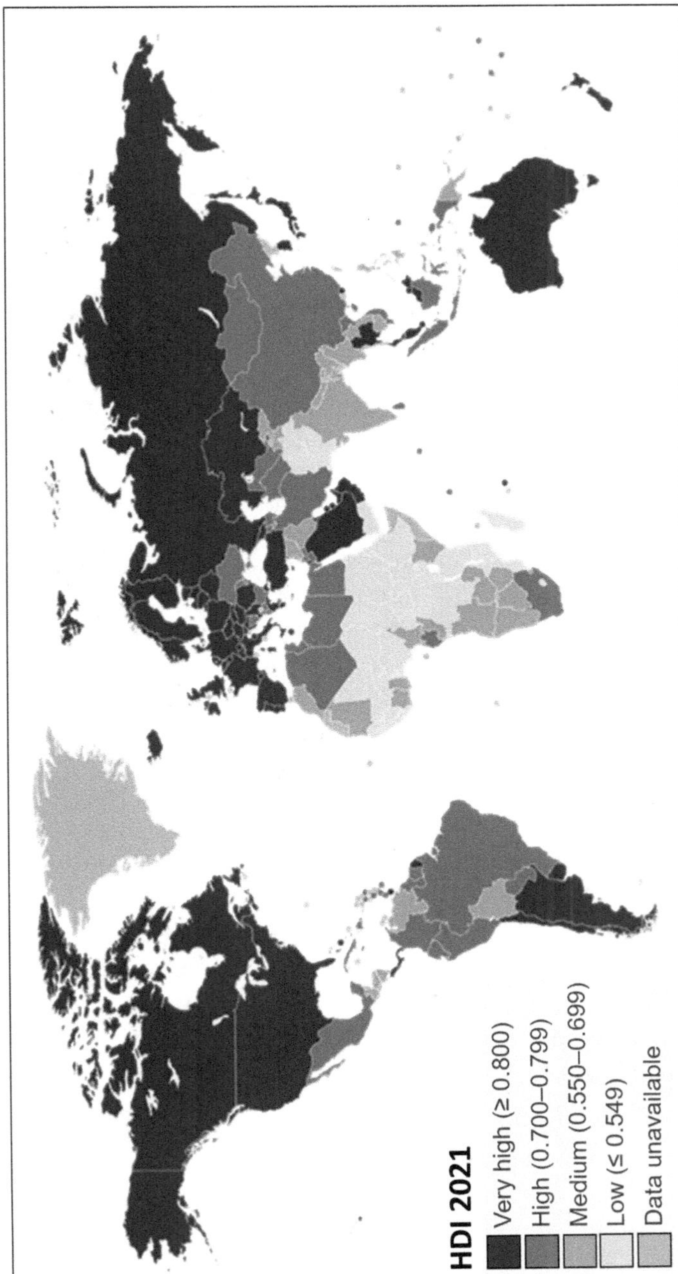

FIGURE 1.6
Classification of developed vs developing countries by HDI.[12]

that almost all the tropical countries are developing countries, and when compared to developed countries such as the Netherlands, Japan, or Korea, which are also prone to the risk of flooding due to seawater level rise, the latter has carried out extensive research and technology development to seek innovative and adaptive solutions to counter the rise of seawater levels via the utilisation of floating solutions.

Nevertheless, there are also plans proposed to counter the rise of seawater level, such as the development of the Maldives floating city (source: www.waterstudio.nl) modelled in the form of a brain coral by the Dutch and the government of Maldives for a next-generation, sea-level, rise-proof floating island consisting of 5,000 housing units.[13] Floating solutions can have a range of applications, including housing, transportation, coastal protection, and energy production. Floating structures, like floating breakwaters and wave barriers, can play a role in reducing wave energy and protecting coastal communities from flooding, as well as improving water quality and supporting fish and bird populations. In terms of coastal protection, floating solutions such as floating breakwaters and wave barriers are considered effective in reducing erosion, wave damage, and flood risks to coastal infrastructure. The use of natural materials and processes can be used in place of, or in conjunction with, the floating structures as nature-based solutions. These interventions can provide multiple benefits beyond coastal protection, such as supporting tourism and fisheries, and promoting biodiversity. Therefore, floating solutions have been gaining attention recently due to the benefits they bring economically, socially, and environmentally, such as the following:

- It is resilient to sea-level rise caused by climate change as the floating module rises and falls with seawater level.
- It does not permanently take up precious sea space, as the floating structure can be easily dismantled and towed away.
- It allows for expansion should a larger platform footprint be needed.
- It can be integrated into existing and future infrastructures, as the structure can be transported and moored close to coastal infrastructure easily.
- It can double up as a coastal breakwater.
- It provides habitats for fishes and birds.
- It is lower in environmental impact than bottom-founded infrastructure, as it does not disrupt the current flows or damage the seabed.
- It adapts easily to the varying nearshore water depth and seabed bathymetry.

It is important to note that there are also some limitations and challenges associated with floating solutions. For example, floating homes and structures can be expensive to construct and maintain, and they may not be suitable for areas with high wave or wind conditions. Additionally, the effectiveness of floating solutions in protecting against rising sea levels will depend on factors such as the specific design and location of the structures, as well as the rate and magnitude of sea-level rise.

Floating solutions could therefore be used to address some of the issues faced by the SIDS. The three major problems pertaining to human migration to SIDS are overpopulation and urbanisation, land-scarcity problems, and climate change that leads to the rise of seawater levels. Details of these issues are described in the subsequent sections.

1.2.1 Overpopulation and Urbanisation

Urbanisation is caused by a combination of factors, including population growth, economic development, and migration. As more and more people move into cities, urban areas are becoming larger and more complex, leading to increased energy consumption, GHG emissions, and land-use changes. The world's cities ranked by population is shown in Figure 1.7, where out of the top 20 countries, 75% are coastal cities, and half are located in the tropical regions. It is projected that more than two-thirds of the world population will be living in urban areas by 2050[14] – that is, the equivalence of seven billion people[14] – and 50% of the largest cities will be tropical coastal cities.[10] It is reported that tropical cities grew by 80,000 people every day compared with 50,000 people outside the tropics over the past 5 years.[15] By 2030, Delhi will overtake Tokyo to be the world's most populous city, and India will add two more megacities – Hyderabad and Ahmedabad – joining Mumbai, Kolkata, Bangalore, and Chennai as megacities status.[10,16] This is a continuous trend in tropical cities, where the size of the cities grows larger over time. By 2030, more megacities will be located along the coastal areas in the tropical regions (see Figure 1.7).

Human migration to these cities has also caused overpopulation problems, as well as strained land usage for housing, infrastructure, and urbanisation. For example, overpopulation in Jakarta has caused it to sink due to heavy groundwater extraction and land subsidence caused by natural soil compaction. These urban areas are particularly vulnerable to the impacts of climate change, such as heat waves, flooding, and air pollution. Urban heat island effect, which is caused by the concentration of buildings and paved surfaces in urban areas, can exacerbate heatwaves and make cities unbearable in summertime. Urban flooding, on the other hand, can cause damage to infrastructure and property, and disrupt the lives of residents. Urban air pollution also has serious health impacts and is becoming a major public health concern.

1.2.2 Land-Scarcity Problems

The current top four cities with population density above 8,000 per square kilometres are Monaco, Macao, Singapore, and Hong Kong (see Figure 1.8).

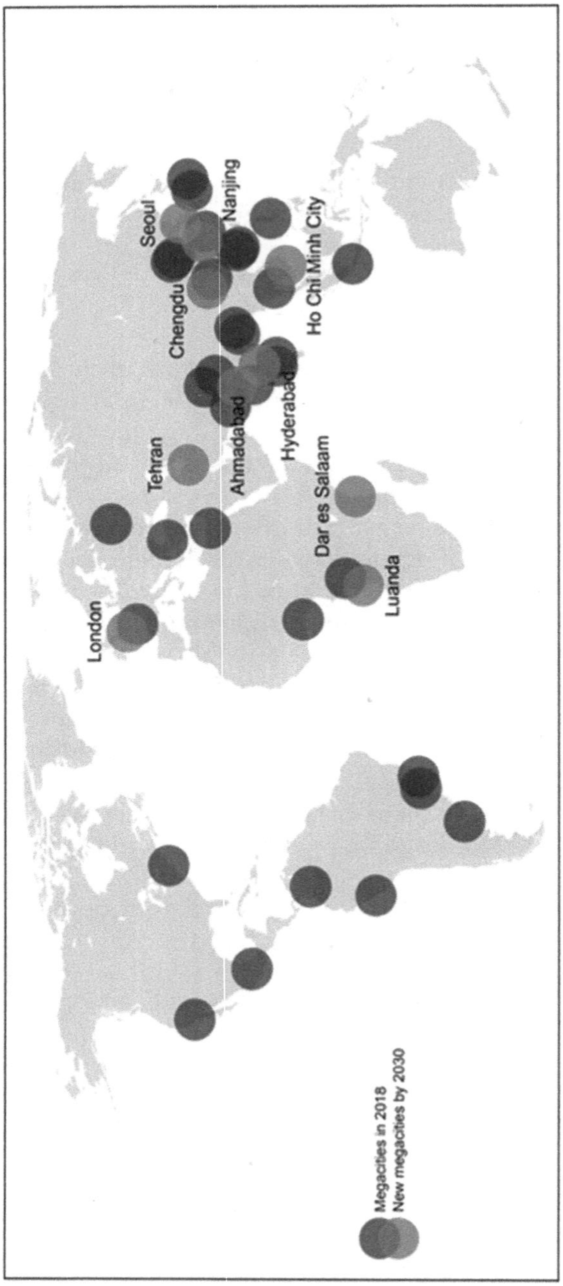

FIGURE 1.7
Megacities of the world in 2018 and 2030.[10,16]

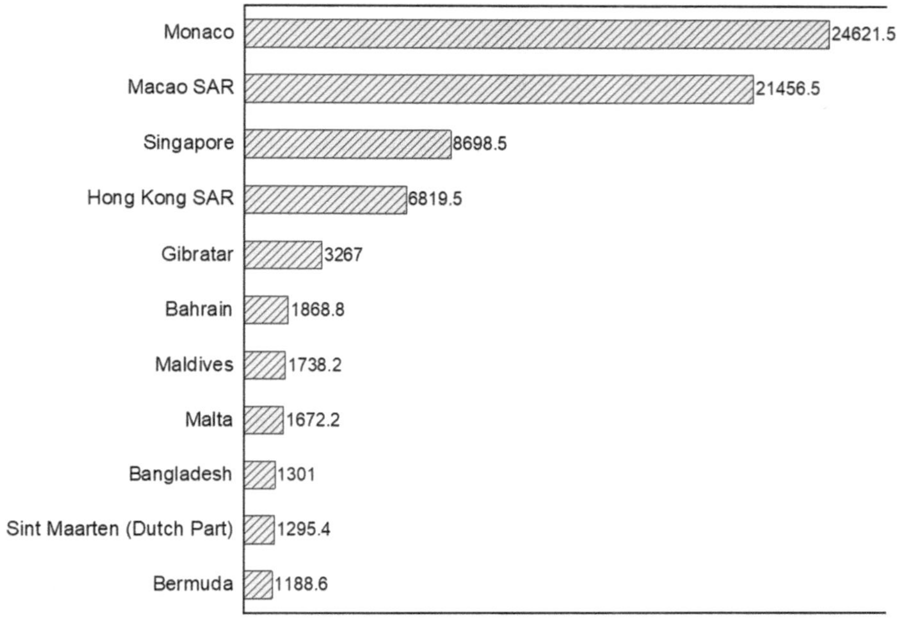

FIGURE 1.8
Countries/cities with highest population density worldwide in 2021 (in inhabitants per square km).[23]

These are land-scarce coastal cities that require more lands for development to catch up with urbanisation and human migration. In addition, some of the top ten countries/cities with highest population density shown in Figure 1.8 are islandic cities, i.e., Bahrain, Maldives, Malta, Sint Marten, and Bermuda, facing high risk of flooding due to seawater level rise. Various innovative methods are sought to tackle the land-scarcity problem. For example, Singapore explores the use of underground spaces (i.e., the Jurong Rock Cavern for oil storage[17]) or subway stations, or developing these spaces for a district cooling system (Marina Bay District Cooling Network[18]). Besides developing underground spaces, high-rise buildings are mushrooming in these land-scarce cities for residential and commercial purposes. To cater for the food security problem, vertical farming techniques are used where plants are grown vertically in stacked layers indoors, typically using artificial lighting and climate control.[19] In addition, rooftop gardens are also used where plants are grown on underutilised rooftops or carparks.[20] For the past decades, the most commonly used method to augment the land mass was via land reclamation. Singapore, for example, has increased its land size by 25% over the past two centuries,[21] and 25% of the land in Hong Kong has been reclaimed.[22] While land reclamation can provide new land for development, it can also have negative impacts, such as the following:

- Loss of natural habitats: Land reclamation can destroy habitats for plants and animals, leading to a loss of biodiversity.
- Loss of coastal ecosystems: Wetlands and other coastal ecosystems can be destroyed during land reclamation, impacting the services these ecosystems provide, such as storm protection, water filtration, and carbon sequestration.
- Disruption of coastal processes: Land reclamation can change the natural coastal processes, such as wave action, sediment transport, and water circulation, which can lead to erosion and flooding.
- Displacement of local communities: Land reclamation can result in the displacement of local communities that have lived on the land for generations.
- Increased air and water pollution: Land reclamation can lead to increased air and water pollution due to the use of heavy equipment and the potential for chemicals used in construction to leach into the surrounding area.
- Increased soil erosion: Land reclamation can make the soil more susceptible to erosion, thus increasing the rate of sedimentation.
- Impacting navigation and navigation safety: Land reclamation can impact navigation and navigation safety by altering the water depths and currents.

The negative impacts on the environment, safety, and ecosystem have highlighted the advantages of floating solutions, which are attractive due to their environmental friendliness, flexibility for expansion, and ability to withstand and adapt to rising seawater levels.

1.2.3 Rise of Seawater Levels

The increase in Earth's atmospheric temperature, driven by escalating GHG, is a contributing factor to climate change. This increase leads to a rise in seawater levels because of a global temperature increase, causing ice sheets and glaciers to melt and causing seawater to expand thermally. According to research conducted by the Intergovernmental Panel on Climate Change (IPCC),[24] the average global sea level has risen by about 1.8 mm per year from 1901 to 2010, with the majority of this increase happening after the 1950s. Human migration to cities and overpopulation are having a direct effect on climate change, as they contribute to global warming, causing a rise in sea levels, and result in more frequent and intense, extreme weather events.

The impacts of sea-level rise on coastal communities and ecosystems can be severe. They can exacerbate existing coastal hazards, such as coastal erosion, storm surges and flooding, which can damage infrastructure, homes, and businesses. A rapid rise in seawater level can also lead to the loss of valuable coastal ecosystems, such as marshes and mangroves, which provide important ecological services, such as storm protection and habitats for wildlife. Rising sea levels and increased frequency of coastal flooding due to climate change have led to population displacement[24] and loss of housing in low-lying coastal areas, particularly in coastal cities and SIDS such as Mumbai, Jakarta, Lagos, and Quito, which coincidently are located in the tropics. It is expected that by 2050, many cities by the sea will face severe disruption to their coastal ecosystems as a result of heat waves, droughts, pluvial floods, tropical cyclones, and marine and coastal erosion.[25] In fact, some of the Maldives Islands and the Indonesia Archipelago have disappeared as a result of rising sea levels caused by global warming.

The Organisation for Economic Co-operation and Development (OECD) has published a list of the largest cities that will be most exposed to coastal flooding in 2070, in terms of population and assets, in which more than half of the top 20 cities are located at tropical climates, contributing to a population of around 76 million people. Out of the top 20 cities ranked in terms of assets exposed to coastal flooding in 2070, the tropical coastal cities contribute to approximately $5.3 trillion of asset exposure in 2070. A comparison between the present-day and future (2070) exposure shows that a 36% increase is projected when taking into account the adverse impact of climate change and socio-economic change.[26]

As floating structures fall and rise with the water level, the floating cities naturally become an enticing alternative solution that is adaptable to seawater level rise. Proposed floating cities, such as those by OCEANIX, the Maldives floating cities, and the Blue-21 Floating Cities (www.blue21.nl), are designed to be self-sustainable such that they are equipped with desalination plants to provide fresh water and renewable energy systems to reduce reliance on fossil fuels, as well as to be flexible for future expansion and relocation.

1.3 Renewable Energy: Wave Energy

A way for a floating city to become self-sufficient is to use renewable sources like wave, tidal, wind, solar, and ocean thermal energy. This book focuses on wave energy, which is a promising form of renewable energy with a great potential sustainable energy source. The ocean has vast energy potential with wave energy estimated at 2.11 TW, tidal energy at 3 TW, and ocean thermal energy around 30 to 90 PWh.[27] Among these ocean energies, wave energy lags in commercialisation compared to its marine energy counterparts but has an estimated total potential of 32 PWh/yr, nearly double the global electricity supply in 2008 of 17 PWh/yr.[27] Experts believe that wave energy is where wind energy was 30 years ago, and the challenges posed by WECs, such as high deployment and operational costs, can be addressed through ongoing research to understand their behaviour in rough seas and to reach optimal design. Wave energy has several advantages, including greater predictability and consistency compared to wind and solar. While all renewable sources are influenced by weather, wave energy tends to be more stable because ocean waves can persist even after winds have calmed, allowing for more reliable energy generation in suitable coastal regions. Additionally, wave energy harvesting technology has a relatively small footprint, causing no damage to land and can be used in offshore locations.

The WEC's development began during the 1970s global oil crisis when extensive work was done on the Salter's duck WEC due to oil shortages.[28] However, the project was abandoned after the oil prices recovered, and there was an underestimation of wave energy's economic viability. Renewable energy, particularly wave energy, regained attention in recent decades due to rising oil prices and concerns about climate change. Extensive research and development on the utilisation of wave energy have been carried out in recent decades, especially for developed countries.[29–34] Several encouraging developments in wave energy farms have shown promising progress recently, such as the WaveRoller WEC deployed by AW-Energy,[35] a new generation of high-efficiency WEC launched by CorPower,[36] and the wave energy farm demonstrated by MOCEAN and AWS Ocean Energy.[37]

There are three common types of WECs operating by oscillatory motion – i.e., the point absorber, terminator, and attenuator WECs, as shown in **Figure 1.9**, where the performance of each of these WECs differs from one another.[31] The point absorber is the simplest type of WEC that usually utilises its heaving motion in generating energy such as the CETO WEC or heaving WECs.[38, 39] On the other hand, the terminator WEC (also known as the flap or

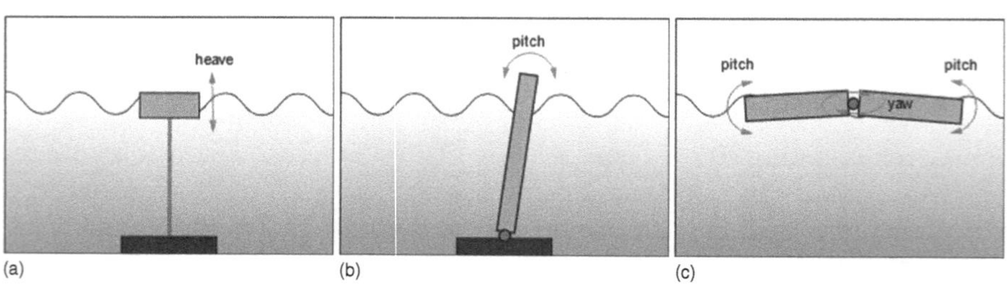

FIGURE 1.9
Oscillatory type of WECs: (a) point absorber, (b) terminator, (c) attenuator.

the oscillating wave surge converter (OWSC)[40]), such as the Oyster WEC, is usually hinged at its bottom to a foundation and pitches with the oscillatory motion of the waves. The attenuator is made of two or more elongated floating bodies connected by hinges, such as the Pelamis WEC or raft-type WEC.[41,42] It generates energy via the pitching and yawing motion when waves hit on it. These technologies convert the kinetic energy of waves into electricity via the power take-off (PTO) system, which can then be fed into the power grid.

The development of WECs has reached the test-bedding stage; however, the adoption of WECs in developing countries is not as widespread as in developed countries. One common misconception is that wave resources in tropical seas are too moderate to be economical. However, wave energy resources available in tropical climates have been shown to be of vast potential for the deployment of wave farms, and there is an increase in energy demand in tropical countries due to urbanisation. The wave energy resources in major oceans in the tropical zone will be presented in Section 1.3.1 and the energy demand in tropical countries in Section 1.3.2.

1.3.1 Wave Energy Resource

Tropical oceans are characterised by relatively warm water temperatures that exceed 20 °C and remain constant throughout the year. This warmth is due to intense solar radiation directly over the equator, which evaporates seawater and creates a mass of warm, humid tropical air that rises and cools as it flows north and south. The Pacific, Atlantic, and Indian Oceans, as well as the South China Sea, which lie in the tropical zones, serve as important ecosystems for the economies, cultures, and societies of about 120 countries.

The tropical seas are considered to be some of the most resource-rich regions for wave energy generation. They are characterised by large swell waves and consistent sea states, which provide ideal conditions for wave energy generation. In particular, the Indian Ocean, the Pacific Ocean, and the Caribbean Sea are known to have high wave energy resources. Figure 1.10 shows the global wave energy resource for the wave front of the coastal zone. Wave energy flux could reach up to 30 kW/m at certain regions, such as the east and west coast of South America and India. The Pacific Islands, which are close to the Tropic of Capricorn, and Southern China cities, which are closer to the Tropic of Cancer, enjoy a higher wave energy resource up to 45 kW/m and wave height of 3 m.

A review on the wave energy resources in the tropical regions by the National Council of Science and Technology in Mexico[44] revealed that although the annual wave power (kW/m) between the Tropics of Cancer and Capricorn is not as higher as their higher latitude counterparts, it is common to find up to 20 kW/m of wave energy resources near the coast and 20–30 kW/m in the intermediate and deep waters. The mean power estimations in the tropical regions were summarised as follows: North-East Brazil

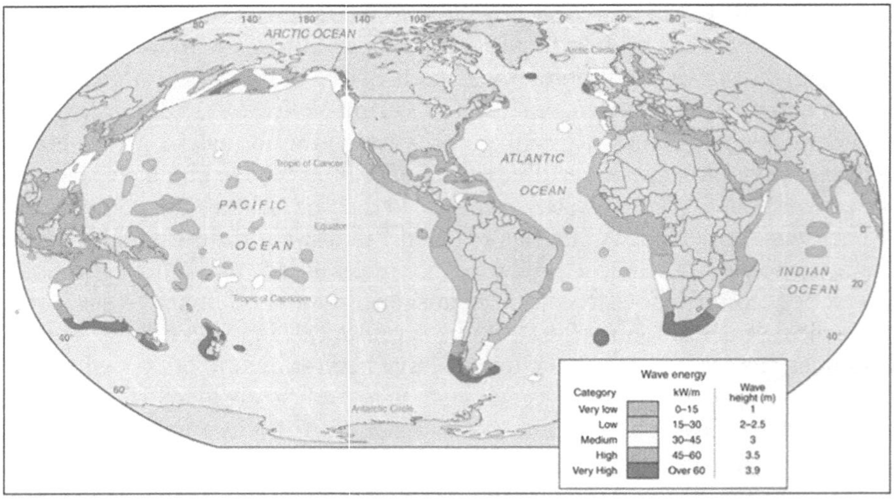

FIGURE 1.10
Global wave energy resource at wave front.[43]

(~2 to 14 kW/m[45]), South China Sea (~5.32kW/m[45]), Indian Coast (~5 to 10 kW/m[46]), and Peninsulas of Malaysia (~6.5 kW/m[47]). The WECs developed for the sub-tropical zone are usually ideal for wave energy resources greater than 15kW/m. However, some of these WECs, as listed in Babarit,[48] could be applicable to the tropical zone with further optimisation of its geometry and PTO control system.

Figures 1.11(a) and 1.11(b) show the yearly percentage in time of power equal or greater than 10 kW/m and 20 kW/m wave energy in the tropical regions.[44] Wave farms could be deployed in regions close to the Pacific and Indian Oceans, which receive high wave energy where the wave power is greater than 10 kW/m at 80% of the time yearly in the Indian and Pacific Oceans. In addition, the wave energy could reach greater than 20 kW/m 50% of the time in regions with increasing power when closer to the southern sub-tropical zone. India has thus begun to explore the use of wave energy as an alternative to fossil fuel. Wave energy resources in the Indian Ocean show a clear seasonal pattern, with the most energetic monsoon (> 20 kW/m) along the Indian coast.[49] A study by the National Institute of Oceanography[50] in India showed that the annual mean wave power is the highest (15 to 20 kW/m) in the southern Bay of Bengal (see **Figure 1.12**). The annual mean wave power is relatively high (~12 kW/m) in the central Arabian Sea and off the southern tip of India. The wave power off the south-east coast of India is the lowest (< 5 kW/m) due to the presence of Sri Lanka. The estimated total annual mean wave power could reach 19.5 GW (171 TWh/y) along the western shelf seas of India and around 8.7 GW (75.5 TWh/y) along the eastern shelf seas.[50] In addition to that, the total incident wave energy resource in India is estimated to be

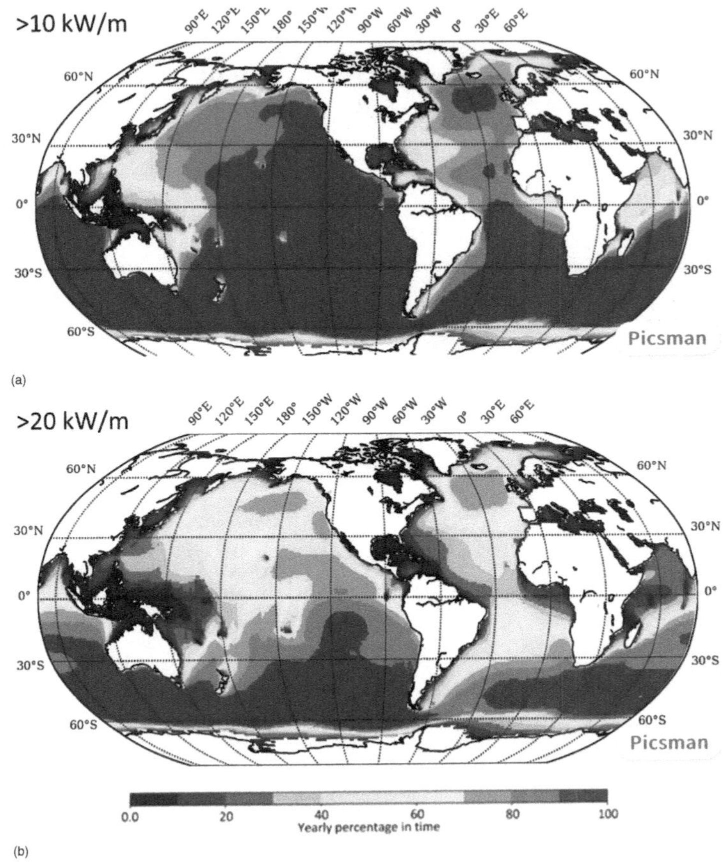

FIGURE 1.11
Yearly percentage in time of power equal or greater than (a) 10 kW/m and (b) 20 kW/m.[44]

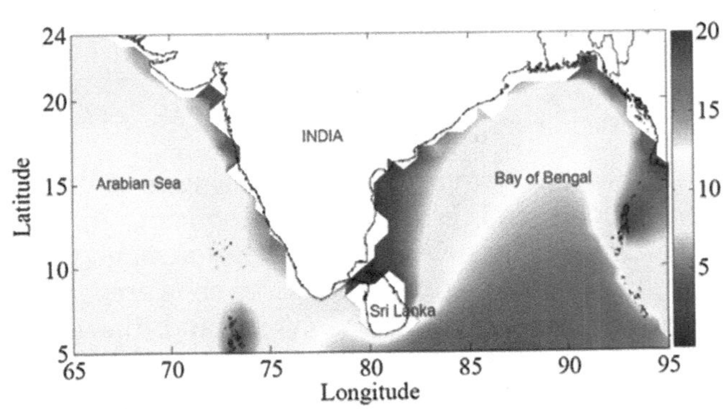

FIGURE 1.12
Distribution of annual mean wave power in the Arabian Sea and Bay of Bengal. The unit of wave power is kW/m.[50]

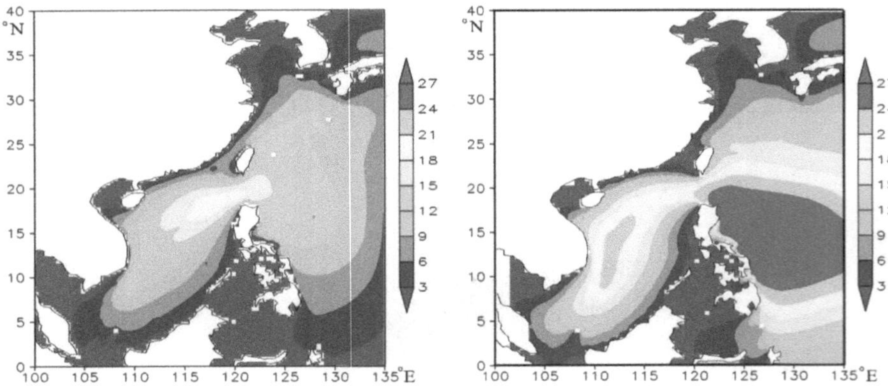

FIGURE 1.13
(a) Multi-year average wave power density and (b) predicted wave power density, in 2015, unit: kW/m.[53]

around 40–60 GW, with attractive locations along the south-west and southern coasts in Maharashtra, Kerala, Karnataka, and Tamil Nadu.[51]

The wave energy resources assessment in the South China Sea carried out by the National Ocean Technology Center in Tianjin showed that the annual average wave energy flux could reach 18 kW/m between the Luzon Strait and south-east of the Indo-China Peninsula.[52] These regions are shared by the Philippines, China, and Vietnam, thus presenting a strategic location with sufficient wave energy potential to house wave farms[53] (see Figure 1.13). However, wave energy close to the shore in certain tropical countries or in the South China Sea could drop below 10kW/h; e.g., the average wave energy density of Malaysian seas facing the South China Sea could be as low as 1.41 kW/m, thereby making wave energy as an alternative source to fossil fuels unattractive.[54] In Section 1.4, the integration of WECs with infrastructure is thus proposed to encourage the utilisation of wave energy in these regions.

1.3.2 Increase of World Energy Demand

World energy demand has been steadily increasing in recent years, driven by population growth and economic development. According to the International Energy Agency (IEA), global energy demand grew by 2.3% in 2018[55, 56] and is projected to continue growing at an average rate of 1.3% per year through 2040.[57] The growth of energy demand directly increases the utilisation of fossil fuels such as oil and natural gas, which are still the primary sources of energy supply for most countries. The BP Statistical Review[58] reported that fossil fuels accounted for around 84% of the world's primary energy demand in 2020. Specifically, oil remains the largest source of primary energy, representing around 36% of the total, followed by coal (27%) and natural gas (24%).

The utilisation of these hydrocarbons (particularly coal, oil, and natural gas) are major contributors to climate change. The burning of fossil fuels releases carbon dioxide (CO_2) and other GHGs (methane, nitrogen oxide (NO_x), sulphur oxide (SO_x), and other industrial chemicals), which trap heat, and contribute to global warming. According to the Netherland Environmental Assessment Agency[59] and the US Environmental Protection Agency,[60] the burning of fossil fuels is the largest source of anthropogenic CO_2 emissions, accounting for about 73% of the total GHG emissions in 2020, in which the IEA[61] estimates that the burning of coal is the largest single source of CO_2 emissions, accounting for around 40% of the total in 2021. This is followed by oil for around 28% and natural gas for around 19.6% of the total. It is worth noting that the production and transportation of hydrocarbons also contribute to emissions. In addition, fugitive emissions of methane (a potent GHG) associated with fossil fuel extraction and transport can represent a significant share of the overall emissions related to fossil fuels.

To mitigate the effects of climate change, it is important to reduce the use of hydrocarbons and transition to cleaner energy sources. Many countries have set targets to reduce their GHG emissions, and there are efforts to develop technologies to capture and store carbon emissions from fossil fuel power plants. One of the UN SGDs,[62] *UN SDG7 Affordable and Clean Energy*, is to ensure everyone has access to affordable, reliable, sustainable, and modern energy. This is further driven by the adverse impact of climate change and government policies, technological advancement, and increasing concern about the environmental impact of fossil fuels. As a result, there is an increased focus on renewable energy and energy transition. Other sources of renewable and sustainable energy have been sought, in line with the UN SDG, such as nuclear power, hydroelectric, wind, solar, and geothermal. The share of renewable energy in total energy consumption has been increasing, particularly for solar and wind power, which are the fastest-growing renewable energy sources.

1.3.3 Challenges

There are however several challenges that need to be addressed to fully realise the potential of wave energy in tropical seas. One of the main challenges is the lack of detailed data and information on wave energy resources in these regions, which limits the ability to accurately assess the potential of wave energy. Additionally, the deployment and maintenance of wave energy devices in tropical seas are challenging; they can be affected by severe weather conditions and ocean currents. The deployment of a large-scale array of WECs could also have a significant impact on the environment, as these areas have a high biodiversity of flora and fauna, such as mangroves, seagrasses, corals, and marine mammals, as well as on the aesthetic value of coastal areas if not designed and installed properly. While extensive research and test-bedding have been carried out on wave energy devices in the developed country, which are targeted for the sub-tropic environment, wave energy technology in the tropical region is still relatively new. Wave energy converters, such as Oyster or Pelamis WECs, developed for the sub-tropical climate might not be efficient to be used for the tropical region. Therefore, more research and development are needed to improve the efficiency and reliability of wave energy systems. Commercial wise, the tropical seas and oceans are important maritime route for shipping activities; hence, the deployment of wave farms in these areas requires proper planning so as not to disturb the economic benefits of the countries. One way to encourage the utilisation of WECs is via the co-location with existing infrastructure that will be described in Section 1.4.

1.4 Integration of WECs with Floating Infrastructure

There are numerous technical and environmental challenges to the deployment of WECs in the tropical region. One main challenge is the high cost of wave energy technology, contributed from the installation, station keeping, and maintenance, which makes it less competitive with other forms of renewable energy. This challenge is coupled with the overpopulation in some of the tropical countries, such as Singapore, the Philippines, and Indonesia. By taking a tiny nation such as Singapore as an example, proper development planning of the city and sea space is needed due to its land-scarcity and urbanisation problems. As Singapore is a global maritime port, thereby making the sea surrounding this island city an important global shipping route, the deployment of WECs in these regions is affected by sea space constraint and marine traffic congestion.

However, the adverse effect of global warming resulting in the rise of seawater level pushes the need for all countries to play a role in mitigating the effect of climate change as described earlier. Therefore, a more efficient way of ocean space utilisation has to be devised to achieve the interests of various stakeholders. Over the years, integrated floating structure with WEC has been proposed to utilise the ocean space more efficiently. By integrating the floating structure with WEC, such a solution is cost-effective and is efficient in resource management, as it optimises usage of the sea space.

1.4.1 Advantages

Proper design planning is necessary when integrating WECs with floating structures to ensure the safety and stability of the integrated systems while optimising the energy output. Some of the benefits of integrating WECs with floating structures are as follows:

i. Increased efficiency: Integrating WECs with coastal infrastructure can increase the overall efficiency of the WECs by using the existing infrastructure as a foundation and by providing a more stable environment for the WECs to operate in.
ii. Reduced costs: Using existing coastal infrastructure as a foundation for WECs can also reduce the cost of building and deploying the WECs. The cost of constructing new foundations can be a significant portion of the total cost of a WEC project.
iii. Additional sources of revenue: Integrating WECs with coastal infrastructure such as ports can provide an additional source of revenue for the ports through the sale of electricity generated by the WECs.
iv. Increase output of grid: Integrating WECs with offshore wind turbines can increase the overall energy output of the combined system.
v. Reducing the environmental impact of wave energy: Integrating WECs with coastal infrastructure can reduce the environmental impact of wave energy, e.g., by reducing the number of materials required to construct the WECs foundation or by reducing the amount of seafloor that needs to be disturbed during installation.
vi. Facilitating the development of wave energy: Integrating WECs with coastal infrastructure can facilitate the development of wave energy by providing a ready-made infrastructure for the WECs.

1.4.2 Examples of Integration of WECs with Floating Structures

There are various concepts of integration of WECs with floating structures. One example is the integration of WECs with breakwaters, which are structures such as ports built along coastlines. Examples of projects for WECs integrated into ports are the SE@Ports[63] and the Eco Wave[64] Projects.

Besides breakwaters, WECs could also be integrated with oil and gas platforms, such as semi-submersibles in order to produce sustainable green energy to power the platform. As such offshore structures are located in offshore deepwater, they are subjected to high sea state with a great amount of wave energy resources, thereby making it feasible for wave farm deployment. The power generation could be further enhanced by using a combination of OWSC and point absorber due to the combination of different resonant periods for the two different WECs.[65] There are also proposal to install WECs to the supporting platform of offshore wind turbines[66], such as by combining spar-type offshore wind turbine (inspired by Hywind) and a WEC (inspired by Wavestar) to increase the energy output form the farm.[67]

As UN@Habitat is considering floating cities as means of adaptation to the rise of seawater level, the world will see more floating solutions being used for human dwelling in the future. A European Union Horizon 2020–funded project, Space@Sea,[68] involving 17 European partners, including engineers, architects, researchers, and industry experts, has been carried out since 2017 to develop a standardised and cost-efficient modular island with low ecological impact. The conceptual design of the floating city is given in Figure 1.14.

FIGURE 1.14
Conceptual design of a floating city for the Space@Sea Project.[69]

As floating cities must be self-sustainable, energy generation from the offshore marine environment has to be sought out. One way is to generate energy via WECs attached to the floating structures. A comprehensive review study on the integration of WECs with floating platforms were carried out to study the feasibility of WECs attached to floating cities.[69] The WECs not only served as devices to generate energy but could also mitigate the response of the platforms arising from structural deformation.[70] Other innovative ways of generating energy for floating cities are, for example, using embedded oscillating water columns (OWCs)[70] (see Figure 1.15a), articulated OWCs (Figure 1.15b), and OWC floating breakwaters (Figure 1.15c).

1.4.3 Challenges

Building floating structures in tropical seas poses a number of challenges that must be considered in order to ensure the safety and functionality

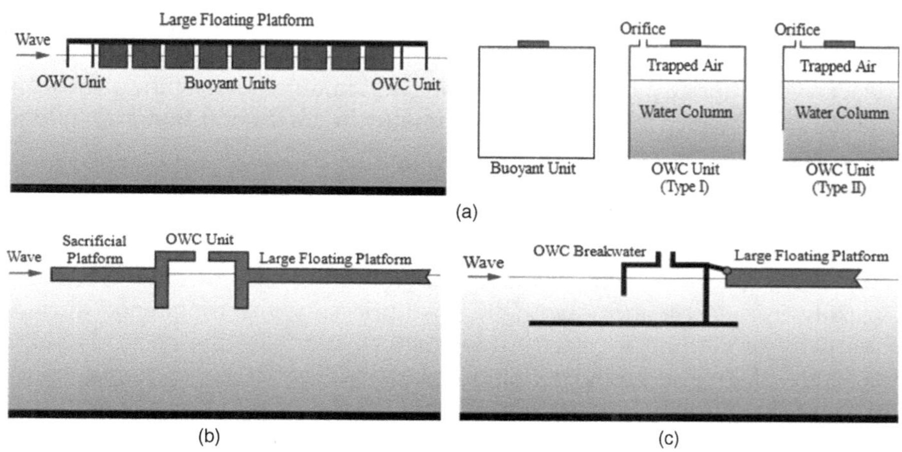

FIGURE 1.15
(a) Embedded OWC, (b) articulated OWC, and (c) OWC floating breakwater attached to floating platforms.[69]

of the structures. These challenges include weather conditions, material degradation, anchoring systems, maintenance, biofouling and environmental impacts.

Weather conditions are one of the main challenges facing the construction of floating structures in tropical seas. Tropical seas are characterised by high winds, waves, and storms that can cause significant damage to floating structures. These conditions also make it difficult to construct and maintain the structures, as high winds and waves can impede the progress of construction and maintenance work, and storms can cause unexpected damage to structures. The design of the floating breakwaters in tropical seas should consider the combined effects of waves, currents, and environmental loads, such as temperature and humidity. The use of durable materials and proper design can greatly improve the structural performance of floating breakwaters in tropical seas.

Material degradation is another major challenge of building floating structures in tropical seas. Floating structures are exposed to a variety of environmental factors in tropical seas, such as high temperatures, humidity, and saltwater, which can lead to accelerated material degradation. This can impact the durability and lifespan of the structure, potentially requiring costly repairs or replacements. To minimise the effects of material degradation, it is important to use materials that are resistant to the corrosion, such as marine-grade steel or concrete.

Anchoring systems also pose a significant challenge for floating structures in tropical seas. The strong currents and waves in tropical seas can make it difficult to anchor floating structures securely. This can increase the risk of structural damage and the potential for a structure to drift away. To address this issue, it is important to use anchoring systems that are specifically designed to withstand the forces exerted by waves and currents in tropical seas. Additionally, regular inspections and maintenance of the anchoring system are essential to ensure that it remains in good condition and able to secure the structure in place.

Biofouling poses a significant challenge for floating structures in tropical climates, where warm, nutrient-rich waters accelerate the growth of marine organisms. The accumulation of biofouling increases weight, affects buoyancy, and accelerates material degradation, particularly through microbiologically influenced corrosion (MIC). Additionally, the roughened surface caused by biofouling increases drag, thereby reducing the hydrodynamic efficiency of WECs. Frequent maintenance and biofouling removal could drive up operational costs and downtime. To mitigate these risks, antifouling coatings can help prevent organism attachment. In addition, ultrasonic and ultra-violet (UV) treatment technologies can be used to disrupt biofilm formation. Regular cleaning and monitoring are therefore essential to prevent excessive accumulation and maintain performance.

Lastly, it is important to consider the environmental impacts of building floating structures in tropical seas. Floating structures in tropical seas can have negative impacts on the surrounding marine ecosystems, such as the destruction of coral reefs and displacement of marine species. It is therefore important to conduct environmental assessments and impact studies prior to construction, and to take steps to minimise the environmental impacts of the structure during construction and operation. The design of floating breakwaters should consider the use of eco-friendly materials in construction to minimise the impact on the environment.

2

Environmental Loads

This chapter focuses on the nature of environmental loading in the tropical sea for the deployment of floating structures and WECs. Three subchapters are introduced:

(i) Understanding water waves: Various types of water waves, essential for understanding ocean dynamics, are introduced. Wind-generated waves, arising from wind-water interactions, and swells, which are long-period waves carrying energy across oceans, are covered. Tsunamis, triggered by underwater earthquakes or volcanic eruptions, are presented, along with tidal waves driven by the gravitational pull of the moon and sun. Breaking waves and their impact on coastal areas are also examined, as well as seiche waves, which are oscillations in enclosed water bodies caused by atmospheric or seismic activity.

(ii) Potential wave theory: Potential wave theory is explored, focusing on key concepts in fluid dynamics and ocean engineering. The Stokes expansion is introduced as a perturbation method for approximating wave motion solutions. Small amplitude wave theory, providing a linear approximation for wave behaviour in shallow water, is covered, followed by Stokes wave theory, which accounts for nonlinear effects in steep waves.

(iii) Wave model: The random wave model for short-term wave analysis is examined. Statistical methods are used to analyse wave height and wave period distributions, with an emphasis on zero-crossing analysis. Spectral analysis is introduced to understand how energy is distributed across different frequencies in the wave spectrum. The complexities of multidirectional seas, where waves approach from various directions, are also addressed.

2.1 Understanding Water Waves

The movement of water waves, shaped by both terrestrial and celestial forces, provides valuable insight into the complex dynamics of the oceans. This chapter explores the wide range of wave phenomena that characterise

DOI: 10.1201/9781003387855-2

marine environments. It examines wind-generated waves, which form the everyday patterns on the ocean's surface, as well as swells that carry energy across vast ocean basins, each revealing distinctive behaviours of the marine system. The chapter also discusses the immense energy of tsunamis, triggered by underwater earthquakes or volcanic activity, and the cyclical nature of tidal waves, governed by the gravitational interactions between the moon and the sun. Additionally, the chapter investigates the impact of breaking waves on coastal landscapes, and the occurrence of seiche waves within enclosed bodies of water, induced by atmospheric or seismic disturbances. Through this analysis, a thorough understanding of wave dynamics is developed, offering deeper insight into the natural forces that shape oceanic environments.

2.1.1 Description of Waves in Tropical Climate

A tropical wave is a type of weather disturbance that is characterised by a trough or low-pressure area in the atmosphere that moves across the tropics. The movement of a tropical wave is accompanied by a shift in wind direction, an increase in humidity, and the development of widespread showers and thunderstorms. The coastal regions in the tropical zone are exposed to a combination of various wave systems, mainly swell and wind-generated water waves.

A swell is a type of ocean wave that travels far from its point of origin and can have a relatively long wavelength. Swells are known for their broad lines and large gaps between each crest, and they can have enough energy to reach heights of over ten metres, as shown in Figure 2.1.

On the other hand, wind-generated waves, also known as surface waves, are due to the friction between the wind and the water, thus creating a wave crest. Wind-generated waves are formed gradually from ripples and grow larger as the friction between air and water stretches the surface, causing

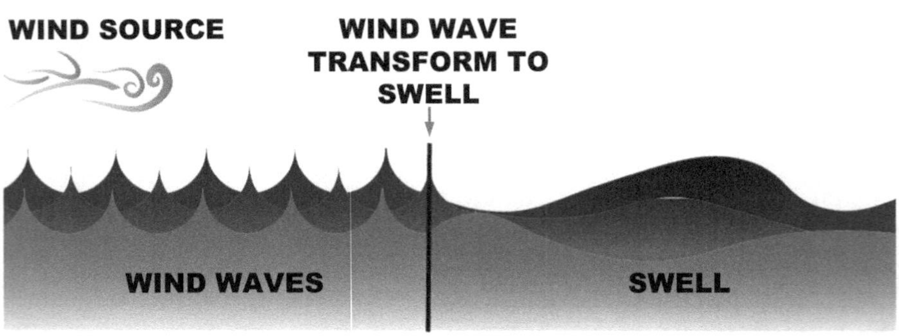

FIGURE 2.1
Wind-generated waves and swell.

waves to form. The amount of energy transferred from the air to the water is proportional to the strength of the wind, so the harder the wind blows, the larger the waves become.

As waves approach the shallow water regions, the velocity of the bottom water particle is slowed down due to friction with the seabed, resulting in a differential velocity between the top and bottom water particle. The wave diverts from the sinusoidal wave profile as the velocity at the top of the water particle increases and finally breaks when it becomes unstable (shown in Figure 2.2). The wave breaks when the wave steepness, defined as the ratio of wave height to wavelength, $\frac{H}{\lambda}$ exceeds 1/7.

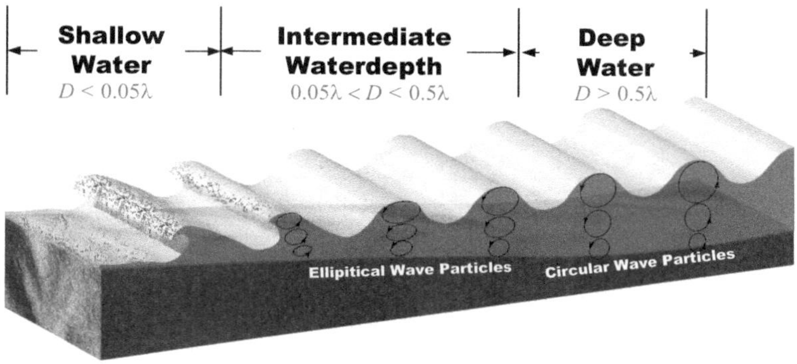

FIGURE 2.2
Waves under different wave depth.

The coastal along the tropical zone are important not only for recreational activities, which are mostly subjected to swell and wind wave, but also important for marina and ports. For the latter, hard structures are built to create sheltered areas for shipping activities. Similarly, coastal protection structures such as breakwaters are also built to protect the shore from further erosion. The construction of marinas, ports and breakwaters creates an enclosed or semi-enclosed zone that is subjected to the occurrence of standing waves, also known as seiche waves (Figure 2.3a). Seiches may disturb shipping by interfering with tides and generating strong reversible currents at the entrances to harbours, or by causing moored vessels to oscillate against their mooring cables and break free. Another phenomenon that may occur is the occurrence of capillary waves (Figure 2.3b). These waves closely resemble ripples in their structure, and the restoring force involved is capillarity, which is the binding force that holds together the water molecules on the ocean's surface.

The characteristics for various waves are summarised in Table 2.1. In the subsequent sections, the formulation for these wave types will be presented.

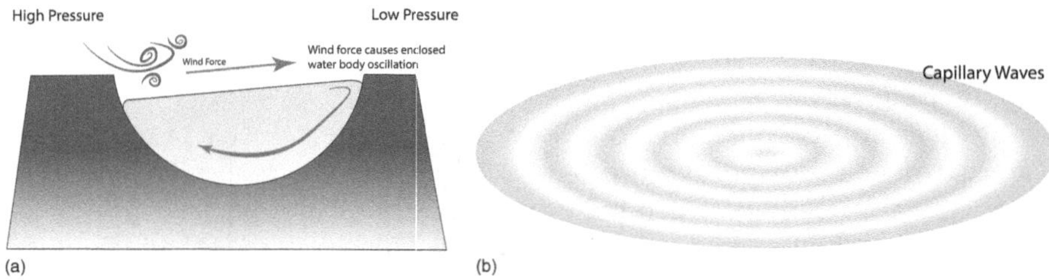

FIGURE 2.3
(a) Seiche and (b) capillary waves.

TABLE 2.1

Characteristics for Different Wave Systems

Wave Type	Typical Wavelength	Disturbing Forces	Restoring Force
Wind-Generated Waves	60–150 m	Ocean Wind	Surface Tension
Swell		Energy below ocean's surface	Gravity
Tsunamis	200 km	Faulting of sea floor, volcanic eruption, landslide	Gravity
Tides	Half the circumference of Earth	Gravitational attraction, Earth's rotation	Gravity
Breaking Waves			
Seiche	Changes according to size of basin/lake/marina	Change in atmospheric pressure, storm surge	Gravity
Capillary Waves	< 2 cm	Wind	Surface tension

2.1.2 Wind-Generated Waves or Surface Waves

The wind-generated waves, also called surface waves, water waves, or wind waves, are gravity waves, where gravity is the main restoring force. They occur as a result of wind blowing over the water's surface. In order to generate wind waves, the wind must be blowing at a distance on an uninterrupted open water, where the distance is known as *fetch*, without significant change in the direction (see Figure 2.4). As the wind blows over the water surface, the speed must be greater than the speed of the wave crest to generate sufficient energy.

As surface waves are wind-driven, they depend significantly on the wind direction and fetch. Wind-generated waves occur in various sizes depending on the wind speed, fetch, duration, and water depth. Figure 2.5 shows an example of the fetch diagram to display the relationship between the wave height with the fetch length. The significant wave height H_s depends on the

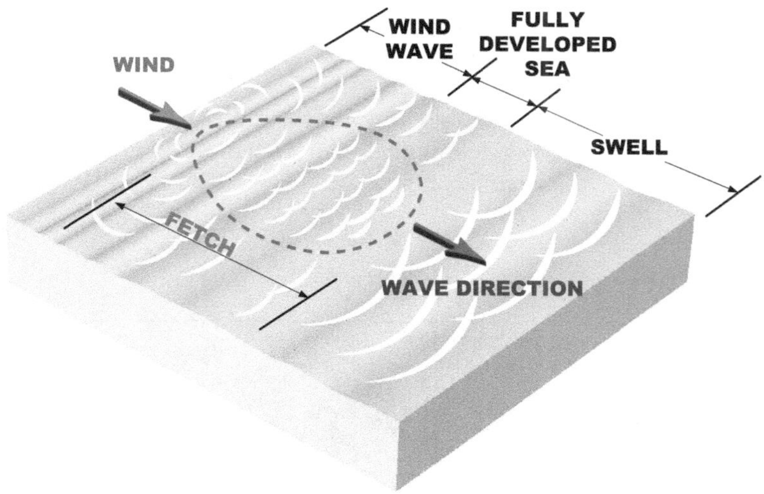

FIGURE 2.4
Generation of wind wave and fetch.

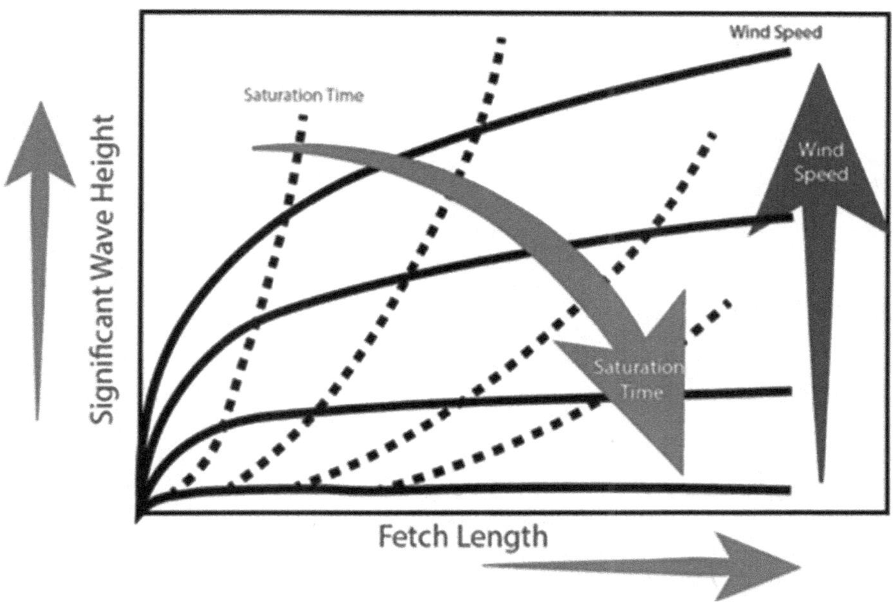

FIGURE 2.5
Fetch diagram.

fetch. The H_s increases with the increase of the fetch and wind speed. The fetch duration also has a significant effect on H_s where longer fetch duration will result in higher H_s.

The mean significant wave height around the globe is shown in Figure 2.6. The H_s in tropical regions is relatively lower as compared to their temperate counterparts and has a mean H_s below 2m.

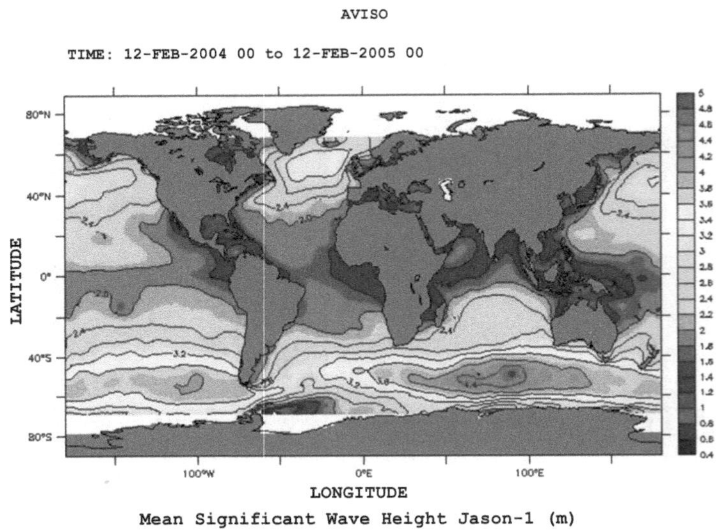

FIGURE 2.6
Global mean significant wave height.

Source: Global Climate Observing System and www.aviso.com.

2.1.2.1 Different Forms of Surface Waves

Depending on water depth and location, the surface waves can appear in different forms, such as breaking waves or seiche waves, that vary depending on the wave period. Another type of wave that has surface tension as the restoring force is the capillary wave. Figure 2.7 shows the various types of waves depending on the wave period. When the waves move out of the fetch area, the wind waves travel thousands of kilometres and are known as swells. Thus, the swells are not affected by the local wind, as they are generated and arrived from elsewhere. The description of swells will be presented in Section 2.1.3.

2.1.2.2 Developing and Fully Developed Sea

The sea can be classified into a developing sea and a fully developed sea. A developing sea refers to the state of the sea with the waves continue to grow larger when energy is continuously supplied to the waves. Up to a point when

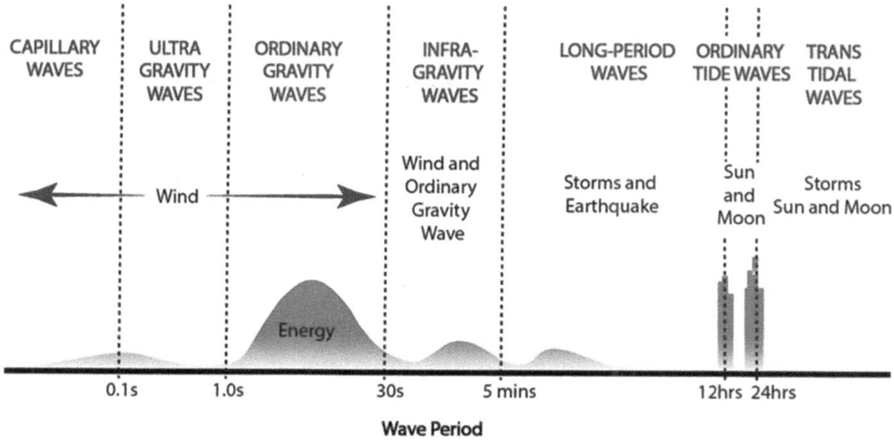

FIGURE 2.7
Classification of spectrum of ocean waves by wave period.

the sea stops growing, the sea state is described as fully developed. The fully developed sea is a state in which the energy input from the local wind to the waves is in balance with the energy transfer among different wave components and with the dissipation of energy caused by wave breaking. The sea state is often reported on the Beaufort scale (see Table 2.2), which ranges from 0 to 12, with the Beaufort Number 0 indicating calm, windless, and waveless conditions, while the Beaufort Number 12 represents a hurricane.

TABLE 2.2
Beaufort Scale

Beaufort Number	Wind Speed (Knots)	Description
0	<1	Calm
1	1–3	Light Air
2	4–6	Light Breeze
3	7–12	Gentle Breeze
4	11–16	Moderate Breeze
5	17–21	Fresh Breeze
6	22–27	Strong Breeze
7	28–33	Near Gale
8	34–40	Gale
9	41–47	Strong Gale
10	48–55	Storm
11	56–63	Violent Storm
12	>64	Hurricane

A fully developed sea is reached when the waves have stopped growing and reached their maximum size for the existing wind speed, duration, and fetch. This state is reached when the energy supplied by the wind is equal to the energy lost in breaking waves. The developing and fully developed sea can be modelled with various random wave model, such as the JONSWAP, Pierson and Moskowitz (PM), and Bretschneider described in Section 2.3.5.

2.1.3 Swells

Swells are long waves created by wind-generated waves that have travelled thousands of kilometres of distance before they break on the shore. Swells may consist of a series of waves with different wavelengths, where the dissipation of wave energy reduces with the increase of wavelengths. Therefore, most swells that arrive at the shore are uniform long waves having the same speed and travel in groups. The distance travelled by the swell can be obtained from the time t of propagation of swells from the source and the wave period T:

$$X_{\text{swell}} = \frac{gT}{4\pi} \cdot t, \tag{2.1}$$

where g is the gravitational acceleration.

Through numerical simulation using WAVEWATCH III, Zheng et al.[71] has identified two pure swell regions with a frequency of occurrence of over 90% found in nearby regions of Indonesia and Cape Verde in the east of the African continent, as shown in Figure 2.8. Figure 2.8 shows that the swell dominated areas (coast waters off Indonesia, the northern Indian Ocean, in the west of central America and Africa, and the western boundary of central South America) are located at tropical regions.

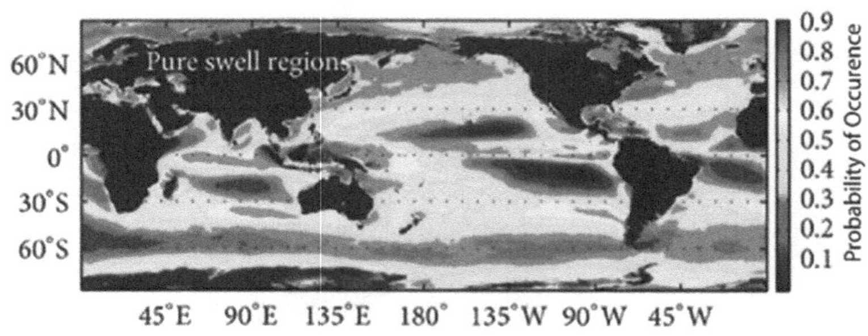

FIGURE 2.8
Pure swell region from WAVEWATCH III simulation.[71]

2.1.3.1 Application of Swells

Swells are important surface waves for surfing activities, as they allow surfers to catch more waves and perform more tricks. Larger and more consistent swells produce better surfing conditions. Very large swells can also be dangerous and challenging, even for experienced surfers. Therefore, it is important for surfers to monitor swell conditions before surfing activities take place. Surfing websites such as Surfline (www.surfline.com) were created to provides surf reports, forecasts, live camera feeds, and other related content for surfers and ocean enthusiasts around the world. An example of a swell travelling from the southern Indian Ocean to Southern California obtained from Surfline is shown in Figure 2.9. The distance travelled by the swell is 10,000 nautical miles (18,520 km, almost half the circumference of Earth!), and the swell travels at a wave period of 20 s. By using Equation (2.1), the swell would take one week to travel from the southern Indian Ocean to Southern California. Along the route, the swell plays an important role in the marine ecosystem by stirring up nutrients and increasing water circulation when passing through Australia, the Indonesian Islands, and the Pacific archipelagos along the route to reach its final destination. Proper understanding of the swell is therefore important, as meteorologists also use data on swell height and frequency to predict the behaviour of weather patterns and make more accurate forecasts, whereas coastal engineers use swell data to make long-term predictions about coastal erosion.

2.1.4 Tsunamis

Tsunamis are a series of ocean waves caused by various underwater geological activities, most commonly earthquakes with a magnitude of 6.5 or above to trigger tsunamis.[72] Tsunamis can also be triggered by volcanic eruptions, icebergs calving, and, in rare cases, meteorites hitting the ocean. Tsunamis form when there is a large, abrupt disturbance at the sea surface, which creates a set of ocean waves which then radiate from the point of origin, as shown in Figure 2.10. Tsunami waves usually start in deep water with a height of less than 1 metre and travel at speeds of 100 kilometres per hour but increase in wave height as they reach shallower waters. The friction between the waves and seabed slows the waves and eventually forces them to increase in height, thereby increasing the energy in the wave and causing detrimental damages to the land.

Some of the notable tsunamis over the past three centuries are listed in Figure 2.11.

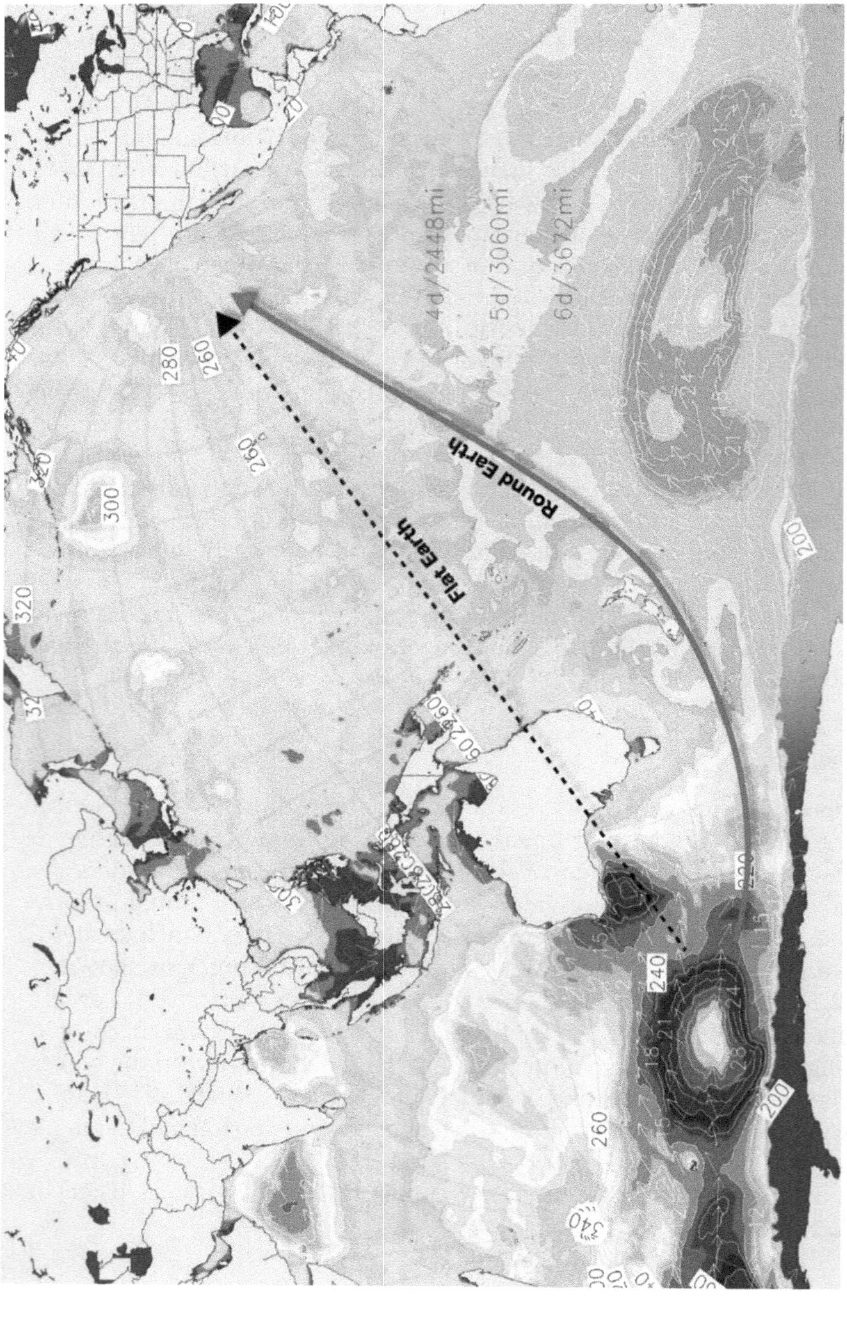

FIGURE 2.9
Swell travelling from the southern Indian Ocean to Southern California.

Source: www.surfline.com.

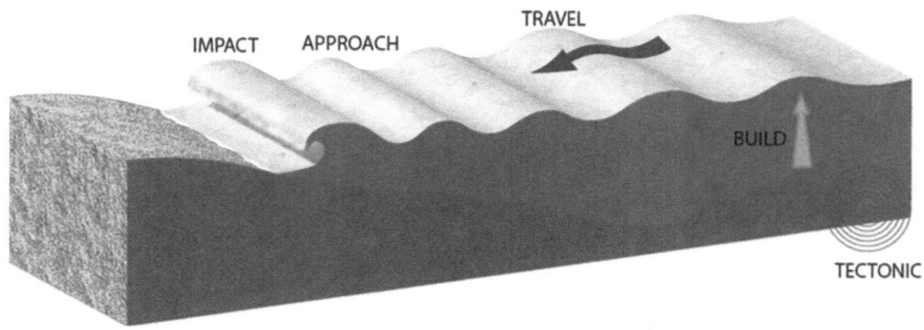

FIGURE 2.10
Formation of tsunamis.

2.1.5 Tides

Tides are the regular rise and fall of ocean waters along coasts where the formation of tides is primarily caused by the gravitational pull of the moon and the sun on Earth (see Figure 2.12). The magnitude and nature of tides vary from place to place and are influenced by factors such as the movement of the moon and changes in the position of the sun and moon relative to Earth.

2.1.5.1 High and Low Tides

As shown in Figure 2.12, the gravitational pull of the moon on Earth causes the oceans to bulge, thus creating high tides. At the same time, the centrifugal force caused by the rotation of Earth raises the water level on the opposite side of the Earth, creating another high tide. The low points between these high tides are where low tides occur.

2.1.5.2 Diurnal, Semi-diurnal, and Mixed Tides

These tides contribute to the diurnal tide in the equator, which occurs when there is only one high tide and one low tide each day. Far away from the equator, there are two high tides and two low tides each day, also known as the semi-diurnal tides. For coastal areas with complex coastlines, there are mixed tides which are similar to the semi-diurnal tides but with the tidal heights varying significantly.

2.1.5.3 Spring and Neap Tides

In addition to the moon, the gravitation pull from the sun will cause the tidal range to change every fortnight. The gravitation pull from the sun causes

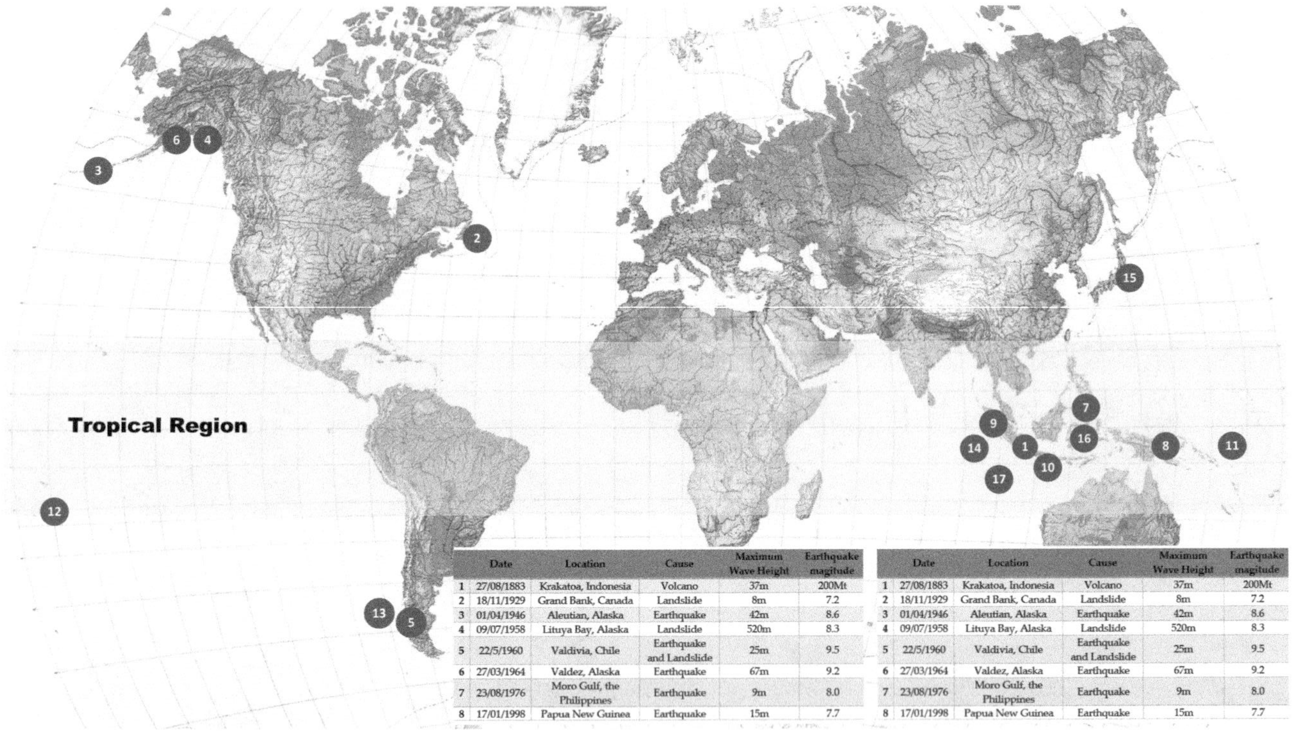

	Date	Location	Cause	Maximum Wave Height	Earthquake magitude
1	27/08/1883	Krakatoa, Indonesia	Volcano	37m	200Mt
2	18/11/1929	Grand Bank, Canada	Landslide	8m	7.2
3	01/04/1946	Aleutian, Alaska	Earthquake	42m	8.6
4	09/07/1958	Lituya Bay, Alaska	Landslide	520m	8.3
5	22/5/1960	Valdivia, Chile	Earthquake and Landslide	25m	9.5
6	27/03/1964	Valdez, Alaska	Earthquake	67m	9.2
7	23/08/1976	Moro Gulf, the Philippines	Earthquake	9m	8.0
8	17/01/1998	Papua New Guinea	Earthquake	15m	7.7

	Date	Location	Cause	Maximum Wave Height	Earthquake magitude
1	27/08/1883	Krakatoa, Indonesia	Volcano	37m	200Mt
2	18/11/1929	Grand Bank, Canada	Landslide	8m	7.2
3	01/04/1946	Aleutian, Alaska	Earthquake	42m	8.6
4	09/07/1958	Lituya Bay, Alaska	Landslide	520m	8.3
5	22/5/1960	Valdivia, Chile	Earthquake and Landslide	25m	9.5
6	27/03/1964	Valdez, Alaska	Earthquake	67m	9.2
7	23/08/1976	Moro Gulf, the Philippines	Earthquake	9m	8.0
8	17/01/1998	Papua New Guinea	Earthquake	15m	7.7

FIGURE 2.11
Notable tsunamis in the past two centuries.

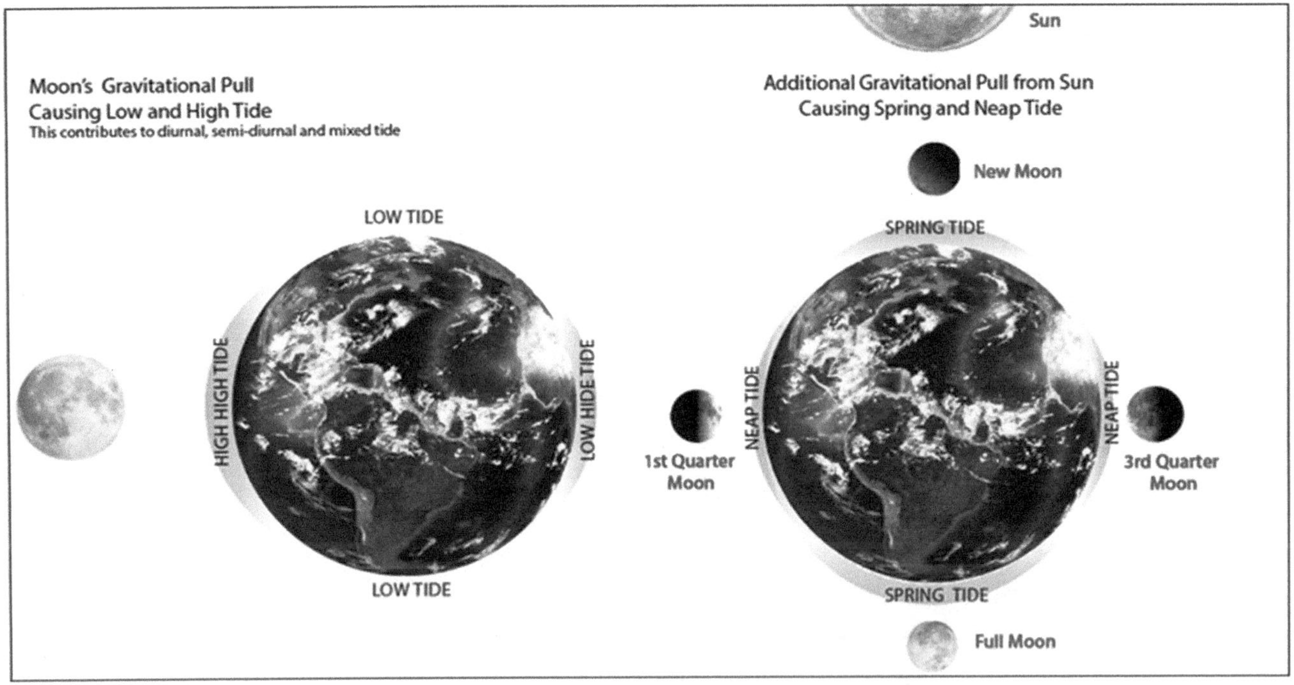

FIGURE 2.12
Gravitation pulls from sun and moon resulting in various tides.

spring tide during the full and new moon and neap tide during the first and third quarter moon, as shown in Figure 2.12. These tides occur due to the gravitational pull from the sun and moon, resulting in large tidal currents which affect the wave propagation energy.

2.1.6 Breaking Waves

Breaking waves are formed when a swell reaches shallow water and slows down, causing the period between the waves to shorten and increase in height. As the waters' depth decreases, the waves get higher and eventually break at roughly 0.7 times the water depth. Breaking waves can be powerful and destructive, eroding shorelines and causing flooding in low-lying areas. They also have significant impact on structures, vessels, seabed sediment movement and facilitate exchanges across the air-sea interface.[73] Breaking waves also produce bubbles, spray, and other effects that can affect marine life. For example, the bubbles generated by breaking waves can help to oxygenate the water, which is important for the survival of fish and other aquatic organisms. Additionally, the turbulence and mixing caused by breaking waves can create nutrient-rich upwelling zones, which support the growth of phytoplankton and other primary producers in the ocean food chain.[74] The type of wave depends on how it breaks, affected by the shape of the sea floor, the size and shape of the wave, and the angle of the approach of the wave. As travelling waves approach the coastal water region, their interaction with the seabed causes the waves to steepen, thereby forming a bulge on the forward face near the crest and capillary waves on the water surface ahead of the toe of the bulge. The waves break when the toe of the bulge moves rapidly down the forward face of the waves.[75] Figure 2.13 shows the various types of breaking waves:

- Spilling Waves: These are the most common type of breaking waves and occur when the wave crest spills down the front of the wave. They typically occur in areas with a gently sloping sea floor.
- Plunging Waves: These waves break suddenly and dramatically, with the crest of the wave curling over and crashing down into the trough. They are often seen at beach breaks and other areas with a steep sea floor.
- Collapsing Waves: These waves occur when a wave breaks on a near-vertical sea wall or another obstacle, causing the wave to collapse in on itself.
- Surging Waves: These waves break without the dramatic curl of plunging waves, instead surging forward with a powerful rush of water. They are typically seen in areas with a very gradual slope.

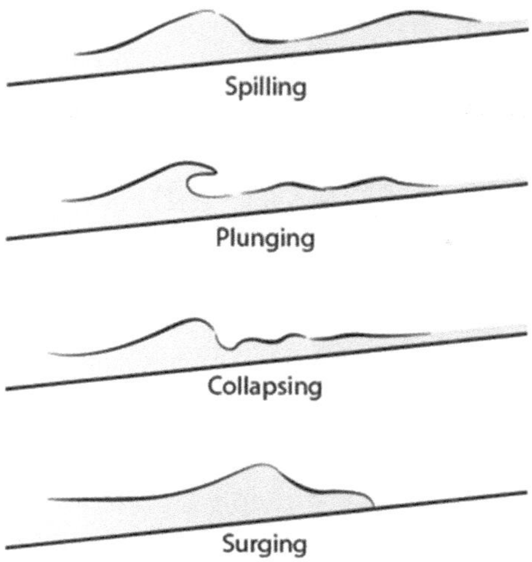

FIGURE 2.13
Types of breaking waves.

- Surging/Spilling Waves: These are a combination of surging and spilling waves, with the front of the wave surging forward while the back spills down the front of the wave. They are typically seen in areas with a moderate slope.

2.1.7 Seiche Waves

Seiche waves are standing waves with longer periods of water-level oscillations, ranging from tens of seconds to several hours depending on the basin's geometry and depth.[76] Seiche waves are formed in partially or fully enclosed basins such as lakes, bays, harbours, caves, and seas with the existence of strong winds or rapid changes in atmospheric pressure to displace the water. The oscillation of seiche waves is simply harmonic where the height of seiche waves depends on the strength of wind and air pressure contrast. Seiche waves have been observed in different water bodies worldwide, such as Lake Erie and Nagasaki Bay. Seiche waves can also be triggered by earthquakes, e.g., the 1959 Hebgen Lake earthquake in Montana triggered seiches in Montana, Wyoming, Idaho, and Alberta, Canada, with the most distant seiche being 545 kilometres from the epicentre.[77] The occurrence of seiche waves as a result of resonance can be useful as a means to enhance the energy generation from wave energy converters. This is shown in detail in Chapter 6.

For a rectangular basin, the longest natural period T_N can be quickly estimated using Merian's formula:[78]

$$T_N = \frac{2\lambda}{\sqrt{gD}}. \tag{2.2}$$

where λ and D are the basin length and water depth, respectively.

2.2 Potential Wave Theory

There are various types of wave theory that exist for modelling ocean waves.[10] The most common being the small amplitude linear wave theory (SAWT), Stokes waves, cnoidal waves, and shallow water waves, where their applicability depends on two ratios, i.e., H/λ and D/λ, as given in Figure 2.14. In this book, the focus is on SAWT in the assessment of fluid-structure interactions of WECs as described later in Chapter 3. A comparison of the wave profile for the various wave theories is shown in Figure 2.15.

2.2.1 Description

In the modelling of waves, it is commonly assumed that the fluid is an ideal fluid which is inviscid and incompressible with its motion irrotational. For an ideal fluid, a velocity potential $\Phi(x,y,z,t)$ exists such that the derivative of the velocity potential would yield the velocity $\mathbf{v}$ of the fluid in the x-, y-, and z-axes given respectively in (2.3a–2.3c), and satisfy the Laplace equation. The potential wave theory is a powerful model in the fluid-structure

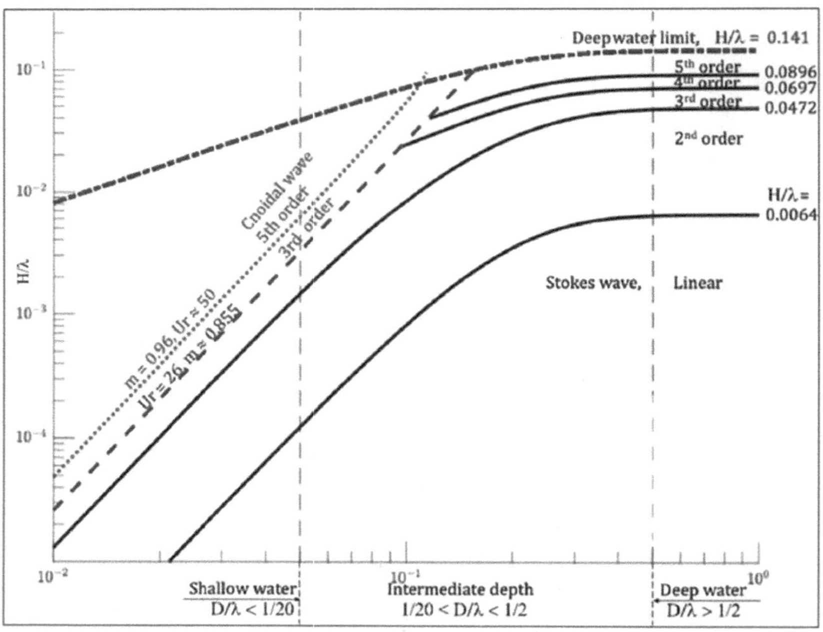

FIGURE 2.14
Applicability of various wave theories.[79,80]

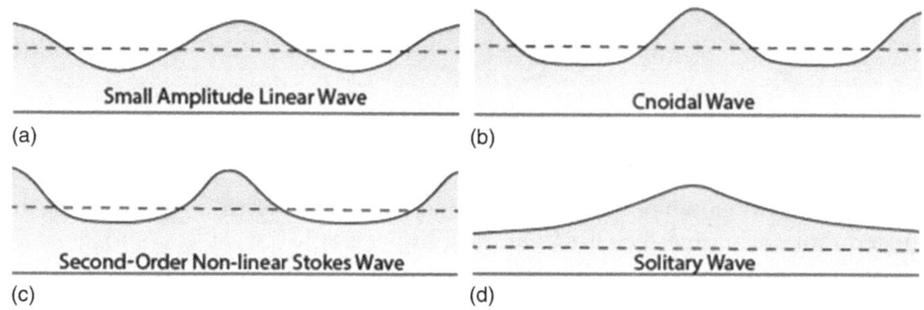

FIGURE 2.15
Wave profile for various wave theories.

modelling that will be presented in Chapter 3. Here, the governing equation and boundary conditions associated with the ideal fluid are presented:

$$v_x = \frac{\partial \Phi(x,y,z,t)}{\partial x}, \tag{2.3a}$$

$$v_y = \frac{\partial \Phi(x,y,z,t)}{\partial y}, \tag{2.3b}$$

$$v_z = \frac{\partial \Phi(x,y,z,t)}{\partial z}, \tag{2.3c}$$

where v_x, v_y and v_z are the velocity in the x, y, and z directions and $\mathbf{v} = (v_x, v_y, v_z)$. It is further assumed that the ideal fluid satisfies the continuity equation given as

$$\frac{\partial \rho}{\partial t} + \nabla \cdot (\rho \mathbf{v}) = 0, \tag{2.4}$$

where $\nabla = \left(\frac{\partial}{\partial x} + \frac{\partial}{\partial y} + \frac{\partial}{\partial z} \right)$. Equation (2.4) describes the conservation of mass, i.e., the mass that enters the system is equal to the rate at which it leaves the system plus the accumulation of mass within the system. Under incompressible fluid conditions, the mass does not change with respect to the time t, i.e., the first term vanishes. Therefore, the continuity equation becomes

$$\nabla \cdot \Phi = 0. \tag{2.5}$$

When we substitute Equations (2.3a) to (2.3b) into Equation (2.5), we arrive at the Laplace equation:

$$\nabla^2\Phi = \left(\frac{\partial^2}{\partial x} + \frac{\partial^2}{\partial y} + \frac{\partial^2}{\partial z}\right)\Phi = 0. \tag{2.6}$$

In addition to the Laplace equation, the velocity potential Φ must also satisfy the boundary conditions at the seabed and the fluid free surface. For simplicity, a two-dimensional wave in the $x-z$ plane will be considered hereafter for the subsequent derivative. The linear assumption is made, i.e., $H \ll \lambda$ and $H \ll D$.

The Bernoulli equation for the inviscid fluid with irrotational flow is first considered. By equating the pressure just beneath the free surface to the atmospheric pressure P_{atm}, we have the Bernoulli equation at the free surface $z = \hat{\eta}$:

$$-\frac{P_{atm}}{\rho} + \frac{1}{2}|\nabla\Phi|^2 + gz + \frac{\partial\Phi}{\partial t} = 0, \quad \text{at} \quad z = \hat{\eta}. \tag{2.7}$$

By rearranging Equation (2.7), the wave elevation $\hat{\eta}$ can be expressed as

$$\hat{\eta} = -\frac{1}{g}\left(\frac{\partial\Phi}{\partial t} + \frac{1}{2}|\nabla\Phi|^2\right), \quad \text{at} \quad z = \hat{\eta}. \tag{2.8}$$

At the free surface, the linearised vertical velocity v_z of the water surface, according to Equation (2.3c), is given as

$$v_z = \frac{\partial\Phi}{\partial z} = \frac{\partial\hat{\eta}}{\partial t} + \frac{\partial\Phi}{\partial x}\frac{\partial\hat{\eta}}{\partial x} + \frac{\partial\Phi}{\partial y}\frac{\partial\hat{\eta}}{\partial y}, \text{where} \quad z = \hat{\eta}. \tag{2.9}$$

Equations (2.8) and (2.9) are, respectively, the dynamic and kinematic boundary conditions of the free surface. These two boundary conditions can be combined by omitting the free surface wave elevation $\hat{\eta}$ term to arrive at the free surface boundary condition:

$$\frac{\partial^2\Phi}{\partial t^2} + g\frac{\partial\Phi}{\partial z} + \left[\frac{\partial}{\partial t}\left(|\mathbf{v}|^2\right) + \frac{1}{2}\mathbf{v}\cdot\nabla\left(|\mathbf{v}|^2\right)\right] = 0 \quad \text{at} \quad z = \hat{\eta}, \tag{2.10}$$

where $\mathbf{v} = \left(\frac{\partial\Phi}{\partial x}, \frac{\partial\Phi}{\partial z}, \frac{\partial\Phi}{\partial z}\right)$. Lastly, the boundary conditions at the seabed are given as

$$\frac{\partial\Phi}{\partial z} = 0, \quad \text{at} \quad z = -D. \tag{2.11}$$

2.2.2 Stokes Expansion

The SAWT is used to model linear waves where the wave height-to-wavelength ratio is significantly large and oscillates in a sinusoidal form. In conditions where the water depth is in intermediate to deepwater condition, following the two ratios H/λ and D/λ given in Figure 2.14, the waves do not oscillate in a sinusoidal form. Rather, the waves are nonlinear and have a steep crest with a flat trough (see Figure 2.15).

The SAWT can be derived by using the Stokes expansion method. Sir George Gabriel Stokes (1819–1903), an Irish physicist and mathematician, has derived a nonlinear representation of these plane waves by using a perturbation series approach, a method now known as the Stokes expansion. In the Stokes expansion technique, the velocity potential Φ and wave elevation $\hat{\eta}$ are expressed as the perturbation series, which involves some "small" parameters that quantify the deviation from the exact solution. In Stokes expansion, the perturbation parameter is ϵ and the perturbation series for Φ and $\hat{\eta}$ are

$$\Phi(x,y,z,t) = \epsilon\Phi^{(1)} + \epsilon^2\Phi^{(2)} + \cdots, \tag{2.12}$$

$$\hat{\eta}(x,y,z,t) = \epsilon\hat{\eta}^{(1)} + \epsilon^2\hat{\eta}^{(2)} + \cdots. \tag{2.13}$$

Here, the derivative of SAWT and Stokes wave theory based on the Stokes expansion will be presented.

2.2.2.1 Derivative of Laplace Equation

Substituting Equation (2.12) into the Laplace equation (2.6) yields

$$\begin{aligned}\left(\epsilon\frac{\partial^2\Phi^{(1)}}{\partial x^2} + \epsilon^2\frac{\partial^2\Phi^{(2)}}{\partial x^2} + \cdots\right) + \left(\epsilon\frac{\partial^2\Phi^{(1)}}{\partial y^2} + \epsilon^2\frac{\partial^2\Phi^{(2)}}{\partial y^2} + \cdots\right)\\ + \left(\epsilon\frac{\partial^2\Phi^{(1)}}{\partial z^2} + \epsilon^2\frac{\partial^2\Phi^{(2)}}{\partial z^2} + \cdots\right) = 0.\end{aligned} \tag{2.14}$$

Collecting the terms associated with ϵ and ϵ^2,

$$\epsilon\left(\nabla^2\Phi^{(1)}\right) + \epsilon^2\left(\nabla\Phi^{(2)}\right) + \ldots = 0. \tag{2.15}$$

This gives us the governing equation, i.e., Laplace equation pertaining to the n th order, expressed as

$$\left(\frac{\partial^2\Phi^{(n)}}{\partial x^2} + \frac{\partial^2\Phi^{(n)}}{\partial y^2} + \frac{\partial^2\Phi^{(n)}}{\partial z^2}\right) = 0 \quad \text{for } n^{\text{th}} \text{ order } \Phi. \tag{2.16}$$

2.2.2.2 Derivative of Boundary Conditions at the Seabed

Similarly, by substituting Equation (2.12) into the boundary condition at the seabed, Equation (2.11) will give us

$$\frac{\partial \Phi^{(n)}}{\partial z} = 0, \quad \text{at} \quad z = -D \text{ for the } n^{\text{th}} \text{ order } \Phi. \tag{2.17}$$

2.2.2.3 Derivative of the Kinematic Free Surface Boundary Condition

We then apply Taylor's theorem to the kinematic free surface boundary condition of (Equation (2.9)) and evaluate at $z = 0$. The Maclaurin series expansion for the boundary condition expressed up to the second order is given as

$$\left(\frac{\partial \hat{\eta}}{\partial t} + \frac{\partial \Phi}{\partial x}\frac{\partial \hat{\eta}}{\partial x} + \frac{\partial \Phi}{\partial y}\frac{\partial \hat{\eta}}{\partial y} - \frac{\partial \Phi}{\partial z}\right)_{z=0} + \hat{\eta}\left[\frac{\partial}{\partial z}\left(\frac{\partial \hat{\eta}}{\partial t} + \frac{\partial \Phi}{\partial x}\frac{\partial \hat{\eta}}{\partial x} + \frac{\partial \Phi}{\partial y}\frac{\partial \hat{\eta}}{\partial y} - \frac{\partial \Phi}{\partial z}\right)\right]_{z=0} + \cdots = 0, \tag{2.18}$$

where $\hat{\eta}$ is the wave elevation.

Then, by using the perturbation series approach, we substitute Equations (2.12) and (2.13) into Equation (2.18):

$$\epsilon\left[\frac{\partial \hat{\eta}^{(1)}}{\partial t} - \frac{\partial \Phi^{(1)}}{\partial z}\right] + \epsilon^2\left[\frac{\partial \hat{\eta}^{(2)}}{\partial t} + \frac{\partial \Phi^{(1)}}{\partial x}\frac{\partial \hat{\eta}^{(1)}}{\partial x} + \frac{\partial \Phi^{(1)}}{\partial y}\frac{\partial \hat{\eta}^{(1)}}{\partial y} - \frac{\partial \Phi^{(2)}}{\partial z}\right] + \cdots = 0. \tag{2.19}$$

2.2.2.4 Derivative of the Dynamic Free Surface Boundary Condition

Lastly, we apply Taylor's theorem to the dynamic free surface boundary condition (Equation (2.8)) and evaluate at $z = 0$. The Maclaurin series expansion for the boundary condition expressed up to the second order is given as

$$\left(\frac{\partial \Phi}{\partial t} + \frac{1}{2}|\nabla \Phi|^2 + g\hat{\eta}\right)_{z=0} + \hat{\eta}\left[\frac{\partial^2 \Phi}{\partial z \partial t} + \frac{1}{2}\frac{\partial}{\partial z}\left(|\nabla \Phi|^2\right)\right]_{z=0} + \cdots = 0. \tag{2.20}$$

Then, by applying the perturbation series approach to Equation (2.20), we substitute Equations (2.12) and (2.13) into Equation (2.20):

$$\epsilon\left[\frac{\partial \Phi^{(1)}}{\partial t} + g\hat{\eta}^{(1)}\right] + \epsilon^2\left[\frac{\partial \Phi^{(2)}}{\partial t} + \frac{1}{2}\left|\nabla\left(\Phi^{(1)}\right)\right|^2 + g\hat{\eta}^{(2)}\right] + \cdots = 0. \tag{2.21}$$

2.2.2.5 Derivative of the Combined (Kinematic + Dynamic) Free Surface Boundary Condition

Similarly, the Maclaurin series expansion can be applied to the free surface boundary condition (2.10). By taking the series up to $n = 2$:

$$\left[\frac{\partial^2 \Phi}{\partial t^2} + g\frac{\partial \Phi}{\partial z} + \frac{\partial}{\partial t}\left(|\nabla\Phi|^2\right) + \frac{1}{2}\nabla\Phi\cdot\nabla\left(|\nabla\Phi|^2\right)\right]_{z=0} + \hat{\eta}\frac{\partial}{\partial z}\left(g\frac{\partial \Phi}{\partial z}\right)_{z=0} + \cdots = 0. \tag{2.22}$$

By using the perturbation series approach, we substitute Equations (2.12) and (2.13) into Equation (2.22) to obtain

$$\epsilon\left[\frac{\partial^2 \Phi^{(1)}}{\partial t^2} + g\frac{\partial \Phi^{(1)}}{\partial z}\right] + \epsilon^2\left[\frac{\partial^2 \Phi^{(2)}}{\partial t^2} + g\frac{\partial \Phi^{(2)}}{\partial z} + \frac{\partial}{\partial t}\left(\left|\nabla\Phi^{(1)}\right|^2\right)\right] + O\left(\epsilon^3\right) = 0, \text{ at } z = 0. \tag{2.23}$$

2.2.2.6 Summary of the Governing Equation and Boundary Conditions

Equations (2.16)–(2.23) give us the Laplace equation, the boundary conditions at the seabed, the dynamic free surface boundary condition, and the combined free surface boundary condition, respectively, for first-order and second-order wave theories. First-order wave theory is also known as SAWT, whereas second-order wave theory is known as second-order nonlinear Stokes waves. The terms related to $\epsilon^{(1)}$ give us the terms related to the SAWT, whereas those related to $\epsilon^{(2)}$ correspond to the second-order Stokes waves. The governing equation and boundary conditions involved in these two wave theories are given as follows:

2.2.2.6.1 *Small Amplitude Wave Theory*

Laplace Equation: $\frac{\partial^2 \Phi^{(1)}}{\partial x^2} + \frac{\partial^2 \Phi^{(1)}}{\partial y^2} + \frac{\partial^2 \Phi^{(1)}}{\partial z^2} = 0$, in fluid domain Ω (2.24a)

BC on Seabed: $\frac{\partial \Phi^{(1)}}{\partial z} = 0$, at $z = -D$ (2.24b)

Kinematic Free Surface BC: $\frac{\partial \Phi^{(1)}}{\partial z} = \frac{\partial \hat{\eta}^{(1)}}{\partial t}$, at $z = 0$ (2.24c)

Dynamic Free Surface BC: $\hat{\eta}_1 = \frac{1}{g}\left(\frac{\partial \Phi^{(1)}}{\partial t}\right)$, at $z = 0$ (2.24d)

Free Surface BC: $\frac{\partial^2 \Phi^{(1)}}{\partial t^2} + g\frac{\partial \Phi^{(1)}}{\partial z} = 0$, at $z = 0$. (2.24e)

2.2.2.6.2 Second-Order Stokes Waves

Laplace Equation: $\frac{\partial^2 \Phi^{(2)}}{\partial x^2} + \frac{\partial^2 \Phi^{(2)}}{\partial y^2} + \frac{\partial^2 \Phi^{(2)}}{\partial z^2} = 0$, in fluid domain Ω (2.25a)

Seabed BC: $\frac{\partial \Phi^{(2)}}{\partial z} = 0$, at $z = -D$ (2.25b)

Kinematic Free Surface BC:

$$\frac{\partial \Phi^{(2)}}{\partial z} = \frac{\partial \hat{\eta}^{(2)}}{\partial t} + \frac{\partial \Phi^{(1)}}{\partial x}\frac{\partial \hat{\eta}^{(1)}}{\partial x} + \frac{\partial \Phi^{(1)}}{\partial y}\frac{\partial \hat{\eta}^{(1)}}{\partial y}, \text{ at } z = 0 \tag{2.25c}$$

Dynamic Free Surface BC:

$$\hat{\eta}^{(2)} = -\frac{1}{g}\left[\frac{\partial \Phi^{(2)}}{\partial t} + \frac{1}{2}\left|\nabla\left(\Phi^{(1)}\right)\right|^2\right], \text{ at } z = \hat{\eta} \tag{2.25d}$$

Combined Free Surface BC:

$$\frac{\partial^2 \Phi^{(2)}}{\partial t^2} + g\frac{\partial \Phi^{(2)}}{\partial z} + \frac{\partial}{\partial t}\left(\left|\nabla \Phi^{(1)}\right|^2\right) = 0, \text{ at } z = 0. \tag{2.25e}$$

2.2.3 Small Amplitude Wave Theory

The SAWT or airy wave theory is the simplest wave theory commonly used to model waves. This theory assumes that the wave as a linear sinusoidal wave (also known as a regular wave) with small wave steepness such that there is no wave breaking, as shown in Figure 2.16, and can be used to model both the surface waves and swells.

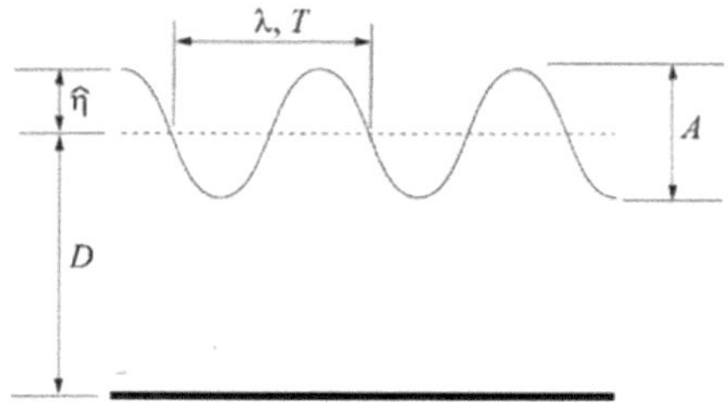

FIGURE 2.16
Regular wave.

Figure 2.16 shows the regular sinusoidal wave in right-hand system coordinates where the wave is propagating in the positive horizontal axis direction (x direction). The surface elevation $\hat{\eta}$ measured from the still water level (SWL), is defined as positive upward therefore the bottom seabed is at $z = -D$. From Figure 2.16, The regular wave has an amplitude of A , wavelength λ, and oscillates with a wave frequency ω for every cycle, which is defined as

$$\omega = \frac{2\pi}{T}, \tag{2.26}$$

where T is the wave period or duration to make one cycle. The propagation of the surface wave elevation can be written as

$$\hat{\eta}^{(1)} = \frac{H}{2}\sin(kx - \omega t), \tag{2.27}$$

where H is the wave height, k the wave number given as $k = \frac{2\pi}{\lambda}$, x is the distance measure along the horizontal axis and t is the time. As waves are made up of a series of wave components, each has a different wave frequency and phase speed. Each wave moves with speed that depends on the wavelength where the wavelength λ is related to the wave frequency ω via the dispersion equation given as

$$\omega = \sqrt{gk \tanh kD}. \tag{2.28}$$

The waves according to Equation (2.28) will move forward with a velocity

$$C = \frac{\lambda}{T}. \tag{2.29}$$

By representing Equation (2.29) with ω and k and by using the wave dispersion relation (Equation (2.28)), the velocity of propagation (known as celerity C) can also be written as

$$C = \frac{\omega}{k} = \sqrt{\frac{g\lambda}{2\pi}\tanh kD} = \frac{gT}{2\pi}\tanh kD. \tag{2.30}$$

Thus, the relationship between λ with the wave period T and water depth D can be expressed as

$$\lambda = CT = \frac{gT^2}{2\pi}\tanh kD. \tag{2.31}$$

2.2.3.1 Group Velocity and Phase Velocity

When the waves move in group, the effect of dispersion causes the waves to move in packets at different speeds. Under such circumstances, the waves move in a distinct phase velocity and group velocity, instead of the wave velocity that is observed in the propagation of individual regular waves.

The group velocity C_g is defined as the changes of wave frequency with respect to the wave number:

$$C_g \equiv \frac{\partial \omega}{\partial k}, \tag{2.32}$$

whereas the phase velocity, which is the rate at which the wave propagates, is given as

$$C_p = \frac{\omega}{k}. \tag{2.33}$$

Equation (2.28) indicates that the longer waves (longer periods) travel faster than shorter waves (shorter periods). Therefore, when groups of waves with different wave frequencies travel together, waves with longer periods, i.e., wavelengths, will lead ahead with time and distance while the shorter wavelengths will lag behind. Due to the interference effect, waves of the same period will group and travel together, while waves with different frequencies may cancel out each other. This results in the formation of wave groups.

By referring to the dispersion equation (2.28), the group velocity is

$$C_g = \frac{\partial \omega}{\partial k} = \frac{C_p}{2}\left(1 + \frac{2kD}{\sinh 2kD}\right). \tag{2.34}$$

The relationship between the C_g and C differs depending on the water depth D, i.e., for deepwater $kD \gg 1$, the deepwater group velocity C_{g0} is given by

$$C_{g0} \cong \frac{C_p}{2} \cong \frac{1}{2}\left(\frac{g}{k}\right)^{\frac{1}{2}}. \tag{2.35}$$

For shallow water $kD \ll 1$, the shallow water group velocity C_{gs} is given by

$$C_{gs} \cong C \cong (gD)^{\frac{1}{2}}. \tag{2.36}$$

Equations (2.35) and (2.36) show that the group wave can never be faster than the carrier wave, with an energy half that of the carrier wave at deep water and equal to the carrier wave at shallow water. As an example, the interference of two waves of different k and ω is shown in Figure 2.17. Each wave is represented by (2.27) as shown in Figure 2.17, where the subscripts 1 and 2 denote the first and second waves, respectively. By superimposing these two equations and representing the equation with the sum-to-product trigonometry identity for cosine function, the wave group equations for elevation $\hat{\eta}_g$ are given as

$$\hat{\eta}_g = 2A\left[\cos\left(\frac{(k_1+k_2)x+(\omega_1+\omega_2)t}{2}\right)\cdot\cos\left(\frac{(k_1-k_2)x-(\omega_1-\omega_2)t}{2}\right)\right], \tag{2.37}$$

$$\hat{\eta}_g = 2A\cdot\left[f_c(x,t)\cdot f_e(x,t)\right], \tag{2.38}$$

where A is the wave amplitude. As such, the propagation of the wave group comprises the envelope wave (modulation) and carrier wave (ripple) as shown in Figure 2.17. The envelope wave or wave group is presented by the function f_e in (2.38) that moves in a lower velocity:

$$C_g = \frac{\omega_1-\omega_2}{k_1-k_2} = \frac{\Delta\omega}{\Delta k} = \frac{\partial\omega}{\partial k}, \tag{2.39}$$

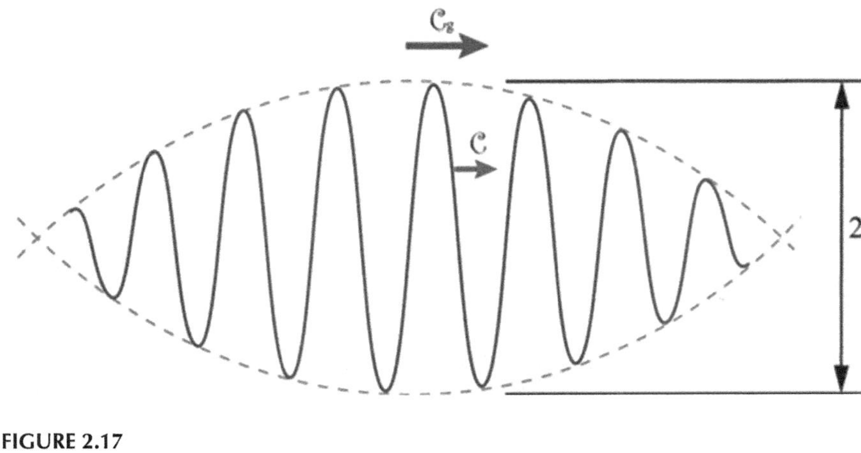

FIGURE 2.17
Wave group.

Note: Wave group elevation is represented by Equation (2.38).

whereas the carrier wave is described by the function f_c that travels in a higher velocity:

$$C_p = \frac{\omega_1 + \omega_2}{k_1 + k_2}. \tag{2.40}$$

Equation (2.39) shows that the phase velocity for the envelope wave is the group velocity under continuous differential case, i.e., when $k_1 \to k_2$. As the group velocity is smaller than phase velocity, i.e., $\frac{\omega_1 - \omega_2}{k_1 - k_2} < \frac{\omega_1 + \omega_2}{k_1 + k_2}$, the envelope wave will be seen moving slower than the carrier wave when travelling in a group, confirming the observation in Equation (2.34).

2.2.3.2 Wave Power in an Irregular Wave (Random Wave)

There are two types of energy contained in a propagating regular wave train, i.e., the kinetic energy T_E and potential energy U_E. Both energies are given as

$$T_E = U_E = \frac{1}{16}\rho g H^2. \tag{2.41}$$

The sum of both energies thus produces the energy density in a period wave train, i.e.,

$$E = \frac{1}{8}\rho g H^2, \tag{2.42}$$

where E is the wave energy, and the unit is in Joule per metre, $\mathrm{J/m^2}$.

The formulation of the rate of power flux across one unit length of wavefront is given from the product of the wave energy E and the group velocity C_g:

$$P_{res} = \bar{E} = EC_g = \frac{1}{2}\rho g |A|^2 C_g = \frac{1}{8}\rho g H^2 \times \frac{\omega}{2k}\left(1 + \frac{2kD}{\sinh 2kD}\right). \tag{2.43}$$

Equation (2.43) has a unit of Watt/m. The wave power P_{res} is usually measured in offshore deepwater condition as

$$P_{res|0} = \overline{E_0} = EC_{g0} = \frac{1}{8}\rho g H^2 \times \frac{1}{2}\left(\frac{g}{k}\right)^{\frac{1}{2}}. \tag{2.44}$$

With the help of the dispersion relation Equation (2.28), the wave number at deep water given as $k = \omega^2/g$ is then substituted into Equation (2.44) to obtain the common expression of the wave energy:

$$P_{res|0} = \overline{E_0} = EC_{g0} = \frac{1}{32\pi}\rho g^2 T H^2. \tag{2.45}$$

Note that Equation (2.45) is for wave power in a regular wave. For random wave conditions in deep water, the wave power $P_{res}|_0^{(irr)}$ is given by

$$P_{res}|_0^{(irr)} = \overline{E_0}^{(irr)} = \frac{1}{64\pi}\rho g^2 T_p H_s^2. \tag{2.46}$$

where T_p is the peak wave period and H_s the significant wave height.

2.2.3.3 Deep Water, Intermediate Water, and Shallow Water

The wave systems can be categorised into deepwater and shallow water waves, which depend on the ratio of wave height-to-wavelength H/λ. The categorisation of the waves is given in Figure 2.2 where the deepwater wave occurs when $H/\lambda > 0.5$, intermediate water $0.05 < H/\lambda < 0.5$, and shallow water $H/\lambda > 0.05$.

As waves such as tsunamis and tides have a relatively longer wavelength λ as compared to the water depth D, they are categorised as shallow water waves. The shape of the wave particle orbital depends on D where a perfectly sinusoidal wave has a circular wave particle (see Figure 2.2) with its horizontal displacement $\hat{\xi}$ and vertical displacement $\hat{\zeta}$ given as

$$\hat{\xi} = -\frac{H}{2}\left[\frac{\cosh k(z+D)}{\sinh kD}\right]\cos(kx - \omega t), \tag{2.47}$$

$$\hat{\zeta} = \frac{H}{2}\left[\frac{\sinh k(z+D)}{\sinh kD}\right]\sin(kx - \omega t). \tag{2.48}$$

The horizontal orbital speed v_x and vertical orbital speed v_z can be obtained by taking the derivatives of Equations (2.47) and (2.48) with respect to the time t, respectively.

$$(v_x)_{orbital} = -\frac{\pi H}{T}\left[\frac{\cosh k(z+D)}{\sinh kD}\right]\sin(kx - \omega t) \tag{2.49}$$

$$(v_z)_{orbital} = \frac{\pi H}{T}\left[\frac{\sinh k(z+D)}{\sinh kD}\right]\cos(kx - \omega t) \tag{2.50}$$

Similarly, the acceleration along the x- and y-axes can also be obtained by taking the derivatives of Equations (2.49) and (2.50) with respect to time t, respectively.

The wave will maintain a perfect sinusoidal shape as long as it is in the deepwater zone, where the seabed has no effect on the wave particle. Under such conditions, the term $kD \gg 1$, so $\tanh kD \approx 1.0$. Similarly, in the shallow water condition, $kD \ll 1.0$ and $\tanh kD \approx kD$ or $\sinh kD \approx kD$. The wave frequency, velocity of propagation, orbital velocities, pressure and energy terms under deep water, intermediate water and shallow water conditions are summarised in Table 2.3.

2.2.3.4 Solutions to the Laplace Equation

There are several methods to solve the Laplace equation, and one well-known method is the method of separable equations. We shall demonstrate the method of separable equations by considering the $x-z$ plane, i.e, the velocity potential is $\Phi_{2D}(x,z,t)$. The solution for the counterpart in the y-axis can be derived easily once the function in the x-axis is solved.

The method of separable equations assumes that the velocity potential $\Phi_{2D}(x,z,t)$ is a product of four different functions; each depends only on the independent variable, i.e., x, z or t. Here, the Φ_{2D} is written as

$$\Phi_{2D}(x,z,t) = X(x)\cdot Z(z)\cdot T(t), \tag{2.51}$$

where $X(x)$, $Z(z)$, and $T(t)$ are functions that depend only on $x, z,$ and t, respectively. As the wave is propagating, the function $T(t)$ can be easily obtained as

$$T(t) = \sin \omega t. \tag{2.52}$$

2.2.3.5 Solution to $Z(z)$

The function $Z(z)$ in Equation (2.51) can be solved from the boundary condition at the seabed (Equation (2.24b)) by assuming a general solution with $Z(z) = \cosh k(z+D)$; i.e.,

$$\Phi_{2D}(x,z,t) = X(x)\cdot B \cosh k(z+D)\cdot \sin \omega t. \tag{2.53}$$

The dynamic free surface boundary condition (Equation (2.8)) is then used to solve for the unknown B by substituting Equations (2.27) and (2.53) into Equation (2.8) at the free surface $z=0$, where $P_a = 0$, to get

$$B = -\frac{gA}{\omega}\cdot\frac{1}{\cosh kD}. \tag{2.54}$$

Therefore, the incident velocity potential (2.53) is now

$$\Phi_{2D}(x,z,t) = -X(x)\cdot\frac{gA}{\omega}\cdot\frac{\cosh k(z+D)}{\cosh kD}\cdot\sin \omega t. \tag{2.55}$$

By substituting Equation (2.55) into a two-dimensional Laplace equation, we obtain the Helmholtz equation:

TABLE 2.3

Summary of Progressive Waves Terms Under Various Water Depth

	Deepwater $D/\lambda > 0.5$	Intermediate Water $0.05 < D/\lambda < 0.5$	Shallow Water $D/\lambda > 0.05$
Wave Elevation		$\hat{\eta} = \frac{H}{2\cos(kx - \omega t)}$	
Velocity of Propagation	$C_0 = \frac{gT}{2\pi} = \sqrt{\frac{g\lambda}{2\pi}}$	$C = \frac{gT}{2\pi}\tanh kD, \sqrt{\frac{g\lambda}{2\pi}\tanh kD}$	$C = \sqrt{gD}$
Wavelength	$\lambda_0 = \frac{gT^2}{2\pi}$	$\lambda = CT = \frac{gT^2}{2\pi}\tanh kD$	
Horizontal Orbital Displacement	$\hat{\xi} = -\frac{H}{2}\left[\frac{\cosh k(z+D)}{\sinh kD}\right]\cos(kx - \omega t)$		
Vertical Orbital Displacement	$\hat{\zeta} = \frac{H}{2}\left[\frac{\sinh k(z+D)}{\sinh kD}\right]\sin(kx - \omega t)$		
Horizontal Orbital Speed	$v_x = -\frac{\pi H}{T}\left[\frac{\cosh k(z+D)}{kD}\right]\sin(kx - \omega t)$	$v_x = -\frac{\pi H}{T}\left[\frac{\cosh k(z+D)}{\sinh kD}\right]\sin(kx - \omega t)$	
Vertical Orbital Speed	$v_z = \frac{\pi H}{T}\left[\frac{\sinh k(z+h)}{kD}\right]\cos(kx - \omega t)$	$v_z = \frac{\pi H}{T}\left[\frac{\sinh k(z+D)}{\sinh kD}\right]\cos(kx - \omega t)$	
Group Velocity	$C_{g0} \approx \frac{C}{2} \approx \frac{1}{2}\left(\frac{g}{k}\right)^{\frac{1}{2}}$	$C_g = \frac{C}{2}\left(1 + \frac{2kD}{\sinh 2kD}\right)$	$C_g \approx C \approx (gD)^{\frac{1}{2}}$
Energy Density	$E = \frac{1}{8}\rho g H^2$		
Wave Power	$P_{res0} = EC_{g0}$	$P_{res} = EC_g$	

$$\frac{\partial^2 \Phi_{2D}}{\partial x} + k^2 \Phi_{2D} = 0. \tag{2.56}$$

2.2.3.6 Helmholtz Equations

The Helmholtz equation (2.56) only involves modelling the plane waves in the x direction. Substitute the function $\Phi_{2D}(x,z,t)$ in the Helmholtz equation (2.56) to get

$$X''(x) - k^2 X(x) = 0. \tag{2.57}$$

Equation (2.57) is a homogeneous second-order linear ordinary differential equation which can be solved by first expressing it in a characteristic equation,

$$r^2 + k^2 = 0, \tag{2.58}$$

where r are the roots for the ordinary differential equation (ODE) given as

$$r_{1,2} = \pm \mathrm{i}k. \tag{2.59}$$

where $\mathrm{i} = \sqrt{(-1)}$ is the imaginary number the general solution for the second-order non-homogeneous partial differential equation (PDE) with two complex roots is given as

$$X(x) = B_1 \cos kx + B_2 \sin kx. \tag{2.60}$$

2.2.3.7 General Solution to Helmholtz Equation

With the angular and radial equation solved, the general solution to the Helmholtz equation (2.56) can be obtained by substituting Equations (2.60) into (2.55):

$$\Phi_{2D}(x,z,t) = \frac{gA}{\omega} \cdot \frac{\cosh k(z+D)}{\cosh kD} \cdot (\cos kx + \sin kx) \cdot \sin \omega t. \tag{2.61}$$

However, Equation (2.61) represents a static wave. For waves to be progressive, the Φ_{2D} has to be written as

$$\begin{aligned} \Phi_{2D}(x,z,t) &= \frac{gA}{\omega} \cdot \frac{\cosh k(z+D)}{\cosh kD} \cdot (\sin kx \sin \omega t - \cos kx \cos \omega t) \\ &= \frac{gA}{\omega} \cdot \frac{\cosh k(z+D)}{\cosh kD} \cdot \sin(kx - \omega t). \end{aligned} \tag{2.62}$$

For the three-dimensional propagation waves, the velocity potential can be obtained by adding the special function $Y(y)$ in the y direction, which has a similar form to $X(x)$; i.e.,

$$\Phi(x,y,z,t) = \frac{gA}{\omega} \cdot \frac{\cosh k(z+D)}{\cosh kD} \cdot \sin\left[k(x+y) - \omega t\right]. \tag{2.63}$$

The dispersion equation (2.28) can be obtained by substituting Equations (2.8) and (2.27) into the first-order kinematic boundary condition (simplified from Equation (2.9)):

$$v_z = \frac{\partial \Phi}{\partial z} = -\frac{\partial \hat{\eta}}{\partial t}. \tag{2.64}$$

By substituting Equations (2.27) and (2.62) into Equation (2.64), we obtain

$$\frac{H}{2} \cdot \frac{gk}{\omega} \cdot \frac{\sinh k(z+D)}{\cosh kD} \cdot \sin(kx - \omega t) = \frac{H}{2} \cdot \omega \cdot \sin(kx - \omega t), \tag{2.65}$$

where $H = 2A$.

Upon simplification and evaluation of z at the free surface, i.e., $z = 0$, the wave dispersion (2.28) can be derived as

$$\frac{gk}{\omega} \cdot \tanh kD \cdot \sin(kx - \omega t) = \omega \cdot \cos(kx - \omega t). \tag{2.66}$$

2.2.4 Stokes Wave Theory

The velocity potential for the Stokes waves can be obtained directly from the combined free surface boundary condition (Equation (2.25e)), which also depends on the velocity potential for the SAWT (Equation (2.63)). Substituting Equation (2.63) into (2.25e), we have

$$\frac{\partial^2 \Phi^{(2)}}{\partial t^2} + g\frac{\partial \Phi^{(2)}}{\partial z} = \omega g k A^2 D \cdot \sin 2(kx - \omega t), \tag{2.67}$$

where

$$D = \frac{gk \cosh 2kD + 5gk - \omega^2 \sinh 2kD}{4\omega^2 \cosh^2 kD}. \tag{2.68}$$

D can be further expressed as

$$D = \frac{gk \cosh 2kD}{4\omega^2 \cosh^2 kD} + \frac{5gk}{4\omega^2 \cosh^2 kD} - \frac{\sinh 2kD}{4 \cosh^2 kD}. \tag{2.69}$$

We can use the wave dispersion relationship $\omega^2 = gk \tanh kD$ where the water depth D can be written as

$$D = \frac{1}{2 \tanh kD} - \frac{\tanh kD}{2} - \frac{1}{4 \sinh kD \cosh kD} + \frac{5}{4 \cosh kD \sinh kD}. \tag{2.70}$$

Upon further simplification, and by expressing D as PDE with $\sinh 2kD = 2\sinh kD \cosh kD$ as the common denominator, we have

$$D = \frac{1}{2}\left[\frac{6}{\sinh 2kD}\right] = \frac{3}{\sinh 2kD}. \tag{2.71}$$

Equation (2.67) is thus given as

$$\frac{\partial^2 \Phi^{(2)}}{\partial t^2} + g\frac{\partial \Phi^{(2)}}{\partial z} = \frac{3gkA^2\omega}{\sinh 2kD}\cdot \sin 2(kx - \omega t). \tag{2.72}$$

By assuming a trial function that satisfies the Laplace equation and boundary condition at the seabed, we have

$$\Phi^{(2)} = B \cdot \cosh 2k(z + D) \cdot \sin 2(kx - \omega t). \tag{2.73}$$

By substituting Equation (2.73) into Equation (2.72),

$$B = -\frac{3}{8}\frac{A^2\omega}{\sinh^4 kD}. \tag{2.74}$$

Substituting Equation (2.74) into Equation (2.73),

$$\Phi^{(2)} = -\frac{3A^2\omega}{8}\cdot\frac{\cosh 2k(z + D)}{\sinh^4 kD}\cdot \sin 2(kx - \omega t). \tag{2.75}$$

The second-order wave elevation can be derived by substituting Equations (2.62) and (2.75) into Equation (2.25d):

$$\hat{\eta}_2 = -\frac{A^2 k}{4}\cdot\frac{\cosh kD}{\sinh^3 kD}\cdot(2 + \cosh 2kD)\cdot \cosh 2(kz - \omega t) - \frac{A^2 k}{2\sinh 2kD}. \tag{2.76}$$

The first term in Equation (2.76) is the oscillatory component of the wave elevation, whereas the second term on the right-hand side is the mean value, which is time independent. The difference between SAWT and Stokes wave is shown in Figure 2.15, modelled using Equations (2.27) and (2.76), respectively.

2.3 Random Wave Model (Short-Term Wave Model)

The SAWT and second-order nonlinear Stoke waves were presented in the previous sections to describe the regular wave in an open sea. Often, in actual sea conditions, the sea is a complex network of waves composed of varying wave frequencies and wave directions. These random waves can be computed by taking the superposition of regular waves of different wave frequencies and phase angles, as modelled in Equation (2.27).

2.3.1 Mean and Standard Deviation

A typical random wave recorded from the sea is shown in Figure 2.18. The characteristic of the random waves can be obtained from the sea state, which involves two important parameters: the significant wave height H_s and the peak wave period T_p. The derivatives of the wave height and wave period will be presented in Sections 2.3.2 and 2.3.3, respectively. In order to obtain the wave sample, as shown in Figure 2.18, sensors can be used to record the wave elevation $\hat{\eta}$ in the actual sea. Post-processing, such as zeroing, has to be carried out before further analysis in order to eliminate any tidal variation. Figure 2.18 thus represents a wave record without tidal effect; i.e., the mean water level is zero.

In the analysis of a single random wave train in a limited duration, the statistical parameters can be used to describe the characteristics of the waves. This group of random wave trains is an independent system where it is assumed that the wave train bears the same characteristic as the average statistic properties of the entire sea, a process known as ergodic. Take for example the unidirectional random waves for a duration of 300 s, as shown in Figure 2.19. The waves can be described using the mean $\bar{\hat{\eta}}$ and standard deviation σ_{j} given as

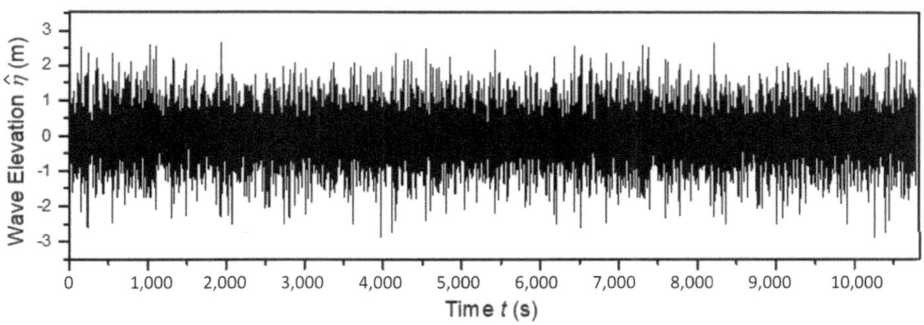

FIGURE 2.18
Typical random wave.

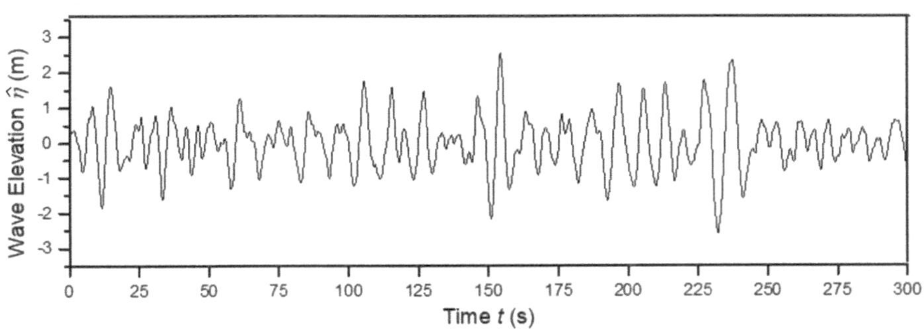

FIGURE 2.19
Unidirectional random waves.

$$\bar{\hat{\eta}} = \frac{1}{N}\sum_{j=1}^{N}\hat{\eta}_j, \tag{2.77}$$

and

$$\sigma = \sqrt{S} = \sqrt{\frac{1}{N}\sum_{j=1}^{N}\left(\hat{\eta}_j - \bar{\hat{\eta}}\right)^2}, \tag{2.78}$$

where S is the variance of $\hat{\eta}$.

2.3.2 Wave Height Distribution

2.3.2.1 Probability Density Function, Gaussian and Rayleigh Distributions

When a continuous random variable such as wave elevation $\hat{\eta}$ is recorded, the probability of occurrence for the variable can be described using the probability distribution function (PDF). According to the central limit theorem, the probability of occurrence for a large random sample will follow a special type of PDF with a bell curve, known as the normal distribution or Gaussian, where the highest probability occurs at its mean value, as shown in Figure 2.20.

The Gaussian function can be expressed in the base form

$$\mathscr{g}(x) = a \exp\left[-\frac{(x-\mathscr{b})^2}{2\sigma^2}\right], \tag{2.79}$$

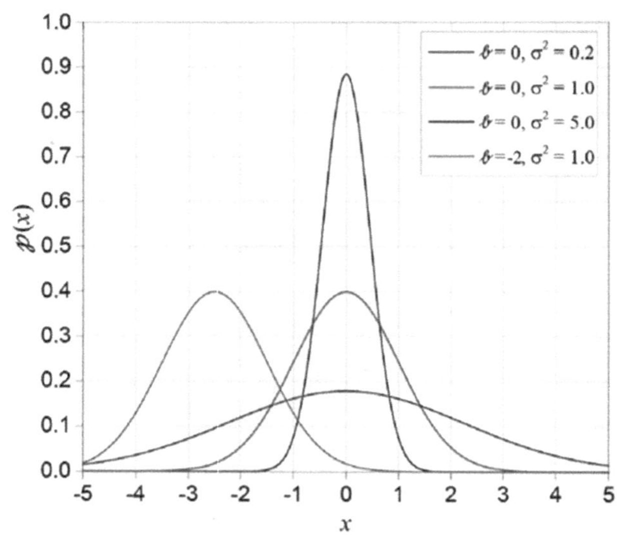

FIGURE 2.20
Gaussian distribution.

where a is the height of the curve's peak given as $a = 1/(\sigma\sqrt{2\pi})$, b the mean value, and σ the standard deviation. By further taking $\hat{\eta}' = \hat{\eta} - \bar{\hat{\eta}}$, i.e., the difference between the wave elevation $\hat{\eta}$ with the mean water level $\bar{\hat{\eta}}$, we have the Gaussian distribution for the wave given as

$$p(\hat{\eta}') = \frac{1}{\sigma\sqrt{2\pi}} \exp\left[-\frac{\hat{\eta}'^2}{2\sigma^2}\right]. \tag{2.80}$$

The standard deviation σ, also known as the shape parameter, plays a role in controlling the shape of the bell curve; i.e., a large σ produces a wide band and flat bell curve, whereas a smaller σ results in a narrow band and steep bell curve.

The exponential terms in Equation (2.80) gives the probability of occurrence for any individual wave $\hat{\eta}'$ that is greater than a specific wave elevation $\hat{\eta}_q$ defined in the Gaussian distribution; i.e., the probability of occurrence of any individual wave elevation $\hat{\eta}'$ greater than a specific wave elevation $\hat{\eta}_q$ defined in the Gaussian distribution is known as the probability of exceedance $q(\hat{\eta}' > \hat{\eta})$:

$$q(\hat{\eta}' > \hat{\eta}_q) = \exp\left[-\frac{\hat{\eta}_q^2}{2\sigma^2}\right]. \tag{2.81}$$

The probability of the occurrence of $\hat{\eta}'$ not greater than $\hat{\eta}_q$ is thus given as

$$p(\hat{\eta}' < \hat{\eta}_q) = 1 - q(\hat{\eta}' > \hat{\eta}_q) = 1 - \exp\left[-\frac{\hat{\eta}_q^2}{2\sigma^2}\right]. \tag{2.82}$$

2.3.2.2 Rayleigh Distribution

In most situations, the wave period of localised wind wave in deep water does not vary significantly. In such cases, the waves are of a narrow frequency band; i.e., energy is confined to a relatively narrow band frequency. Instead of following a normally distributed curve, the distribution of the wave elevation is skewed and can be described by a Rayleigh distribution. The shape of a Rayleigh distribution is given in Figure 2.21.

The Rayleigh distribution can be expressed in the base form

$$p(x) = \frac{x}{\sigma^2} \exp\left[-\frac{1}{2}\frac{x^2}{\sigma^2}\right]. \tag{2.83}$$

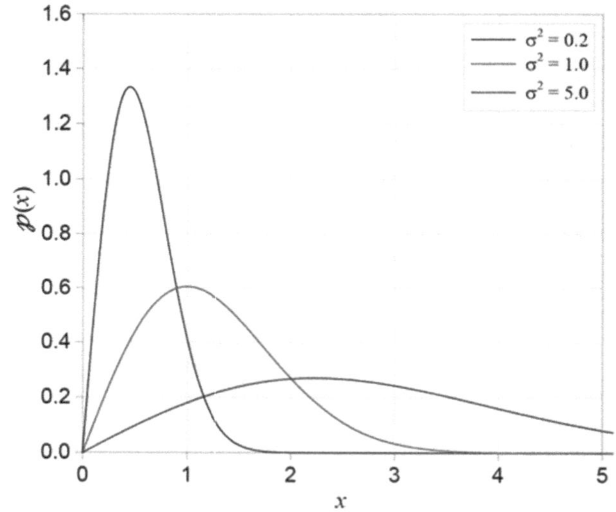

FIGURE 2.21
Rayleigh distribution.

Similar to Equation (2.80), the standard deviation/shape parameter σ controls the shape of the skewed curve where the distribution gets narrower and steeper with a smaller σ.

In an ocean wave, it is common to express the wave elevation in terms of wave height H, i.e., $x = H$ in Equation (2.83). The Rayleigh distribution of the wave height is thus given as

$$p(H) = \frac{H}{4\sigma^2} \exp\left[-\frac{H^2}{8\sigma^2}\right]. \tag{2.84}$$

The probability of exceedance for individual wave height H' that is greater than a specific wave height H_q, for the Rayleigh distribution can be written as

$$q(H' > H_q) = \exp\left[-\frac{H_q^2}{8\sigma^2}\right], \tag{2.85}$$

and the probability of occurrence for the Rayleigh distribution can be written as

$$p(H' < H_q) = 1 - exp\left[-\frac{H_q^2}{8\sigma^2}\right], \tag{2.86}$$

where $P(H' < H_q)$ is also known as the cumulative distribution function (CDF).

2.3.2.3 Significant Wave Height

Here, we introduce the significant wave height H_q, which is the mean wave height of the highest q th of the waves. Traditionally, the H_q can be obtained from first sorting the wave height record in descending order and then taking the average of the highest q th of the wave record. Mathematically, it can be obtained by

$$H_q = \frac{\int_{H_Q}^{\infty} H\, p(H' > H_q) \cdot dH}{\int_{H_Q}^{\infty} p(H' > H_q) \cdot dH} = \frac{1}{4\sigma^2} \frac{\int_{H_Q}^{\infty} H^2 \exp\left(-\frac{H^2}{8\sigma^2}\right) \cdot dH}{Q(H' > H_q)}, \tag{2.87}$$

where the H_q, the value that any individual wave height H' that would exceed in the probability of exceedance q is obtained from Equation (2.86) as

$$H_q = 2\sqrt{2m_0 \ln\left(\frac{1}{q}\right)} = 2\sigma\sqrt{2\ln\left(\frac{1}{q}\right)}. \tag{2.88}$$

The most common q in (2.87) is $q = \frac{1}{3}$, which, when substituted into the equation, is given as $H_{\frac{1}{3}}$, also commonly referred to as the significant wave height H_s. Numerical evaluation of Equation (2.87) gives us

$$H_s = H_{\frac{1}{3}} \approx 4\sigma \approx 4\sqrt{m_0}. \tag{2.89}$$

Mathematically, this shows that H_s is about *four times the standard deviation* σ *of the surface elevation* – or equivalently *four times the square root of the zeroth-order moment* m_0 *of the wave spectrum.* The zeroth-order moment m_0 is obtained by taking the area under the curve of a wave spectrum. Other significant H_q are

$$\begin{array}{ll} H_{\frac{1}{10}} \cong 5.09\sigma \cong 1.27H_s & H_{\frac{1}{100}} \cong 6.67\sigma \cong 1.67H_s \\ H_{\frac{1}{2}} \cong 3.55\sigma \cong 0.63H_s & H_{rms} \cong 2\sqrt{2}\sigma \cong \frac{2}{\sqrt{2}H_s} \end{array}. \tag{2.90}$$

Using the relationship of $\sigma = \sqrt{m_0}$ in Equation (2.86) and $H_{rms} = 2\sqrt{2}\sigma$ in Equation (2.90), the Rayleigh distribution in (2.84) can also be written as

$$p(H) = \frac{H}{4m_0} \exp\left[-\frac{H^2}{8m_0}\right] = \frac{2H}{H_{rms}^2} \exp\left[-\left(\frac{H}{H_{rms}}\right)^2\right], \tag{2.91}$$

where H_{rms} is the root mean square wave height, which is related to the average wave energy $\overline{E_0}$:

$$H_{rms}^2 = \int_0^{\infty} p(H) \cdot H^2 \cdot dH = \frac{8}{\rho g} \overline{E_0}, \tag{2.92}$$

Returning to Equation (2.91), the shape of the Rayleigh distribution depends on m_0, i.e., the area under the curve of a wave spectrum or the H_{rms}. A narrow-banded wave spectrum will have a smaller m_0; therefore, it produces a smaller H_{rms}, resulting in a narrower and steeper wave distribution.

2.3.3 Wave Period (Zero-Crossing Analysis)

Besides the wave height H_s presented in Section 2.3.2, the wave period is also essential in order to define a sea state. From a typical wave record with zero mean water level shown in Figure 2.19, it is possible to group the waves in accordance with their wave height and wave period. The wave period T is defined as the duration required for one wave to make a full cycle. As shown in Figure 2.22, five different waves are recorded, where each wave can be described by the H and T. However, sometimes small waves such as ripples which are insignificant to the response of floating bodies are also recorded (see Figure 2.22a). In order to record only the significant waves, the zero-crossing method must be applied.

The zero-crossing method only records a wave that has crossed the mean water level, which is zero in the wave sample considered in Figure 2.19. There are two ways the zero-crossing analysis can be performed, i.e., the zero down-crossing (Figure 2.22b) and the zero up-crossing (Figure 2.22c) wave height H. The former takes the wave height as the vertical distance between the maximum and minimum water levels that lie between two subsequent zero down-crossings, whereas the latter is for two subsequent zero up-crossings.

2.3.4 Spectral Analysis

As mentioned in the beginning of Section 2.3, the periodic random waves, assumed to be stationary and ergodic, can be expressed mathematically as the superposing of regular waves, where each wave has its own amplitude and wave frequency, as shown in Figure 2.23.

Therefore, it is possible to use the Discrete Fourier Transform (DFT) to break the finite random waveform into an alternate representation characterised by the sum of sine and cosine functions. The DFT allows the transformation of the waves in the time domain to the frequency domain. This reveals some of the important characteristics of the wave through frequency components and can be used for filtering out unwanted noise in the wave.

If there are N number of samples in the finite random wave, then the j th individual wave component in frequency domain $\hat{\eta}_j(\omega)$ that makes up the time-series random waves can be expressed as

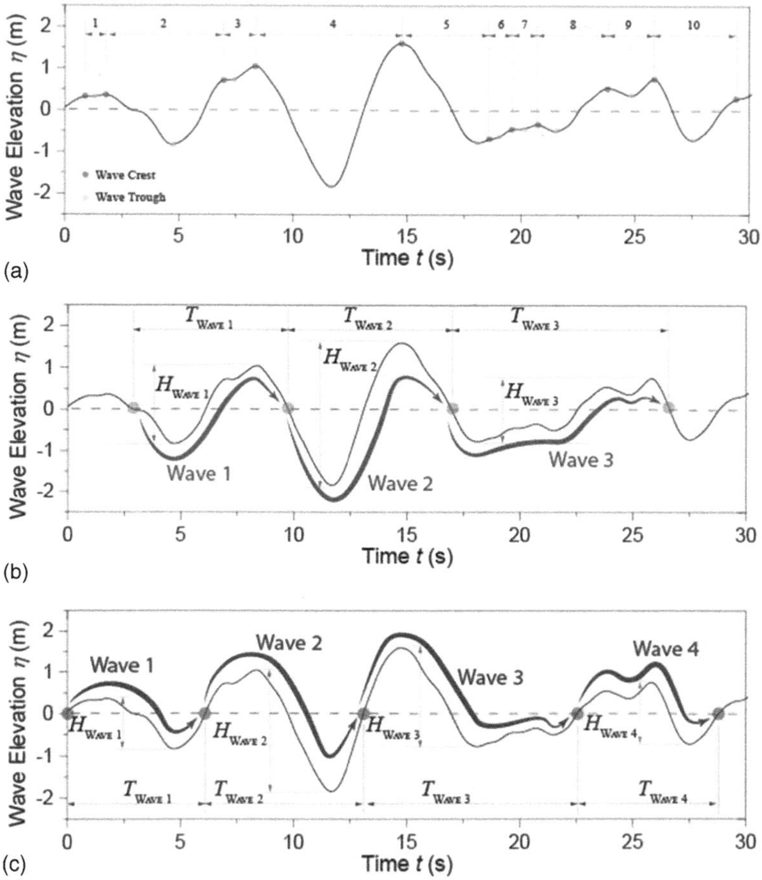

FIGURE 2.22
Various methods to measure wave height and wave period.

$$\hat{\eta}_j(\omega) = \sum_{n=0}^{N-1} \hat{\eta}_n(t) \cdot \exp\left[-\mathbb{i}\left(\frac{2\pi j}{N}\right)n\right], \tag{2.93}$$

where $\hat{n}_n$ is the wave elevation recorded for each sample in the time domain.

Conversely, the time-series random wave can be obtained from $\hat{\eta}_j(\omega)$ by using the inverse Discrete Fourier Transform (iDFT).

$$\hat{\eta}_n(t) = \frac{1}{N} \sum_{n=0}^{N-1} \hat{\eta}_j(\omega) \cdot \exp\left[\mathbb{i}\left(\frac{2\pi}{N}\right)n\right]. \tag{2.94}$$

Equations (2.93) and (2.94) represent the DFT pair.

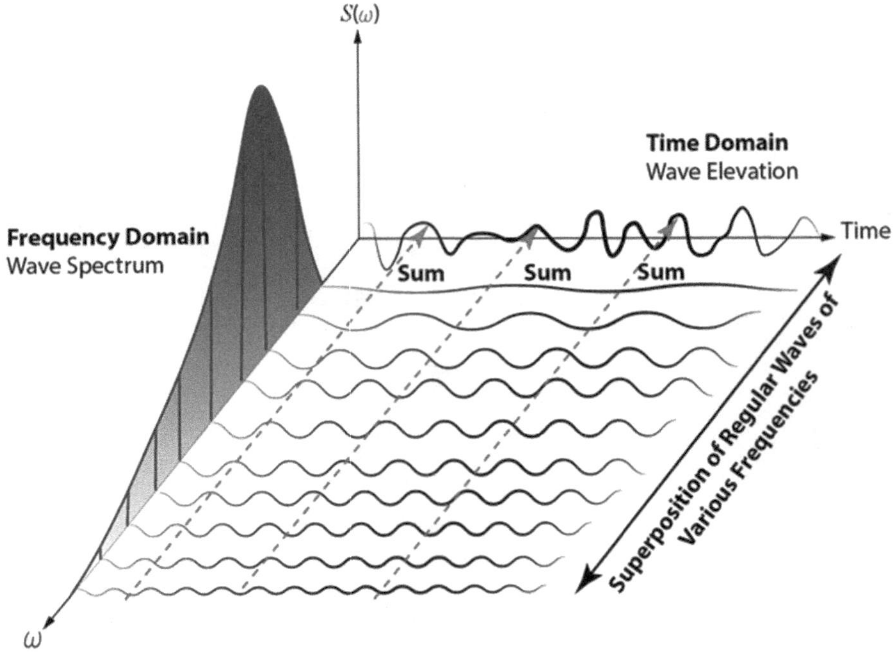

FIGURE 2.23
Superposition of waves (reproduced from Faltinsen[81]).

It should be noted here that $\eta_j(\omega)$ obtained from Equation (2.93) is a complex number due to the Euler term:

$$\exp\left[-\mathrm{i}\left(\frac{2\pi j}{N}\right)n\right] = \cos\left[-\mathrm{i}\left(\frac{2\pi j}{N}\right)n\right] + \mathrm{i}\sin\left[-\mathrm{i}\left(\frac{2\pi j}{N}\right)n\right]. \tag{2.95}$$

Therefore, $\hat{\eta}_j(\omega)$ can also be written as

$$\hat{\eta}_j(\omega) = \sum_{n=0}^{N-1}\left\{A_{j,n} + \mathrm{i}B_{j,n}\right\}, \tag{2.96}$$

where $A_{j,n} = \hat{\eta}_n \cdot \cos\left[-\mathrm{i}\left(\frac{2\pi j}{N}\right)n\right]$ and $B_{j,n} = \hat{\eta}_n \cdot \sin\left[-\mathrm{i}\left(\frac{2\pi j}{N}\right)n\right]$.

The magnitude of $\eta_j(\omega)$ for each n is thus

$$\left|\hat{\eta}_j(\omega)\right| = \sqrt{A_j^2 + B_j^2}, \tag{2.97}$$

and the phase angle is given as

$$\theta_j = \tan^{-1}\left(\frac{B_j}{A_j}\right). \tag{2.98}$$

The wave components obtained from Equation (2.93) can be plotted on the frequency spectrum, as shown in Figure 2.24. The dominant wave frequencies can be observed from the frequency spectrum, thereby allowing us to filter out frequencies which are insignificant to the design. Once the frequencies are filtered, they can be transformed back into the time-series waves by using iDFT.

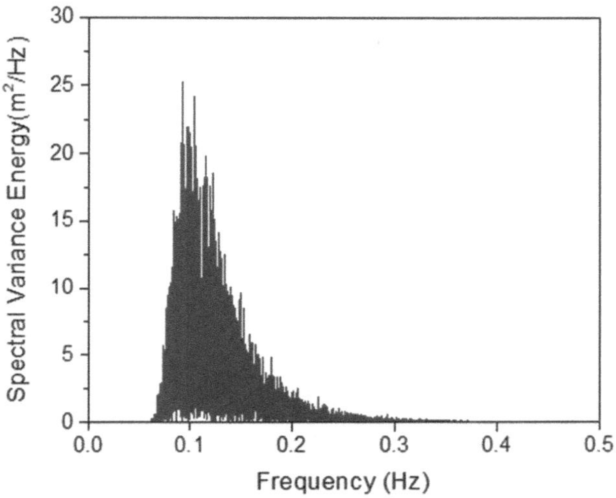

FIGURE 2.24
Frequency spectrum of waves.

The wave energy distribution can be calculated from

$$E(f) = \rho g \int_0^\infty S(f) \cdot df, \tag{2.99}$$

where f is the frequency of the wave in Hertz (Hz) and $S(f)$ is the wave spectrum to be described in the next section.

2.3.5 Wave Spectrum (Unidirectional Sea)

When time-series wave records are unavailable, it is useful to represent the sea state by using the wave spectrum. A comprehensive wave spectrum for the random sea is covered by Goda.[82] There are various types of wave spectra, depending on when the sea is a fully developed or developing sea.

2.3.5.1 Fully Developed Sea

One of the earliest wave spectra for the fully developed sea was proposed by Phillips (1958), where the spectral shape $S_P(\omega)$ is a function of ω^{-5} given as

$$S_P(\omega) = \frac{\alpha_P}{2\pi^4} g^2 \omega^{-5}, \tag{2.100}$$

where α_p is the Phillip constant given as $\alpha_p = 0.0074$.

Pierson and Moskowitz[83] extended the Phillips wave spectrum by adding a low-frequency filter over the complete frequency range:

$$S_{PM}(\omega) = S_p \cdot E_{PM}, \tag{2.101}$$

where

$$E_{PM} = \exp\left[-\frac{5}{4}\left(\frac{\omega}{\omega_p}\right)^{-4}\right]. \tag{2.102}$$

Therefore, (2.101) can be written as

$$S_{PM}(\omega) = \frac{\alpha_{PM}}{(2\pi)^4} \cdot \frac{g^2}{\omega^5} \exp\left[-\frac{5}{4}\left(\frac{\omega_p}{\omega}\right)^4\right], \tag{2.103}$$

where α_{PM} is commonly taken as $\alpha_{PM} = 0.0081$.

Another type of wave spectrum that can be used to describe tropical storm waves is the one-sided Bretschneider (BS) wave spectrum, also known as the two-parameter PM wave spectrum, given as

$$S_{BS}(\omega) = 5\pi^4 \frac{H_s^2}{T_p^4} \cdot \frac{1}{\omega^5} \cdot \exp\left[-\frac{20\pi^4}{T_p^4} \cdot \frac{1}{\omega^4}\right]. \tag{2.104}$$

2.3.5.2 Fetch-Limited Developing Sea

A team of researchers in the Joint North Sea Wave Observation Project (JONSWAP) realised that the wave spectrum is never fully developed.[84] The waves continue to develop through nonlinear wave-wave interactions, even for very long times and distances. In view of this, an extra artificial factor was added to the PM spectrum to improve the fit to their measurements. The JONSWAP spectrum is thus a PM spectrum multiplied by an extra peak enhancement factor given as

$$S_{JONSWAP}(\omega) = S_{PM}(\omega) \cdot \gamma_{PM}^{c}, \tag{2.105}$$

where the peak enhancement factor γ_{PM}^{c} ranging from 1 to 7 depending on weather conditions and ocean region. The typical γ_{PM}^{c} value is around 3.3 for most applications and c depends on the parameter σ given as[82]

$$c = \exp\left(-\frac{\omega - \omega_p}{2\sigma^2\omega_p^2}\right), \tag{2.106}$$

where,

$$\sigma = \begin{cases} 0.07 & \text{for}\, \omega < \omega_p \\ 0.09 & \text{for}\, \omega > \omega_p \end{cases}. \tag{2.107}$$

The JONSWAP wave spectrum can be adjusted to represent waves in West Africa and offshore Brazil. Compared to the BS spectrum shown in Figure 2.25, the latter has a greater frequency bandwidth to represent the characteristics of waves in the tropical sea.

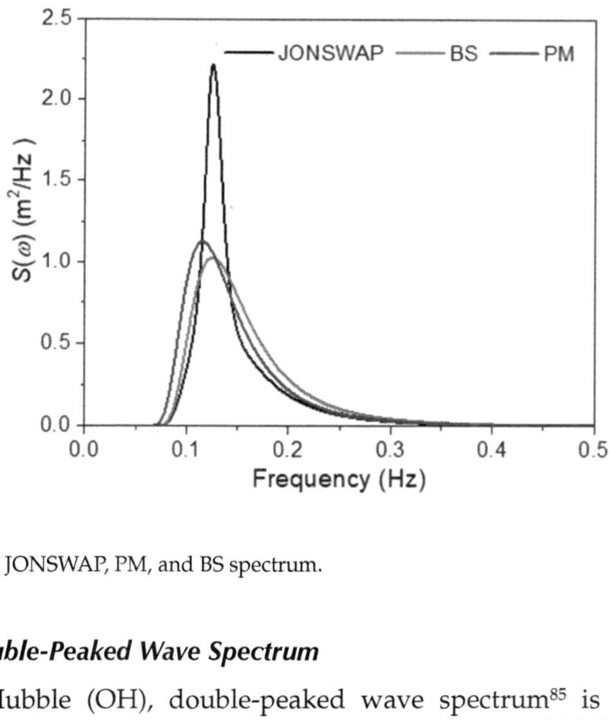

FIGURE 2.25
Comparison of JONSWAP, PM, and BS spectrum.

2.3.5.3 *Double-Peaked Wave Spectrum*

The Ochi-Hubble (OH), double-peaked wave spectrum[85] is given by a combination of two unidirectional wave spectra of low and high frequencies denoted, respectively, by subscript $j = 1$ and 2:

$$S_{OH}(\omega) = \sum_{j=1}^{2} \frac{\frac{1}{4}\left[\frac{4\lambda_j + 1}{4}\omega_{p,j}^4\right]^{\lambda_j}}{\Gamma(\lambda_j)} \frac{H_{s,j}^2}{\omega^{4\lambda_j+1}} \exp\left[-\left(\frac{4\lambda_j + 1}{4}\right)\left(\frac{\omega_{p,j}}{\omega}\right)^4\right], \tag{2.108}$$

where $\Gamma(\lambda_j)$ is the gamma function of the spectral shape parameter λ_j. The OH can be used to represent the sea state in the tropical climate, which comprises the swell and wind wave. Here, a smaller value of $\lambda_1 = 1$ corresponds to a wider spectrum (swell), whereas a larger value of $\lambda_2 = 2$ corresponds to a narrower spectrum (developing sea), as shown in Figure 2.26.

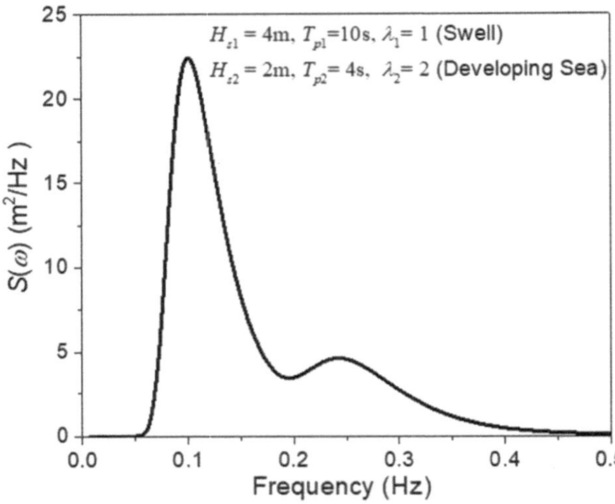

FIGURE 2.26
OH wave spectrum.

2.3.6 Multidirectional Sea

The aforementioned wave spectrum is unidirectional, assuming that waves approach from one direction. The actual sea is made of a combination of waves coming from varying directions. To model the sea state correctly, the multidirectional wave spectrum $S(\omega,\theta)$ can be derived by multiplying the wave spectrum $S(\omega)$ with a spreading function $d(\theta)$ as given in Equation (2.110):

$$S_I(\omega,\theta) = \sum_{j=1}^{N} S(\omega)\cdot d_j(\theta), \text{where } I = \text{PM, BS, JS or OH,} \tag{2.109}$$

where the spreading function $d(\theta)$ is given by Duarte et al.:[86]

$$d_j(\theta) = \frac{\sqrt{\pi}\,\Gamma(s+1)}{2\theta_{\max}\Gamma\left(s+\frac{1}{2}\right)}\cdot\left\{\cos\left[\frac{\pi\left(\theta-\overline{\theta}_j\right)}{2\theta_{\max}}\right]\right\}^{2s} \text{ for } -\theta_{\max} < \left(\theta-\overline{\theta}_j\right) < \theta_{\max}, \tag{2.110}$$

where $\bar{\theta}_j$ is the mean wave direction, s the wave spreading parameter, and $\theta_{\max} = \pi / 2$ (2.111). Note that the subscript j in (2.111) depends on the upper bound of the summation N in (2.110), where $N = 1$ for BS or JS and $N = 2$ for OH. The gamma function Γ ensures that

$$\int_{-\frac{\pi}{2}}^{\frac{\pi}{2}} d(\theta) \cdot d\theta = 1. \tag{2.111}$$

3

Principles of WEC Arrays

This chapter begins by exploring the fundamental concept of simple harmonic motion and its solution, establishing a foundation for understanding wave energy converter's (WEC) dynamics. It progresses to the motion of a WEC, presenting the optimal power take-off (PTO) damping and power capture under regular waves, as well as in unidirectional and multidirectional seas. The performance of a WEC deployed in an array configuration is analysed, with a focus on its interaction factor, known as the q-factor. The chapter delves into advanced methods for solving the velocity potential, utilising model expansion and the boundary element method (BEM). These solutions are critical for accurately predicting wave-structure interactions and optimising the design and performance of WEC systems. Through a detailed examination of these topics, the chapter provides a comprehensive understanding of WEC technology and its application in harnessing ocean energy.

3.1 Rigid Body Motion Versus Hydroelastic Response

The design of a floating structure requires a special branch of engineering known as naval architecture that traditionally involves the design, development, and construction of boats and ships to meet the requirements of prospective owners and operators. During the 20th century, naval architecture continued to evolve with the development of new technologies such as computer-aided design (CAD), computational fluid dynamics (CFD), and finite element analysis (FEA). These technologies enabled naval architects to design and construct ships and offshore structures with greater accuracy and efficiency than ever before.

In the design stage, it is always important to determine the motion of the floating body under the influence of environmental loadings such as wind, wave, and current, as mentioned in Chapter 2, in order to measure how well the structure responds under different sea states, where the analysis objective is to minimise the motion to a reasonable threshold under the various sea states as stipulated in the classification requirement. The solutions to the motion of floating bodies can be obtained analytically or numerically depending on the complexity of the structure's geometry. The BEM is by

 DOI: 10.1201/9781003387855-3

far the most commonly used numerical technique in assessing the motion of floating bodies due to its computational efficiency and ability to predict the motion with reasonable accuracy. The earliest solution from using the BEM was given by John,[87,88] in which he used the Green's function within a boundary integral formulation to solve for the wave scattering from floating bodies. A detail description of the linear wave theory was published by Wehausen and Laiton[89] in their remarkable review article "Surface Waves". This review article contains benchmark solutions for wave-structure interaction problems[90–95].

However, earlier works by these researchers only considered the floating structure as a rigid body. Over the centuries, ships have increased in size tremendously due to military and commercial needs. For example, the size of a container ship has grown steadily in recent years, with many vessels now capable of carrying more than 20,000 twenty-foot equivalent units (TEUs). The motion of ships with increasing length should no longer be considered as rigid because the structure deforms significantly under wave action. Such flexible motion of a floating body under hydrodynamic force is termed *hydroelasticity* and has a significant effect on the structural design. One of the earliest documentations on hydroelasticity was published by Bishop and Price[96] specifically for ship-like structure.

As mentioned in Chapter 1, the very large floating structure (VLFS) has gained tremendous interest in recent decades as one of the future solutions to creating land for land-scarce countries. This encouraged a new research area in hydroelasticity in the early 1990s. Among the pioneers working on the hydroelastic theory of VLFS are Ertekin and Riggs,[97] Suzuki et al.,[98,99] Yago and Endo,[100] Kashiwagi,[101,102] Utsunomiya et al.,[103] and Ohmatsu,[104,105] to name a few. The development of hydroelastic theory should also be attributed to Meylan and Squires[106] and Meylan[107,108] who studied ice-flow problems, but these problems were similar to VLFS problems.

The global energy demand has indirectly influenced the advancement of floating structures since the exploration of offshore oil and gas began in the late 19th century in the United States. As the exploration went deeper and further from the shore, floating oil and gas platforms were built to cater for the environmental changes, thereby seeing new types of floating structure such as semi-submersible, spar, mobile offshore drilling unit (MODU), tension leg platform (TLP), and drillship being built. As the world is venturing into the 21st century, the demand for energy continues to rise, and at the same time, a more sustainable energy has to be sought to phase out the dependency on fossil fuels. This causes the popularity of WECs such as the Pelamis and Oyster WECs. The design of WECs is however different from the conventional ship and offshore structures, as the WECs are usually designed for maximum motion under wave action in order to optimise the wave energy generation.

In this chapter, the mathematical formulation for the hydrodynamic responses of WECs and the hydroelastic response of VLFS will be presented. Section 3.2 presents the various motions of the floating body, focusing on generic harmonic motion; Section 3.3 covers the governing equation for solving the WEC motion, followed by Section 3.4 on the WECs in array; Section 3.5 presents the solution to the velocity potential, and Section 3.6 shows the BEM for solving the governing equation. Lastly, Section 3.7 shows some examples of WECs in regular and irregular waves in the tropical climate.

3.2 Harmonic Motion

3.2.1 Simple Oscillator

The motion of a floating body can be described by a simple harmonic motion oscillating at a frequency ω and can be modelled by a simple spring-mass-damper (SMD) system, as shown in Figure 3.1. The mass m of the SMD is supported by a spring system with a constant k and a damping c. We will be showing the derivative of the equation of motion of the simple oscillator using two methods, i.e., the direct equilibrium method and the principle of virtual work. The latter can be used to derive the governing equation of a complex structure, as will be shown later for the VLFS.

3.2.2 Direct Equilibrium Method (D'Alembert's Principle)

The equation of motion for the SMD system with one degree of freedom (dof) is given by

$$ma + cv + kx = F, \tag{3.1}$$

where F is the external force acting on the mass m, a the acceleration, v the velocity, and x the displacement of the mass from the equilibrium position, as shown in Figure 3.1.

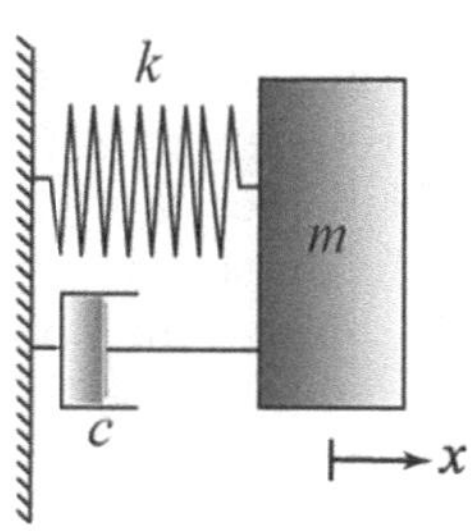

FIGURE 3.1
Simple SMD system.

By expressing the acceleration a and velocity v in terms of displacement x, Equation (3.1) reduces to a second-order linear univariate equation

$$m\frac{d^2x}{dt^2}+c\frac{dx}{dt}+kx=F. \tag{3.2}$$

3.2.3 Principle of Virtual Work

The direct equilibrium method is suitable for less complex structures such as the simple SMD system shown in Figure 3.1. For complex structures, the principle of virtual work (PVW) can be utilised to obtain the governing equation, as well as the boundary conditions associated with the system. The PVW as described in Appendix A is used to derive the equation of motion (Equation (3.2)) of the same SMD in Section 3.2.2. The action functional for the PVW according to Hamilton's principle is given as

$$\delta\Pi=\int_{t_i}^{t_f}\delta(T_E-U_E)dt+\int_{t_i}^{t_f}(\delta W_{nc})dt=0, \tag{3.3}$$

where T_E is the total kinetic energy of the system, U_E the potential energy (including strain energy and potential of conservative external force), W_{nc} the work done by nonconservative forces, and δ the variation taken during the indicated time interval. For the SMD given in Figure 3.1, the kinetic energy, potential energy, and work done are given by

$$T_E=\frac{1}{2}m\left(\frac{dx}{dt}\right)^2, \tag{3.4a}$$

$$U_E=\frac{1}{2}kx^2, \tag{3.4b}$$

$$\delta W_{nc}=F\delta x-c\frac{dx}{dt}\delta x. \tag{3.4c}$$

Substituting Equation (3.4a–c) into Equation (3.3), we obtain the action function

$$\delta\Pi=\int_{t_i}^{t_f}\delta\left(\frac{1}{2}m\left(\frac{dx}{dt}\right)^2-\frac{1}{2}kx^2\right)\cdot dt+\int_{t_i}^{t_f}\left(F\delta x-c\frac{dx}{dt}\delta x\right)\cdot dt=0. \tag{3.5}$$

Applying the variation to the velocity and displacement,

$$\delta\Pi = \int_{t_i}^{t_f} \left(m\frac{dx}{dt}\frac{d(\delta x)}{dt} - kx\delta x + F\delta x - c\frac{dx}{dt}\delta x \right) \cdot dt = 0. \tag{3.6}$$

Equation (3.6) can now be solved by using integration by parts:

$$\delta\Pi = \int_{t_i}^{t_f} \left[-m\frac{dx}{dt} - c\frac{dx}{dt} - kx + F \right] \cdot \delta x \cdot dt + \left[m\frac{d^2x}{dt^2} \right]_{t_i}^{t_f} = 0,$$

$$\delta\Pi = \int_{t_i}^{t_f} \left[-m\frac{dx}{dt} - c\frac{dx}{dt} - kx + F \right] \cdot \delta x \cdot dt + m\frac{d}{dt^2}\left(x_{tf} - x_{ti} \right) = 0. \tag{3.7}$$

According to Hamilton's principle, the variation δx in the first term vanishes at the limits of integration:

$$\int_{t_i}^{t_f} \left[-m\frac{dx}{dt} - c\frac{dx}{dt} - kx + F \right] \cdot \delta x \cdot dt = 0, \delta x \to 0. \tag{3.8}$$

Therefore, we arrive at the equation of motion of the SMD system:

$$-m\frac{dx}{dt} - c\frac{dx}{dt} - kx + F = 0, \tag{3.9}$$

which upon arranging, is the same as Equation (3.2).

3.2.4 Solutions to the Free Oscillation System

We first consider the free oscillation of the SMD system where the force F in the right-hand side of Equation (3.2) is disregarded. Therefore, we arrive with a homogeneous equation; i.e., $m\frac{d^2x}{dt^2} + c\frac{dx}{dt} + kx = 0$. Here, we introduce the critical damping parameter $c_c = 2\sqrt{km} = 2m\omega_N$, which is the damping value that results in the quickest possible return to equilibrium without oscillation after disturbance. Note that ω_N is the natural frequency of the system.

By taking the ratio of the actual damping c to the critical damping c_c, we have the damping ratio $\zeta_c = \frac{c}{c_c}$. The damping ratio ζ_c can be used to classify whether the system is underdamped, critically damped, or overdamped, as will be explained later.

The homogeneous Equation (3.9) can then be written as

$$\frac{d^2x}{dt^2} + 2\zeta_c \omega_N \frac{dx}{dt} + \omega_N^2 x = 0. \tag{3.10}$$

The homogeneous Equation (3.10) can be solved by first obtaining the roots $r_{1,2}$ of its characteristic equation (refer to the method of characteristic equation shown in solving the Helmholtz equation in Chapter 2). The roots for the characteristic equation are thus given by

$$r_{1,2} = -\omega_N \left(\zeta_c \pm \sqrt{\zeta_c^2 - 1} \right), \tag{3.11}$$

where the natural frequency ω_N of the floating body is given by

$$\omega_N = \frac{1}{2\pi} \sqrt{\frac{k}{m}}. \tag{3.12}$$

For homogeneous second-order ODE, the solution for r can take three different forms depending on the discriminant $(\zeta_c^2 - 1)$ or the square of the damping ratio ζ_c^2 value, as shown in Table 3.1. The plots for the oscillator under different ζ_c values are provided in Figure 3.2. The normalised amplitude is used in Figure 3.2 by dividing the amplitude x with the maximum of the oscillator so that the oscillation has a maximum amplitude of 1.0. As the damping ratio expresses the relationship between the actual damping c with the critical damping c_c, where $c = c_c$ (i.e., $\zeta_c^2 = 1$), it can be seen in Figure 3.2 that the displacement $x(t)$ returns to its equilibrium without any oscillation at the shortest time. For $c > c_c$

TABLE 3.1

Solution to Homogeneous Equation for SMD

	Roots Characteristics	Solution $x(t)$	System Classification
$\zeta_c^2 > 1$	Two real and distinct	$x(t) = B_1 e^{-\omega_N\left(\zeta_c - \sqrt{\zeta_c^2 - 1}\right)t} + B_2 e^{-\omega_N\left(\zeta_c + \sqrt{\zeta_c^2 - 1}\right)t}$	Overdamped system
$\zeta_c^2 > 1$	One real	$x(t) = (B_1 + B_2 t) e^{-\zeta_c \omega_N t}$	Critically damped system
$0 < \zeta_c^2 < 1$	Two complex and distinct	$x(t) = e^{-\omega_N \zeta_c t} \left\{ \begin{array}{l} B_1 \sin\left[\left(\sqrt{1 - \zeta_c^2}\right)\omega_N t\right] \\ + B_2 \cos\left[\left(\sqrt{1 - \zeta_c^2}\right)\omega_N t\right] \end{array} \right\}$	Underdamped system
Special case: $\zeta_c^2 = 0$	Two complex and distinct	$x(t) = B_1 \sin(\omega_N t) + B_2 \cos(\omega_N t)$	Undamped system

Note: B_1 and B_2 are the unknown coefficients to be solved from the boundary conditions.

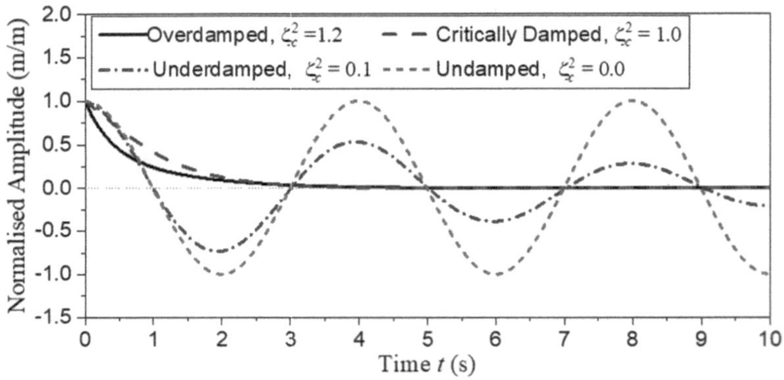

FIGURE 3.2
Oscillator with different damping system.

(i.e., $\zeta_c^2 > 1$), the system has a similar characteristic as the critically damped system but with its $x(t)$ approaching equilibrium over a longer period of time, whereas an underdamped system with $0 < c < c_c$ (i.e., $0 < \zeta_c^2 < 1$) will oscillate with a gradually decreasing amplitude until it eventually comes to a rest. For the purpose of practicality, ζ_c^2 or c_c is taken to be positive (i.e., greater than or equal to 0). In a special case where $\zeta_c^2 = 0$ or $c_c = 0$, we have the undamped system where the displacement will oscillate indefinitely.

3.2.5 Solutions to Forced Oscillation

A floating body on water under wave action can be assumed to be a forced oscillation system where the wave exciting force $F(x,t)$ is taken to be oscillating at the wave frequency ω; i.e.,

$$F(x,t) = f(x)\cos\omega t, \tag{3.13}$$

where $f(x)$ is the force in the spatial domain. Substituting Equation (3.13) into Equation (3.2) and expressing it in terms of ζ_c and ω_N, we have

$$\frac{d^2x}{dt^2} + 2\zeta_c\omega_N\frac{dx}{dt} + \omega_N^2 x = f\cos\omega t. \tag{3.14}$$

Equation (3.14) is a second-order non-homogeneous equation where the solution can be decomposed into the complimentary solution x_c and the particular solution x_p:

$$x = x_c + x_p \tag{3.15}$$

The complimentary solution x_c can be obtained by first solving the homogeneous solution for Equation (3.15), i.e., solving Equation (3.10), and will not be repeated here. The complimentary solution x_c given in Table 3.1 depends on the damping ratio ζ_c^2. The practical real engineering problem for floating bodies has an oscillating motion that decays with time; i.e., $\zeta_c^2 > 0$. Therefore, the complimentary solution x_c, as a function of time, is also known as the transient response.

The particular solution x_p, on the other hand, requires an assumption of a general particular solution given as

$$x_p = B_3 \cos\omega t + B_4 \sin\omega t, \tag{3.16}$$

where B_3 and B_4 are the unknown coefficients. By substituting Equation (3.16) into Equation (3.14), we thus arrive with

$$\begin{aligned} &\left[B_3\left(-\omega^2 + \omega_N\right) + 2B_4\omega\omega_N\zeta_c\right]\cos\omega t \\ &+\left[B_4\left(-\omega^2 + \omega_N\right) - 2B_3\omega\omega_N\zeta_c\right]\sin\omega t = f\cos\omega t. \end{aligned} \tag{3.17}$$

Comparing the left-hand side (LHS) and right-hand side (RHS) of Equation (3.17), two equations can be obtained; i.e.,

$$\begin{aligned} &B_3\left(-\omega^2 + \omega_N^2\right) + 2B_4\omega\omega_N\zeta_c = f, \\ &B_4\left(-\omega^2 + \omega_N^2\right) - 2B_3\omega\omega_N\zeta_c = 0. \end{aligned} \tag{3.18}$$

By solving Equation (3.18), simultaneously, B_3 and B_4 are given as

$$\begin{aligned} B_3 &= \frac{f\left(\omega_N^2 - \omega^2\right)}{\omega^4 + \omega_n^2 + 2\omega^2\omega_N\left(-1 + 2\omega_n\zeta_c^2\right)}, \\ B_4 &= \frac{2f\omega\omega_N\zeta_c}{\omega^4 + \omega_N^2 + 2\omega^2\omega_N\left(-1 + 2\omega_N\zeta_c^2\right)}. \end{aligned} \tag{3.19}$$

Upon substituting Equation (3.19) into Equation (3.16), the particular solution is given as

$$x_p = f\frac{\left(\omega_N^2 - \omega^2\right)}{\left(\omega^2 - \omega_N^2\right)^2 + \left(2\omega\omega_N\zeta_c\right)^2}\left[\cos\omega t + \frac{2\omega\omega_N\zeta_c}{\left(\omega_N - \omega^2\right)}\sin\omega t\right]. \tag{3.20}$$

Here, we note that the terms in the square brackets are a trigonometry equation in the form of

$$B_1\cos\omega t + B_2\sin\omega t = \mathbf{A}\cos\left(\omega t - \phi\right), \tag{3.21}$$

where the magnitude $\mathbf{A} = \sqrt{A_1^2 + A_2^2}$ and phase angle $\phi = \tan^{-1}\left(\frac{A_2}{A_1}\right)$ with A_1 and A_2 being constants. Using this identity, the square bracket terms in Figure (3.20) are thus

$$\left[\cos\omega t + \frac{2\omega\omega_N\zeta_c}{\left(-\omega^2 + \omega_N\right)}\sin\omega t\right] = \mathbf{A}\cdot\cos\left(\omega t - \phi\right), \tag{3.22}$$

where

$$\mathbf{A} = \left[\frac{\left(\omega_N^2 - \omega^2\right)^2 + \left(2\omega\omega_N\zeta_c\right)^2}{\left(-\omega^2 + \omega_N\right)^2}\right] \text{and } \phi = \tan^{-1}\left(\frac{2\omega\omega_N\zeta_c}{\omega_N - \omega^2}\right). \tag{3.23}$$

The particular solution (3.20) is thus given as

$$x_p = \frac{f\cos\left(\omega t - \phi\right)}{\sqrt{\left(\omega^2 - \omega_N^2\right)^2 + \left(2\omega\omega_N\zeta_c\right)^2}}. \tag{3.24}$$

By manipulating Equation (3.24), the particular solution can be written as

$$x_p = \Lambda\left(\omega_N\right)\cdot\frac{f}{\omega}\cos\left(\omega t - \phi\right), \tag{3.25}$$

where the magnitude $\Lambda\left(\omega_N\right)$, also known as the dynamic amplification factor, is given as

$$\Lambda\left(\omega_N\right) = \frac{1}{\sqrt{\left[1 - \left(\frac{\omega_N}{\omega}\right)^2\right]^2 + 4\left(\frac{\omega_N}{\omega}\right)^2\zeta_c^2}}. \tag{3.26}$$

The particular solution x_p has an oscillating motion that does not decay with time. Such motion is said to be in steady state, and its magnitude depends on the term $\Lambda\left(\omega_n\right)$. Plotting only the magnitude term $\Lambda\left(\omega_n\right)$ in Figure 3.3, the magnitude depends on the ratio $\left(\frac{\omega_N}{\omega}\right)$, where the amplification of the maximum $\Lambda\left(\omega_N\right)$ occurs when $(\omega_N / \omega) = 1$. This is when the wave frequency ω coincides with the natural frequency ω_N of the system, a phenomenon

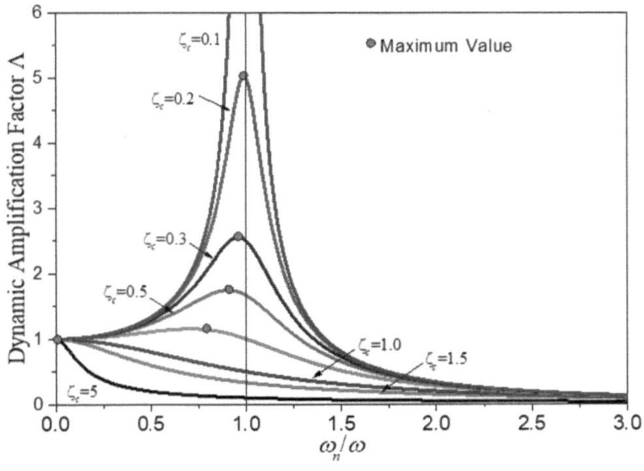

FIGURE 3.3
Dynamic amplification factor Λ.

known as resonance. For marine and offshore structures, the main objective is to design the structure such that the ω_N is away from the wave frequency of the sea state to avoid amplification of the magnitude. However, for some WECs, the amplification of the magnitude of the displacement is required in order to maximise the wave energy generation. This also poses some structural issues for WECs that are designed to generate energy via resonance effect. Figure 3.3 shows that inclusion of damping ζ_c in the model results in a change in the natural frequency ω_N of a floating body where ω_N reduces with the increase of ζ_c. Therefore, it is important to include the damping in assessing the performance of the floating structure. The damping c can be obtained from Equation (3.27) via free decay motion in an experimental test or by using the numerical method (e.g., using CFD for obtaining the viscous damping for floating body).

$$c = \frac{2\kappa k}{\omega_n}, \tag{3.27}$$

where the non-dimensional damping, known as the logarithmic decrement κ, is given by the natural logarithm of the ratio of the amplitudes of any two successive peaks, given as[109]

$$\kappa = \frac{1}{2\pi} \ln \frac{z_k - z_{k+1}}{z_{k+2} + z_{k+3}}. \tag{3.28}$$

Note that $\zeta_c = \frac{c}{c_c}$ where $c_c = 2m\omega_N$.

3.3 Motion of WEC

In this chapter, the governing equation for a generic WEC subjected to regular and irregular waves is derived. The WEC has a relatively simple geometry where its equation of motion can be obtained by using the direct equilibrium method described in Section 3.2.2, whereas the water is modelled as an ideal fluid by assuming irrotational motion.

Before delving into the governing equation, it is necessary to define the degrees of freedom (dofs) of a freely floating body. For a freely floating body in a three-dimensional space defined by the cartesian coordinate, as shown in Figure 3.4, there exists six dofs, i.e., surge, sway, heave, roll, pitch, and yaw. The surge, sway and heave are the translation motions along the x-, y-, and z-axes, respectively, whereas the roll, pitch and yaw are the rotational motions along the x-, y-, and z-axes, respectively. However, for seakeeping, the significant dofs only involve the heave $\mathbf{w}$, roll $\boldsymbol{\theta}_x$, and pitch $\boldsymbol{\theta}_z$.

3.3.1 Water Equation

Figure 3.5 shows a floating body with the origin of the global cartesian coordinate system (x-y-z axes) located at the free surface $z = 0$. The floating body is surrounded by fluid domain Ω. The coordinate system follows the right-hand rule with the z-axis pointing upwards. The surface boundary at the seabed, free surface, wetted surface of floating body, and infinity is represented by S_B, S_F, S_W, and $S_{\pm\infty}$, respectively.

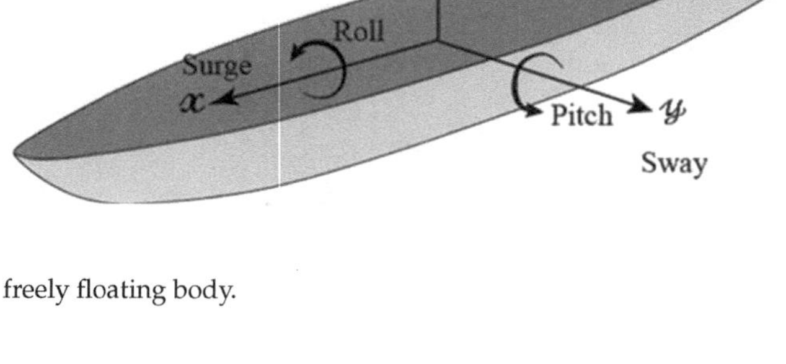

FIGURE 3.4
Six dofs for a freely floating body.

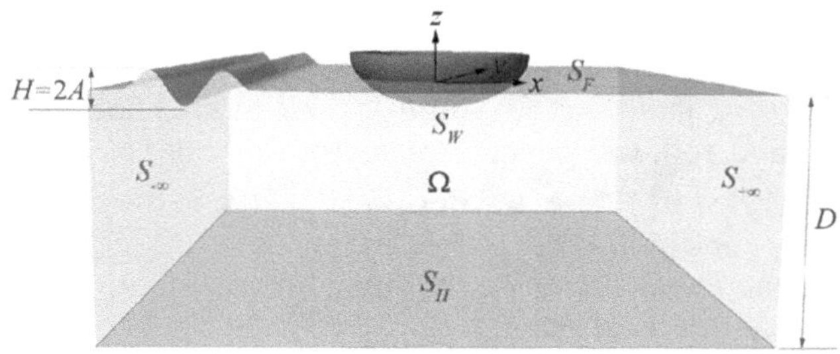

FIGURE 3.5
Mathematical domain.

The SAWT is assumed. As presented in Section 2.2, the fluid motion can be represented by a velocity potential $\Phi(x,y,z,t)$ such that $\Phi(x,y,z,t)$ satisfies the Laplace Equation (2.25a), together with the boundary conditions on the seabed Equation (2.25b) and on the free surface (2.25e). For simplicity, the velocity potential for the SAWT in this chapter will simply be denoted as Φ instead of $\Phi^{(1)}$.

In addition to these boundary conditions, the velocities of the fluid and floating body have to satisfy the following boundary conditions:

$$\frac{\partial \Phi}{\partial n} = \frac{\partial \mathbf{W}}{\partial t}, \quad \text{on} \quad S_W, \tag{3.29}$$

where $\mathbf{W}(x,y,z,t) = (\mathbf{w} \ \theta_x \ \theta_y)$. It is to note that Equation (3.29) implies that the floating body is always in contact with the water such that it obeys the non-wave breaking condition.

3.3.2 Complex Plane

As summarised in Table 3.1, there are situations where the roots to the solution of the second-order ODE that governs the motion of a harmonic oscillator are complex numbers. In fact, most of the real oscillation systems in real life application, such as WEC or VLFS, are underdamped systems. Therefore, it would be easier to express the motion of a WEC in complex plane. With this, it is thus assumed that the velocity potential $\Phi(x,y,z,t)$, floating body motion $\mathbf{W}(x,y,z,t)$ and external force $\mathbf{F}(x,y,z,t)$ can be represented in the complex domain as

$$\Phi(x,y,z,t) = \text{Re}\{\varphi(x,y,z)e^{\mathrm{i}\omega t}\}, \tag{3.30}$$

$$\mathbf{W}(x,y,z,t) = \text{Re}\{W(x,y,z)e^{\mathrm{i}\omega t}\}, \tag{3.31}$$

$$\mathbf{F}(x,y,z,t) = \text{Re}\{F(x,y,z)e^{\mathrm{i}\omega t}\}, \tag{3.32}$$

where φ and W are the spatial velocity potential and floating body response, respectively, and $e^{i\omega t} = (\cos\omega t + i\sin\omega t)$ is the Euler formula. $W(x,y,z) = (w,\theta_x,\theta_y)$ with w, θ_x, and θ_y represent the spatial motion corresponding to heave, roll, and pitch, respectively. It should be noted that the Φ, $\mathbf{W}$, and $\mathbf{F}$ follow the same time function, i.e., $e^{i\omega t}$ due to the assumption made in the boundary condition Equation (3.29).

By using Equations (3.30) and (3.31), the Laplace equation and boundary conditions are summarised as follows:

$$\text{Laplace Equation}: \nabla^2\varphi = 0, \text{ in fluid domain } \Omega, \tag{3.33a}$$

$$\text{BC on Seabed}: \frac{\partial\varphi}{\partial z} = 0, \quad \text{at} \quad z = -D, \tag{3.33b}$$

$$\text{Free Surface BC: } \omega^2\varphi + g\frac{\partial\varphi}{\partial z} = 0, \quad \text{at} \quad z = 0, \tag{3.33c}$$

$$\text{BC on Wetted Surface}: \frac{\partial\varphi}{\partial \mathbf{n}} = i\omega W, \quad \text{on} \quad S_w, \tag{3.33d}$$

where $\mathbf{n} = (n_x, n_y, n_z, n_{x'}, n_{y'}, n_{z'})$ is the unit vector normal to the body boundary for the six dofs. $(n_x, n_y, n_z) = (\hat{i}, \hat{j}, \hat{k}) = \mathbf{n}_t$ and $(n_{x'}, n_{y'}, n_{z'}) = \mathbf{x} \times \mathbf{n}_t = \mathbf{n}_r$, where $\mathbf{x} = (x,y,z)$; $\hat{i}$, $\hat{j}$, and $\hat{k}$ are the unit vectors in the x-, y-, and z-axes, respectively. Thus, for seakeeping motion with heave, roll, and pitch,

$$\boldsymbol{n}_r = \mathbf{x} \times \boldsymbol{n}_t = \begin{pmatrix} n_z & n_{x'} & n_{y'} \end{pmatrix} = \left(y\cdot\hat{k} - z\cdot\hat{j}, \quad z\cdot\hat{i} - x\cdot\hat{k}, \quad x\cdot\hat{j} + y\cdot\hat{i} \right). \tag{3.34}$$

The scattered and radiated potential must also satisfy the radiation condition,

$$\frac{\partial\varphi_k}{\partial r} = i\,k\varphi_k, \; kr \to \infty \text{ for } k = D \text{ or } S, \tag{3.35}$$

where D is the diffracted waves and S the scattered waves introduced later in Section 3.3.4. The radiation condition requires that as the φ_S and φ_R propagate away from the source, its value will approach zero as $r \to \pm\infty$. Therefore, $\varphi \to \varphi_I$ as $r \to \pm\infty$.

3.3.3 Hydrostatic Coefficients

The equation of motion of the floating body can be derived by using the direct equilibrium method described in Section 3.2.2. By using Equations (3.30)–(3.32), the equation of motion (3.2) of the WEC with mass m and mass moment of inertia about x-axis I_x and mass moment of inertia about y-axis I_y for the heave, roll, and pitch can be written as below.

$$\left(-\omega^2 m + \mathrm{i}\omega c_z + k_z\right) \cdot w = F_z, \tag{3.36a}$$

$$\left(-\omega^2 I_x + \mathrm{i}\omega c_{x'} + k_{x'}\right) \cdot \theta_y = M_x, \tag{3.36b}$$

$$\left(-\omega^2 I_y + \mathrm{i}\omega c_{y'} + k_{y'}\right) \cdot \theta_y = M_y, \tag{3.36c}$$

where c and k are the viscous damping and hydrostatic stiffness, respectively. Here, the subscripts x, y, and z associated with c and k are used to denote values corresponding to their translational motions, whereas subscripts x', y', and z' are used to denote values related to their corresponding rotational motions. The vertical external force, external moment about the x-axis, and external moment about the y-axis acting on the floating body are given by F_z, M_x, and M_y; i.e., $F = \left(F_z, M_x, M_y\right)$.

When a body with mass m is floating on a water surface, its weight is supported by the buoyancy force, which is analogous to a spring system with linear constant k_z given as

$$k_z = \rho g A_{wp}, \tag{3.37}$$

where A_{wp} is the free surface water plane area of the floating body. For example, a vertically floating cylindrical body with radius r_a and length L_a has $A_{wp} = \pi r_a^2$, whereas the same cylinder placed horizontally with draft d_a has $A_{wp} = 2L_a r_a$. It should be noted that Equation (3.37) has the same unit of a spring constant, i.e., $\frac{N}{m}$. This buoyancy force acts as a linear restoring force for the floating body to return it to its equilibrium position when any linear external forces are removed.

Similarly, when a rotating floating body is subjected to a moment regarding the x-axis M_x or y-axis M_y, the rotational restoring force due to the rotational spring stiffness about the x-axis $k_{x'}$ and about the y-axis $k_{y'}$ will return the floating body to its equilibrium position; i.e.,

$$k_{x'} = \rho g \nabla \cdot GM_{xx}, \tag{3.38a}$$

$$k_{y'} = \rho g \nabla \cdot GM_{yy}, \tag{3.38b}$$

where GM_{xx} and GM_{yy} are the metacentric height of the floating body about the x-axis and y-axis, respectively. ∇ is the volume displaced by the floating body. Here, the rotational motion is affected by the mass moment of inertia I_x and I_y instead of the mass m. For all cases, the viscous damping due to the fluid viscosity that affects the heave, roll, and pitch motions can be modelled by c_z, $c_{x'}$, and $c_{y'}$, respectively.

3.3.4 Hydrodynamic Force and Components of Velocity Potential

Besides the hydrostatic force k, a floating body under wave action is subjected to hydrodynamic force **F**. Mathematically, this force can be derived from hydrodynamic pressure by integrating the pressure $\mathbf{P} = \rho \frac{\partial \Phi}{\partial t}$ over the wetted surface S_w of the floating body:

$$\mathbf{F}(x,y,z,t) = \int_{S_w} \rho \frac{\partial \Phi(x,y,z,t)}{\partial t} \cdot dS_W, \tag{3.39}$$

or

$$\mathbf{F}(x,y,z,t) = F(x,y,z) \cdot e^{i\omega t}, \tag{3.40}$$

where

$$\mathrm{F}(x,y,z) = i\omega\rho \int_{S_W} \varphi(x,y,z) \cdot dS_W. \tag{3.41}$$

The hydrodynamic forces can be decomposed into three components, i.e., the exciting force due to incident wave, the forces due to scattered wave, and the radiated force.[110] Therefore, the velocity potential can be further decomposed as

$$\varphi(x,y,z) = \varphi_I(x,y,z) + \varphi_S(x,y,z) + \varphi_R(x,y,z), \tag{3.42}$$

or

$$\varphi(x,y,z) = \varphi_D(x,y,z) + \varphi_R(x,y,z), \tag{3.43}$$

where $\varphi_D = \varphi_I + \varphi_S$ is the diffracted velocity potential, φ_I the incident velocity potential and φ_R the radiated velocity potential. It is noted that φ_I is already derived in Equation (2.55). The three-dimensional φ_I expressed in complex plane is[89]

$$\varphi_I(x,y,z) = \frac{gA}{\omega} \cdot \frac{\cosh k(z+D)}{\cosh kD} \cdot \exp\left[ik(x\cos\theta + y\sin\theta)\right]. \tag{3.44}$$

The scattered wave φ_S is the wave that is being diverted away from the direction of the incident wave. It is usually accompanied by a change in wave amplitude when the incident wave φ_I encounters some disturbance such as hitting on a floating object or when there is a change of seabed bathymetry.

The scattered wave has to satisfy the condition that

$$\frac{\partial \varphi_I}{\partial \mathbf{n}} = -\frac{\partial \varphi_S}{\partial \mathbf{n}}. \tag{3.45}$$

Following the boundary conditions of Equations (3.33d) and (3.45) and using Equation (3.29), the boundary condition for the φ_D and φ_R on S_w is thus

$$\frac{\partial \varphi}{\partial \mathbf{n}} = \begin{cases} \mathrm{i}\omega W & \text{for } \varphi_R \\ 0 & \text{for } \varphi_D \end{cases}. \tag{3.46}$$

Substituting Equation (3.43) into the hydrodynamic force F (3.40),

$$\mathrm{F}(x,y,z) = \mathrm{i}\omega\rho \int_{S_w} \varphi_D \cdot \mathbf{n} \cdot dS + \mathrm{i}\omega\rho \int_{S_w} \varphi_R \cdot \mathbf{n} \cdot dS. \tag{3.47}$$

The first term in the RHS of Equation (3.47) is the excitation force:

$$F_E = \mathrm{i}\omega\rho \int_{S_w} \varphi_D \cdot \mathbf{n} \cdot dS, \tag{3.48}$$

whereas the second term is the force contribute to the added mass $\mathcal{A}$ and radiated damping $\mathcal{B}$.

3.3.5 Equation of Motion with PTO Damping Force

As the WEC generates energy from the oscillating motion, a PTO system is thus needed to transform the kinetic energy into electricity. This PTO is commonly modelled using a damping coefficient B_{pto} as an external restoring force:

$$F_{pto} = -\mathrm{i}\omega B_{pto} W, \tag{3.49}$$

where the negative sign indicates that the F_{pto} is a restoring force acting in the opposite direction of the hydrodynamic force.

By using Equations (3.30) and (3.31), Equations (3.36a) to (3.36c) can be written in a more compact form with three additional terms – the added mass $\mathcal{A}(\omega)$, radiated damping $\mathcal{B}(\omega)$, and PTO damping B_{pto}:

$$\left[-\omega^2 (M + \mathcal{A}) + \mathrm{i}\omega\left(C + \mathcal{B} + B_{pto}\right) + K\right] \cdot W = F_E, \tag{3.50}$$

where $W(x,y,z)=\{w \quad \theta_x \quad \theta_y\}^{\mathrm{T}}$, $M = diag(m \quad I_x \quad I_y)$, $C = \mathrm{diag}(c_z \quad c_{x'} \quad c_{y'})$, and $K = \mathrm{diag}(k_z \quad k_{x'} \quad k_{y'})$. The excitation force F_E comprises the forces in the z-direction and moment about the x- and y-axes. Here, the viscous damping C is ignored due to the inviscid fluid assumption.

The floating body response is thus

$$\mathrm{W} = \frac{F_E}{-\omega^2(M+\mathcal{A})+\mathrm{i}\omega(\mathcal{B}+B_{pto})+K}, \tag{3.51}$$

or

$$\mathrm{W} = \frac{F_E}{(M+\mathcal{A})(-\omega^2+\omega_N^2)+\mathrm{i}\omega(\mathcal{B}+B_{pto})}. \tag{3.52}$$

In order for the WEC to have the maximum heave response (to generate the most energy), the determinator in Equation (3.51) has to be zero; i.e.,

$$-\omega^2(M+\mathcal{A})+\mathrm{i}\omega(\mathcal{B}+B_{pto})+K=0. \tag{3.53}$$

3.3.6 Optimal PTO Damping

The optimal PTO damping $(B_{pto})_{opt}$ can be obtained by solving Equation (3.53):

$$(B_{pto})_{opt} = \frac{\omega^2(M+\mathcal{A})-K}{i\omega} - \mathcal{B}. \tag{3.54}$$

The magnitude of the $(B_{pto})_{opt}$ is thus

$$\left|(B_{pto})_{opt}\right| = \sqrt{\left[\frac{\omega^2(M+\mathcal{A})-K}{\omega}\right]^2 + \mathcal{B}}, \tag{3.55}$$

or

$$\left|(B_{pto})_{opt}\right| = \sqrt{\left[\frac{(\omega_n^2-\omega^2)(M+\mathcal{A})}{\omega}\right]^2 + \mathcal{B}^2}. \tag{3.56}$$

The value for $\left(B_{pto}\right)_{opt}$ depends on the $\mathcal{A}(\omega)$ and $\mathcal{B}(\omega)$ values that change with the frequency ω. This means that the optimal values of the PTO damping are different for various wave frequencies. For a complex structure, modelling a WEC with varying B_{pto} value for different wave frequencies is not practical. Therefore, a constant $\left(B_{pto}\right)_{opt}$ is usually assumed to enhance the computational time in the simulation. It is common to take the $\mathcal{A}$ and $\mathcal{B}$ at small wave frequency, i.e., large wave period, as the constant $\left(B_{pto}\right)_{opt}$. An example of the performance of a generic WEC by using the aforementioned constant $\left(B_{pto}\right)_{opt}$ is given in Section 3.7.

3.3.7 Power Capture

The captured power or absorbed power P_a of a WEC can be calculated from the WEC responses W given by

$$P_a = \frac{1}{2}\omega^2 \cdot B_{pto} \cdot |W|^2. \tag{3.57}$$

Substituting Equations (3.52) and (3.55) into Equation (3.57),

$$P_a = \frac{1}{2}\omega^2 \cdot B_{pto} \cdot \frac{|F_E|^2}{(M+\mathcal{A})\left(-\omega^2+\omega_N^2\right)+\mathrm{i}\omega\left(\mathcal{B}+B_{pto}\right)}. \tag{3.58}$$

Dividing the power P_a with the rate of power flux given in Equation (3.58) gives us the capture width CW in unit length:

$$CW = \frac{P_a}{\bar{E}} = \frac{\omega^2 B_{pto}}{\rho g |A|^2 C_g} \cdot \frac{|F_E|^2}{(M+\mathcal{A})^2\left(-\omega^2+\omega_N^2\right)^2+\omega^2\left(\mathcal{B}+B_{pto}\right)^2}. \tag{3.59}$$

By multiplying Equation (3.59) with the wave number k, we have the capture width ratio (CWR) given as

$$CWR = \frac{k}{C_g}\frac{\omega^2 B_{pto}}{\rho g |A|^2} \cdot \frac{|F_E|^2}{(M+\mathcal{A})^2\left(-\omega^2+\omega_N^2\right)^2+\omega^2\left(\mathcal{B}+B_{pto}\right)^2}. \tag{3.60}$$

The CW is maximum when the denominator is small. This can be achieved when resonance occurs, i.e., $\omega_n = \omega$, and assuming that $\mathcal{B} = B_{pto}$,

$$(CWR)_{max} = \frac{1}{4}\frac{k}{\rho g C_g |A|^2} \cdot \frac{|F_E|^2}{\mathcal{B}}. \tag{3.61}$$

Therefore, the maximum power $(P_a)_{max}$ is given by

$$(P_a)_{max} = \frac{1}{8} \cdot \frac{|F_E|^2}{\mathcal{B}} \tag{3.62}$$

Example 3.1

The equation of motion (3.50) can be solved analytically for simple geometry like a cylindrical point absorber using the equation shown in Section 3.3.7 for the P_a, CW and CWR. Consider a cylindrical point absorber WEC (PA-WEC) with a diameter $\varnothing = 5$ m and a draft $d_{wec} = 0.25$ m floating in a water depth $D = 25$ m.

The added mass $\mathcal{A}$ is assumed to be a constant value; i.e.,

$$\mathcal{A} = \frac{\rho g \pi \varnothing^2}{4}, \tag{3.63}$$

and the radiated damping $\mathcal{B}$ can be derived from Haskind's relation:

$$\mathcal{B} = \frac{k}{8\pi\rho g C_g |A|^2} \int_0^{2\pi} |F_D(\theta)|^2 \cdot d\theta, \tag{3.64}$$

where the evaluation of the integral in (**3.64**) is

$$\int_0^{2\pi} |F_D(\theta)|^2 \cdot d\theta = 2\pi \times |F_D(\theta)|^2. \tag{3.65}$$

Substituting Equations (3.63) and (3.64) into Equation (3.61), we obtain the $(CWR)_{max}$:

$$(CWR)_{max} = 1. \tag{3.66}$$

The maximum power $(P_a)_{max}$ is given by

$$(P_a)_{max} = \frac{1}{2k} \cdot \rho g C_g |A|^2. \tag{3.67}$$

3.3.8 Unidirectional and Multidirectional Sea

For unidirectional irregular wave, the average heave response $\overline{w}$ generated over the range of wave frequency ω considered can be obtained from the response spectrum expressed by[111]

$$S_{res}(\omega)\big|_I = \int_{\omega} |w(\omega)|^2 \cdot S_I(\omega) \cdot d\omega \quad \text{for} \quad I = \text{PM, BS, JS or OH.} \tag{3.68}$$

The average heave $\overline{w}$ is thus given by

$$\overline{w}(\omega)\big|_I = 4\sqrt{S_{res}\big|_I} \quad \text{for} \quad I = \text{PM, BS, JS or OH.} \tag{3.69}$$

The response spectrum of the absorbed power can be written as

$$S_{res}^{(P_a)}(\omega)\Big|_I = \int_{\omega} |P_a(\omega)|^2 \cdot S_I(\omega) \cdot d\omega \quad \text{for} \quad I = \text{PM, BS, JS or OH,} \tag{3.70}$$

and the average power $\overline{P_a}$ is thus given by

$$\overline{P_a}(\omega)\big|_I = 4\sqrt{S_{res}^{(P_a)}\big|_I} \quad \text{for} \quad I = \text{PM, BS, JS, or OH.} \tag{3.71}$$

Similarly, $\overline{P_a}$ over multidirectional sea is given as

$$\overline{P_a}(\omega,\theta)\big|_I = 2\int_{\theta}\int_{\omega} |P_a(\omega)|^2 \cdot S_I(\omega,\theta) \cdot d\omega \, d\theta \quad \text{for} \quad I = \text{PM, BS, JS or OH.} \tag{3.72}$$

Similarly, $S_I(\omega,\theta)$ is the multidirectional wave spectrum given in Section 2.3.6.

3.4 WECs in Array

The WECs are commonly arranged in an array in order to enhance the energy generation from the wave farm. It is necessary to quantify the influence of one WEC to another when assessing the performance of the WEC array.

3.4.1 Equation of Motion

For different numbers N_{wec} of WECs, the equation of the motion of the body a due to the body b is written as[31,112]

$$\left\{-\omega^2\left[M+\mathcal{A}_{aa}\right]+\mathrm{i}\omega\left[B_{pto}+\mathcal{B}_{aa}\right]+K\right\}\cdot W_a-\sum_{\substack{b=1\\b\neq a}}^{N_{wec}}\left[\omega^2\mathcal{A}_{ba}-\mathrm{i}\omega\mathcal{B}_{ab}\right]\cdot W_b=\left(F_E\right)_a. \quad (3.73)$$

3.4.2 Interaction Factor (q-Factor) in a Regular Wave

In order to quantify the interaction between devices, Budal[113] defines the q-factor to facilitate the discussion on the performance of the array. For regular waves, this is written as

$$q=\frac{P_T}{N_{wec}\times\left(P_a\right)_0}=\frac{\sum_{n=1}^{N_{wec}}\left(P_a\right)_n}{N_{wec}\times\left(P_a\right)_0}, \quad (3.74)$$

where P_T is the total absorbed power, $\left(P_a\right)_n$ is the absorbed power by the nth number of WEC, and $\left(P_a\right)_0$ the absorbed power of an isolated WEC in the open sea.

Equation (3.74) is used as a performance evaluator for the array where a constructive interaction is denoted by a value greater than 1.0 and a destructive interaction when smaller than 1.0. Thus, a constructive interaction informs the engineers that the wave interaction due to wave diffraction and radiation has a positive effect to the wave arrays in terms of energy generation, whereas a destructive interference implies that there is cancellation of waves due to wave superposing which results in low energy generation. Thus, it is the job of the engineers to ensure that the constructive interference can be achieved, and this can be done via optimisation, as will be explain in Chapter 7.

3.4.3 Interaction Factor (q-Factor) in Irregular Wave

Similar to the regular wave, the q-factor for irregular wave can be computed by

$$\bar{q}=\frac{\left(\overline{P_a}\right)_T}{N_{wec}\times\left(\overline{P_a}\right)_0}=\frac{\sum_{n=1}^{N_{wec}}\left(\overline{P_a}\right)_n}{N_{wec}\times\left(\overline{P_a}\right)_0}, \quad (3.75)$$

where $\left(\overline{P_a}\right)_T$ is the total average power, $\overline{P_a}$ the average power as given in Equations (3.71) and (3.72) for unidirectional wave and multidirectional sea, respectively. Similarly, the subscript n denotes the power for the nth WEC and subscript 0 denotes power for an isolated WEC in the open sea.

3.5 Solution to Velocity Potential

The evaluation of *CWR*, *CW* and P_a requires the knowledge of diffracted and radiated potentials. In this section, evaluation of diffracted and radiated potential using Green's identity, which results in the boundary integral equation (BIE), will be presented. This is followed by the introduction of Haskind's relation, which allows the evaluation of exciting force from the radiated potential. Lastly, the numerical scheme based on the BEM used to solve for the boundary integral in the boundary value problem (BVP) will be presented. In order to derive the BIE, we first present the modal expansion method and free surface Green's function in Sections 3.5.1 and 3.5.2.

3.5.1 Modal Expansion Method

Equation (3.46) for the radiated wave shows that the velocity potential φ and the floating body motion W depend on each other, and therefore we have a coupled problem. This is to say that if boundary condition Equation (3.46) is imposed on the equation of motions (Equation (3.36)) via the force F on the RHS of the equation, we will end up with two unknows for each equation, which is unsolvable. In order to solve such a coupled problem, the equation of motion can be decoupled by using the modal expansion method.

According to the modal expansion method proposed by Newman,[90] the radiated velocity potential $\varphi_R(x, y, z)$ can be expanded by a series of the products of the complex amplitudes of the body oscillatory motion in its three dofs W_l and the corresponding unit-amplitude radiation potentials φ_l, given by

$$\varphi_R = \mathrm{i}\omega \sum_l W_l \cdot \varphi_l \text{ for } l = z, x' \text{ and } y', \tag{3.76}$$

where the complex motion of the floating body W_l can be written in unit vector form:

$$W_l = W \cdot \mathbf{n} = n_l = \begin{pmatrix} w \cdot n_z & \theta_x \cdot n_{x'} & \theta_y \cdot n_{y'} \end{pmatrix} \text{for } l = z, x' \text{ and } y'. \tag{3.77}$$

Therefore, the alternative form of Equation (3.76) is

$$\varphi_R = \mathrm{i}\omega \sum_l W_l \cdot \varphi_l \cdot \mathbf{n} = \mathrm{i}\omega \sum_l W_l \cdot \varphi_l \cdot n_l \text{ for } l = z, x' \text{ and } y'. \tag{3.78}$$

3.5.2 Free Surface Green's Function

Green's function was first developed by the British mathematician George Green (1793–1841),[114,115] where he derived a function that is taken to be the fundamental solution of the Laplace equation.

Green's function G has a property that it satisfies the Laplace equation,[116]

$$\nabla^2 G(\mathbf{x},\boldsymbol{\xi}) = \delta(\mathbf{x}-\boldsymbol{\xi}), \text{in } \Omega, \tag{3.79}$$

where $\delta(\mathbf{x}-\boldsymbol{\xi})$ is the Dirac delta function which denotes solution exists when $\mathbf{x} \neq \boldsymbol{\xi}$. The solution to the second-order PDE for Green's function Equation (3.79) has a fundamental solution in the form of

$$G(\mathbf{x},\boldsymbol{\xi}) = -\frac{1}{\|\mathbf{x}-\boldsymbol{\xi}\|} = -\frac{1}{R}, \tag{3.80}$$

where $\mathbf{x} = (x,y,z)$ is the location of the field and $\boldsymbol{\xi} = (\xi,\eta,\zeta)$ is the location of the source point. $R = \|\mathbf{x}-\boldsymbol{\xi}\|$ is the distance between the source and field points. The objective of $\|\mathbf{x}-\boldsymbol{\xi}\|$ is to evaluate the influence of the sources $\boldsymbol{\xi}$, when distributed on surfaces, has on its field $\mathbf{x}$.

Linton[117] has derived a Green function, known as the free surface Green's function, which satisfies the boundary conditions

$$\frac{\partial G(\mathbf{x},\boldsymbol{\xi})}{\partial z} = 0, \quad \text{at} \quad z = -D, \tag{3.81a}$$

$$\frac{\partial G(\mathbf{x},\boldsymbol{\xi})}{\partial z} = \frac{\omega^2}{g} G(\mathbf{x},\boldsymbol{\xi}), \quad \text{at} \quad z = 0. \tag{3.81b}$$

The free surface Green's function is given as[117]

$$G(\mathbf{x},\boldsymbol{\xi}) = -\sum_{j=0}^{\infty} \frac{K_0\left(k_j|\mathbf{x}-\boldsymbol{\xi}|\right)}{2\pi C_j} \cos\left[k_j(z+D)\right]\cos\left[k_j(\zeta+D)\right], \text{at } z=0, \tag{3.82}$$

where K_0 is the second kind Bessel function of order zero and k_j the positive real roots of the dispersion equation $-k_j \tan k_j D = \frac{\omega^2}{g}$, where $j \leq 0$. The zeroth root k_0 is related to the wavenumber k by $k_0 = -ik$, and the constant C_j is given by

$$C_j = \frac{D}{2}\left(1 + \frac{\sin 2k_j D}{2k_j k}\right). \tag{3.83}$$

3.5.3 Green's Theorem: Green's Second Identity

Green's second identity states that for two scalar functions ϕ_1 and ϕ_2 defined on some region Ω, where both functions are twice differentiable,

$$\int_{\Omega}\left(\phi_1\nabla^2\phi_2-\phi_2\nabla^2\phi_1\right)d\Omega=\oint_{S}\left(\phi_1\frac{\partial\phi_2}{\partial\mathbf{n}}-\frac{\partial\phi_1}{\partial\mathbf{n}}\phi_2\right)dS \tag{3.84}$$

where S is the surface of the domain Ω.

3.5.4 Boundary Integral Equation

Referring to Figure 3.5, S encompasses the surface at the seabed S_B, the free water surface S_F, the wetted surface of the floating body S_W, and the surface at infinite boundary $S_{\pm\infty}$. We have to define the function ϕ_1 equals to the free surface Green's function $G(\mathbf{x},\boldsymbol{\xi})$ and by replacing ϕ_2 with the velocity potential φ, the BIE for the Laplace equation is given by

$$\int_{\Omega}\left(G\nabla^2\varphi-\varphi\nabla^2G\right)\cdot d\Omega=\oint_{S\supset\{S_B+S_F+S_{\pm\infty}+S_W\}}\left(G\frac{\partial\varphi}{\partial\mathbf{n}}-\frac{\partial G}{\partial\mathbf{n}}\varphi\right)\cdot dS. \tag{3.85}$$

It is easy to proof that by substituting the boundary conditions for φ and $G(\mathbf{x},\xi)$, i.e., Equations (3.33b), (3.33c), (3.33d), (3.81a), and (3.81b), we have

$$0=\int_{S\supset\{S_B+S_F+S_{\pm\infty}+S_W\}}\left(G\frac{\partial\varphi}{\partial\mathbf{n}}-\frac{\partial G}{\partial\mathbf{n}}\varphi\right)\cdot dS. \tag{3.86}$$

As described by the Dirac delta function $\delta(\mathbf{x},\boldsymbol{\xi})$ in Equation (3.79), the solution for G does not exist when $\mathbf{x}=\boldsymbol{\xi}$ due to singularity. Therefore, we can decompose the surface S into $S_{W'}$ and S_ϵ, where S_ϵ is a small circular of radius ϵ around the source point $\mathbf{x}$, and $S_{W'}$ is the surface of the domain without S_ϵ. This is shown in Figure 3.6, where, for simplicity, the surface $S_{w'}$ will be denoted as S_w. With this, Equation (3.86) can be re-written as

$$0=\int_{S_w}\left(G\frac{\partial\varphi}{\partial\mathbf{n}}-\frac{\partial G}{\partial\mathbf{n}}\varphi\right)\cdot dS_W+\int_{S_{\pm\infty}}\left(G\frac{\partial\varphi}{\partial\mathbf{n}}-\frac{\partial G}{\partial\mathbf{n}}\varphi\right)\cdot dS_{\pm\infty}+\int_{S_\epsilon}\left(G\frac{\partial\varphi}{\partial\mathbf{n}}-\frac{\partial G}{\partial\mathbf{n}}\varphi\right)\cdot dS_\epsilon. \tag{3.87}$$

If we substitute the fundamental solution for G from (3.80), i.e., $G=-\frac{1}{R}$, into (3.87), this yields

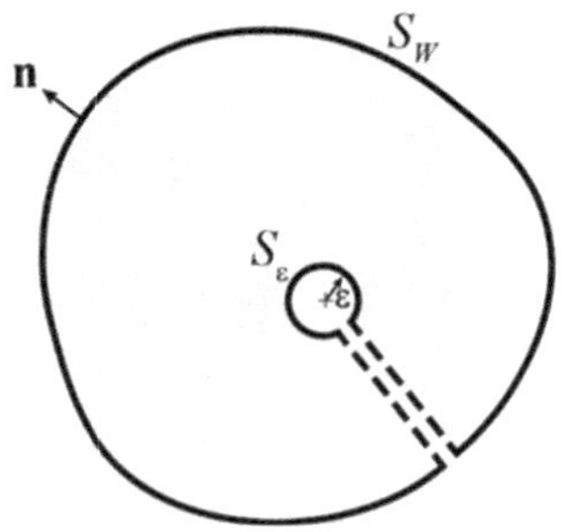

FIGURE 3.6
Discretisation of Ω into $S_W \cup S_\epsilon$.

$$0 = \int_{S_w + S_{\pm\infty}} \left(G\frac{\partial \varphi}{\partial \mathbf{n}} - \frac{\partial G}{\partial \mathbf{n}}\varphi \right) \cdot dS + \int_{S_\epsilon} \left[-\frac{1}{R}\frac{\partial \varphi}{\partial \mathbf{n}} - \frac{\partial}{\partial \mathbf{n}}\left(-\frac{1}{R} \right)\varphi \right] \cdot dS_\epsilon \tag{3.88}$$

and

$$0 = \int_{S_w + S_{\pm\infty}} \left(G\frac{\partial \varphi}{\partial \mathbf{n}} - \frac{\partial G}{\partial \mathbf{n}}\varphi \right) \cdot dS + \int_{S_\epsilon} \frac{1}{R^2}\left(R\frac{\partial \varphi}{\partial \mathbf{n}} - \varphi \right) \cdot dS_\epsilon . \tag{3.89}$$

The term $\frac{\int dS_\epsilon}{R^2}$ in Equation (3.89) is the solid angle $C(\theta)$, therefore giving us

$$0 = \left(\epsilon\frac{\partial \varphi}{\partial \mathbf{n}} - \varphi \right) \cdot C_\theta(\theta) + \int_{S_W + S_{\pm\infty}} \left(G\frac{\partial \varphi}{\partial \mathbf{n}} - \frac{\partial G}{\partial \mathbf{n}}\varphi \right) \cdot dS \tag{3.90a}$$

or

$$0 = \epsilon\frac{\partial \varphi}{\partial \mathbf{n}} - \frac{\varphi}{\epsilon^2} C_\theta(\theta)\epsilon^2 + \int_{S_W + S_{\pm\infty}} \left(G\frac{\partial \varphi}{\partial \mathbf{n}} - \frac{\partial G}{\partial \mathbf{n}}\varphi \right) \cdot dS. \tag{3.90b}$$

Upon simplification and taking the small circular radius ϵ to be an infinitesimal small value, i.e., $\epsilon \to 0$, Equation (3.90b) becomes

$$C_\theta(\theta) \cdot \varphi_k(\mathbf{x}) = \int_{S_w} \left[G(\mathbf{x},\boldsymbol{\xi})\frac{\partial \varphi_k(\mathbf{x})}{\partial \mathbf{n}} - \frac{\partial G(\mathbf{x},\boldsymbol{\xi})}{\partial \mathbf{n}}\varphi_k(\mathbf{x}) \right] \cdot dS_W, k = R \text{ or } D, \tag{3.91}$$

where R represents the radiated wave and D the diffracted wave.

The solid angle $C_\theta(\theta)$ is a three-dimensional angular volume that measures the amount of field of view covered by an object from a particular point of observation. It is defined analogously to a plane angle in two dimensions and is measured in steradians (sr)[114–116]. Therefore, $C_\theta(\theta)$ has a value of 4π steradian when the source point is located in the fluid domain Ω and a value of 2π steradian when located at the wetted surface of a floating body, i.e.,

$$C_\theta(\theta) = \begin{cases} 4\pi & \text{in } \Omega \\ 2\pi & \text{on } S_W \end{cases}. \tag{3.92}$$

3.5.5 Solution to φ_D and φ_R

By substituting Equations (3.76) and (3.77) into Equation (3.91) and imposing the boundary condition Equation (3.46) on the wetted surface S_W,

$$\varphi_k(\mathbf{x}) = \begin{cases} \int\limits_{S_w} \left[\mathrm{i}\omega \cdot G(\mathbf{x},\xi) \cdot W(\mathbf{x}) \cdot \mathbf{n} - \dfrac{\partial G(\mathbf{x},\xi)}{\partial \mathbf{n}} \cdot \varphi_k(\mathbf{x}) \right] \cdot dS & \text{for } k = R \\ \int\limits_{S_w + S_{\pm\infty}} \left[G(\mathbf{x},\xi) \dfrac{\partial \varphi_k(\mathbf{x})}{\partial \mathbf{n}} \right] \cdot dS & \text{for } k = D \end{cases}. \tag{3.93}$$

Note that the term on RHS in Equation (3.93) for φ_R is evaluated only over S_w, as the radiated potential satisfies the radiated boundary condition Equation (3.35) at $S_{\pm\infty}$.

To evaluate the RHS term in Equation (3.93) for φ_D over $S_{\pm\infty}$, it can be rewritten as

$$\int\limits_{S_{\pm\infty}} \left(G \frac{\partial \varphi_D}{\partial \mathbf{n}} - \frac{\partial G}{\partial \mathbf{n}} \varphi_D \right) \cdot dS = \int\limits_{S_{\pm\infty}} \left(G \frac{\partial \varphi_I}{\partial \mathbf{n}} - \frac{\partial G}{\partial \mathbf{n}} \varphi_I \right) \cdot dS + \int\limits_{S_{\pm\infty}} \left(G \frac{\partial \varphi_S}{\partial \mathbf{n}} - \frac{\partial G}{\partial \mathbf{n}} \varphi_S \right) \cdot dS \tag{3.94}$$

As the scattered potential satisfies the radiated boundary condition Equation (3.86), Equation (3.94) becomes

$$\int\limits_{S_{\pm\infty}} \left(G \frac{\partial \varphi_D}{\partial \mathbf{n}} - \frac{\partial G}{\partial \mathbf{n}} \varphi_D \right) \cdot dS = \int\limits_{S_{\pm\infty}} \left(G \frac{\partial \varphi_I}{\partial \mathbf{n}} - \frac{\partial G}{\partial \mathbf{n}} \varphi_{In} \right) \cdot dS. \tag{3.95}$$

Using the same technique as in Equation (3.87), i.e., by assuming a small circular S_ϵ of radius ϵ around the source point $\mathbf{x}$ and upon evaluation,

$$\int\limits_{S_{\pm\infty}} \left(G \frac{\partial \varphi_D}{\partial \mathbf{n}} - \frac{\partial G}{\partial \mathbf{n}} \varphi_D \right) \cdot dS = C_\theta(\theta) \cdot \varphi_I, \text{where } C_\theta(\theta) = 4\pi. \tag{3.96}$$

The solid angle $C_\theta(\theta) = 4\pi$ because φ_I is evaluated at the fluid domain Ω. Therefore, Equation (3.93) becomes

$$C(\theta)\cdot\varphi_k(\mathbf{x})=\begin{cases}\int\limits_{S_w}\left(\mathrm{i}\omega\cdot G(\mathbf{x},\xi)\cdot W(\mathbf{x})\cdot\mathbf{n}-\frac{\partial G(\mathbf{x},\xi)}{\partial\mathbf{n}}\varphi_k(\mathbf{x})\right)\cdot dS & \text{for } k=R\\ \int\limits_{S_w}\left(G(\mathbf{x},\xi)\frac{\partial\varphi_k(\mathbf{x})}{\partial\mathbf{n}}-\frac{\partial G(\mathbf{x},\xi)}{\partial\mathbf{n}}\varphi_k(\mathbf{x})\right)\cdot dS+4\pi\varphi_I(\mathbf{x}) & \text{for } k=D\end{cases},$$

$$\text{where } C(\theta)=2\pi. \tag{3.97}$$

By rearranging Equation (3.97) and imposing the boundary condition Equation (3.46) on S_w for φ_S and φ_D, we have

$$\left[2\pi+\int\limits_{S_w}\left(\frac{\partial G(\mathbf{x},\xi)}{\partial\mathbf{n}}\right)\cdot dS\right]\cdot\varphi_k(\mathbf{x})=\begin{cases}\mathrm{i}\omega\int\limits_{S_w}G(\mathbf{x},\xi)\cdot W(\mathbf{x})\cdot\mathbf{n}\cdot dS & \text{for } k=R\\ 4\pi\varphi_I(\mathbf{x}) & \text{for } k=D\end{cases} \tag{3.98}$$

or

$$\varphi_k(\mathbf{x})=\left[2\pi+\int\limits_{S_w}\left(\frac{\partial G(\mathbf{x},\xi)}{\partial\mathbf{n}}\right)\cdot dS\right]^{-1}\times\begin{cases}\mathrm{i}\omega\int\limits_{S_w}G(\mathbf{x},\xi)\cdot W(\mathbf{x})\cdot\mathbf{n}\cdot dS & \text{for } k=R\\ 4\pi\varphi_I(\mathbf{x}) & \text{for } k=D\end{cases}. \tag{3.99}$$

The diffracted potential φ_D allows the evaluation of excitation force F_E in Equation (3.48), and the radiated potential φ_R contributes to the added mass and radiated damping described in the Section 3.5.7.

With that, we have successfully decoupled the equation of motion (Equation (3.50)), which now only has one unknown W. The complex amplitude for the motion can thus be solved easily for this algebraic equation.

3.5.6 Haskind's Relation

Equation (3.48) shows that in order to derive the excitation force F_E, the knowledge of the scattered potential is necessary. By substitute $\varphi_D=\varphi_I+\varphi_S$ into (3.48), we have

$$F_E=\mathrm{i}\omega\rho\int\limits_{S_w}\varphi_D\cdot\mathbf{n}\cdot dS=\mathrm{i}\omega\rho\int\limits_{S_w}(\varphi_I+\varphi_S)\cdot\mathbf{n}\cdot dS. \tag{3.100}$$

Haskind[118] has formulated an equation that relates the hydrodynamic radiated damping of the body to the excitation force F_E, now known as the

Haskind's relation, which allows the solution to the excitation force from the radiated potential without needing to solve the scattered potential.

We know that the unit-normal vector to a surface is given by

$$\mathbf{n} = \frac{\nabla \varphi_I}{|\nabla \varphi_I|} = \frac{\partial \varphi_I}{\partial \mathbf{n}}. \tag{3.101}$$

Substituting Equation (3.101) into Equation (3.100) gives us

$$F_E = \mathrm{i}\omega\rho \int_{S_w} (\varphi_I + \varphi_S) \cdot \frac{\partial \varphi_I}{\partial \mathbf{n}} \cdot dS = \mathrm{i}\omega\rho \int_{S_w} \left(\varphi_I \cdot \frac{\partial \varphi_I}{\partial \mathbf{n}} + \varphi_S \cdot \frac{\partial \varphi_I}{\partial \mathbf{n}} \right) \cdot dS \tag{3.102}$$

Imposing Green's second identity for φ_S and φ_I in the form of (3.84) gives us

$$0 = \int_{S_w} \left(\varphi_S \cdot \frac{\partial \varphi_I}{\partial \mathbf{n}} - \frac{\partial \varphi_S}{\partial \mathbf{n}} \cdot \varphi_I \right) dS \rightarrow \int_{S_w} \left(\varphi_S \cdot \frac{\partial \varphi_I}{\partial \mathbf{n}} \right) \cdot dS = \int_{S_w} \left(\frac{\partial \varphi_S}{\partial \mathbf{n}} \cdot \varphi_I \right). \tag{3.103}$$

With the boundary condition Equation (3.45), Equation (3.102) becomes

$$F_E = \mathrm{i}\omega\rho \int_{S_w} \left(\varphi_I \cdot \frac{\partial \varphi_I}{\partial \mathbf{n}} - \varphi_I \cdot \frac{\partial \varphi_I}{\partial \mathbf{n}} \right) \cdot dS = \mathrm{i}\omega\rho \int_{S_w} \left(\varphi_I \cdot \mathbf{n} - \varphi_I \frac{\partial \varphi_I}{\partial \mathbf{n}} \right) \cdot dS, \tag{3.104}$$

which is the well-known Haskind's relation that allows the evaluation of F_E from the incident and radiated potentials.

3.5.7 Added Mass and Radiated Damping

3.5.7.1 *Added Mass $\mathcal{A}$*

Added mass refers to the inertia added to a system due to the acceleration or deceleration of a body. In such a situation, some of the volume surrounding the floating body has to be moved to allow the body movement. This phenomenon occurs because the object and surrounding fluid cannot occupy the same physical space simultaneously. The added mass is modelled similar to the mass and results in additional force/inertia acting in the opposite direction of the excitation force/moment F_E (i.e., Newton's third law of motion). Therefore, this additional term is significant, as it has a profound effect on the motion of the floating body, i.e., larger added mass/inertia may reduce the seakeeping motion. In addition, added mass/inertia can also change the natural frequency of the floating body, i.e.,

$$(\omega_N)_z = \sqrt{\frac{k_z}{m + (m_a)_z}} = \sqrt{\frac{\rho g \cdot A_{wp}}{m + (m_a)_z}}, \tag{3.105a}$$

$$(\omega_N)_{x'} = \sqrt{\frac{k_{x'}}{I_x + (I_a)_x}} = \sqrt{\frac{\rho g \nabla \cdot GM_{xx}}{I_x + (I_a)_x}}, \tag{3.105b}$$

$$(\omega_N)_{y'} = \sqrt{\frac{k_{y'}}{I_y + (I_a)_y}} = \sqrt{\frac{\rho g \nabla \cdot GM_{yy}}{I_y + (I_a)_y}}, \tag{3.105b}$$

where are $\mathcal{A} = \left[(m_a)_z \quad (I_a)_x \quad (I_a)_y \right]$. The added mass for a relatively simple geometry can be obtained analytically, as shown in Table 3.2.

3.5.7.2 Radiated Damping $\mathcal{B}$

Under the absent of incident wave, when a floating body is being pushed down into the water or rotated to a certain angle and released, waves will be created as the floating body oscillates and eventually returns to its equilibrium position. The waves are observed to radiate away from the source, i.e., the floating body, as shown in Figure 3.7. This radiated wave can be said to dissipate the energy created from the source and damp the motion of the floating body, thereby is termed the radiated damping, or added damping.

The numerical solution to potential wave theory can be used to solve for the radiated damping (and added mass), as will be shown in Section 3.6.3. It is to be noted that when comparing radiated damping with viscous damping, the former has a stronger influence over the motion of the structure, as viscous damping can be negligible small when the structure gets bigger, i.e., large structure length-to-wavelength ratio, which is the case for most WEC and VLFS. Having said that, the disregard of viscous damping may cause unrealistic motion during the occurrence of resonance.

TABLE 3.2

List of Added Mass for Simple Geometry

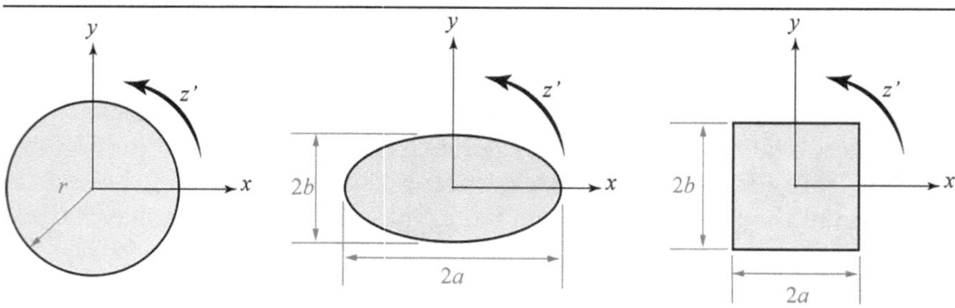

Circle	Ellipse	Square
$(m_a)_x = (m_a)_y = 4\rho r^2$ $(I_a)_{z'} = 0$	$(m_a)_x = \rho\pi b^2$ $(m_a)_y = \rho\pi a^2$ $(I_a)_{z'} = \rho\left(a^2 - b^2\right)$	$(m_a)_x = (m_a)_y = 1.51\pi\rho a^2$ $(I_a)_{z'} = 0.234\pi\rho a^4$

FIGURE 3.7
Radiated wave.

$\mathcal{A}$ and $\mathcal{B}$ from φ_R

The added mass and radiated damping for floating body of complex geometry can be determined from the radiated potential φ_R (3.99). By using the expression for φ_R in (3.78), we have

$$\varphi_R = \mathrm{i}\omega \textstyle\sum_l W_l \cdot \varphi_l \text{ for } l = z, x' \text{ and } y' \tag{3.106}$$

The unit-amplitude radiated potential φ_l has a unit metre and is given by

$$\varphi_l = \left[2\pi + \int_{S_w} \frac{\partial G}{\partial \mathbf{n}} \cdot dS_W \right]^{-1} \cdot \int_{S_w} G \cdot \mathbf{n} \cdot dS_W. \tag{3.107}$$

The added mass $\mathcal{A}$ and radiated damping $\mathcal{B}$ is related to φ_R by the following relationship.

$$\mathcal{A} - \frac{1}{\omega}\mathcal{B} = \rho \int_{S_w} \mathbf{n} \cdot \varphi_l \cdot dS \tag{3.108}$$

or

$$\mathcal{A} = \mathrm{Re}\left\{ \rho \int_{S_w} \mathbf{n} \cdot \varphi_l \cdot dS \right\} \tag{3.109a}$$

$$\mathcal{B} = \mathrm{Im}\left\{ \omega\rho \int_{S_w} \mathbf{n} \cdot \varphi_l \cdot dS \right\} \tag{3.109b}$$

3.5.8 Wave Elevation

The wave elevation surrounding the floating body can be derived from Equation (2.8) for the SAWT, re-written as

$$\hat{\eta} = -\frac{1}{g}\left(\frac{\partial \varphi}{\partial t} + \frac{1}{2}\left|\nabla \varphi\right|^2 \right), \quad \text{at} \quad z = \hat{\eta}, \tag{3.110}$$

where $\varphi = \varphi_D + \varphi_R$, which can be derived from Equation (3.97) by taking $C(\theta) = 4\pi$,

$$C_{\theta(\theta)} \cdot \varphi_k(\mathbf{x}) = \begin{cases} \int\limits_{S_w} \left[\mathrm{i}\omega \cdot G(\mathbf{x},\boldsymbol{\xi}) \cdot W(\boldsymbol{\xi}) \cdot \mathbf{n} - \dfrac{\partial G(\mathbf{x},\boldsymbol{\xi})}{\partial \mathbf{n}} \varphi_k(\boldsymbol{\xi}) \right] \cdot dS & \text{for} \quad k = R \\ \int\limits_{S_w} \left[G(\mathbf{x},\boldsymbol{\xi}) \dfrac{\partial \varphi_k(\boldsymbol{\xi})}{\partial \mathbf{n}} - \dfrac{\partial G(\mathbf{x},\boldsymbol{\xi})}{\partial \mathbf{n}} \varphi_k(\boldsymbol{\xi}) \right] \cdot dS + 4\pi \varphi_I(\mathbf{x}) & \text{for} \quad k = D \end{cases} . \tag{3.111}$$

Rearranging Equation (3.111),

$$\varphi_k(\mathbf{x}) = \begin{cases} \dfrac{\mathrm{i}\omega}{4\pi} \int\limits_{S_w} \left[G(\mathbf{x},\boldsymbol{\xi}) \cdot W(\boldsymbol{\xi}) \cdot \mathbf{n} - \dfrac{\partial G(\mathbf{x},\boldsymbol{\xi})}{\partial \mathbf{n}} \varphi_k(\boldsymbol{\xi}) \right] \cdot dS & \text{for} \quad k = R \\ \varphi_I(\mathbf{x}) - \dfrac{1}{4\pi} \int\limits_{S_w} \left[-\dfrac{\partial G(\mathbf{x},\boldsymbol{\xi})}{\partial \mathbf{n}} \varphi_k(\boldsymbol{\xi}) \right] \cdot dS & \text{for} \quad k = D \end{cases} , \tag{3.112}$$

where $\varphi_R(\xi)$ and $\varphi_D(\xi)$ are the velocity potential on S_W obtained from (3.99).

Example 3.2
Consider a generic cylindrical PA-WEC with a diameter $\varnothing$ and draft d_{wec} as shown in Figure 1.10(a). The added mass and radiated damping of the PA-WEC in the heave direction are plotted in Figure 3.8. Here, five different PA-WECs of different dimensions are further considered to study the effect of diameter and draft has on the added mass and radiated damping:

$$\varnothing = 2\,\text{m and } d_{wec} = 0.25\text{ m}$$

$$\varnothing = 3\,\text{m and } d_{wec} = 0.25\text{ m}$$

$$\varnothing = 4\,\text{m and } d_{wec} = 0.25\text{ m}$$

$$\varnothing = 2\,\text{m and } d_{wec} = 0.25\text{ m}$$

$$\varnothing = 2\,\text{m and } d_{wec} = 0.25\text{ m}$$

Figure 3.8 shows $\mathcal{A}$ and $\mathcal{B}$ increase with the diameter $\varnothing$ and draft d_{wec} of the WEC. In other words, the increase in the mass will result in the increase of $\mathcal{A}$ and $\mathcal{B}$, which in turn will have a profound effect on the motion of the WEC. Also, the inclusion of $\mathcal{A}$ of the WEC reduces the natural frequency ω_N of the WEC, as described in Equation (3.12), where ω_N decreases with the increase of $\mathcal{A}$.

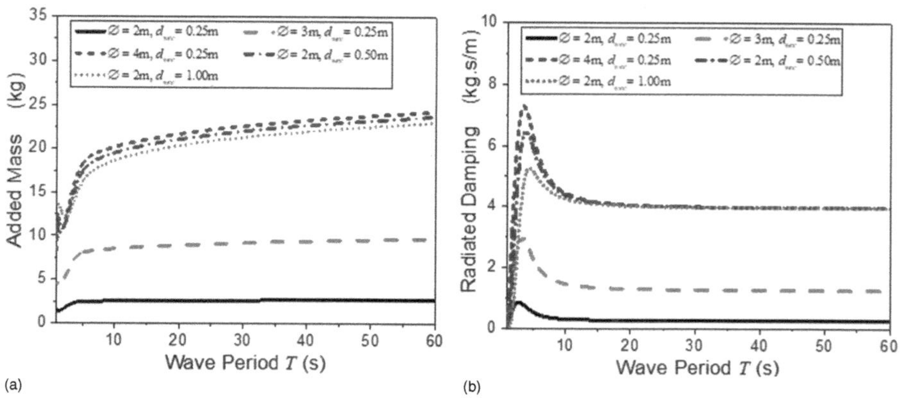

FIGURE 3.8
Comparison of (a) added mass $\mathcal{A}$ (b) radiated damping $\mathcal{B}$ for PA-WEC under different diameter $\varnothing$ and draft d_{wec}.

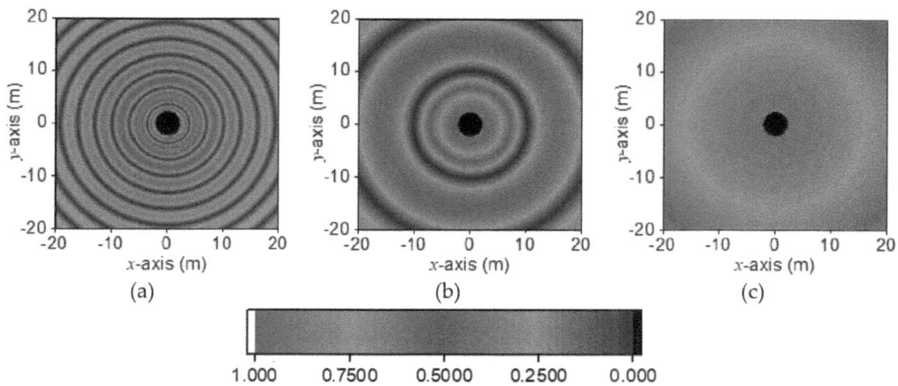

FIGURE 3.9
Radiated waves from PA-WECs for wave period (a) $T = 2$s (b) $T = 4$s (c) $T = 8$s. $\varnothing = 4$m and $d_{wec} = 1.00$m.

According to Equation (3.108), $\mathcal{A}$ and $\mathcal{B}$ depend on the unit-amplitude radiated velocity potential φ_l of the fluid. The radiated potential measures how energy in a floating body is dissipated via the wave radiated away from the floating body, as shown in Figure 3.7. The rate of wave dissipation varies with wavelength (i.e., wave periods or wave frequencies), as shown in Figure 3.9, where the radiated wave elevation computed from φ_l according to Equation (3.109) is plotted. As the radiated wave is generated when a floating body is being pushed down into the water or rotated to a certain angle and released under the absent of incident wave, the radiated potential and thus $\mathcal{A}$ and $\mathcal{B}$ are independent of the wave direction.

Next, the radiated waves from three different types of WECs, namely the point absorber, the terminator and attenuator, are investigated. The schematic diagrams of the terminator and attenuator are shown in Figure 3.10(b) and (c), respectively.

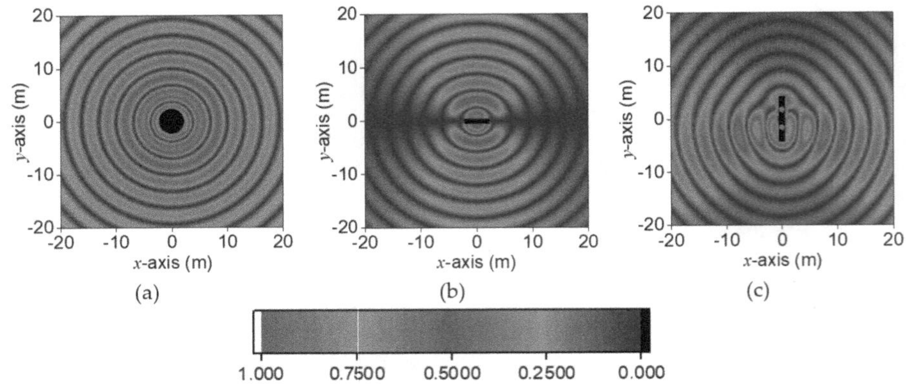

FIGURE 3.10
Radiated Waves from (a) point absorber (b) terminator and (c) attenuator WECs.

Figure 3.10 presents the radiated waves of the same wavelength generated from the three different types of WECs. The terminator considered here is a rectangular flap-type WEC with a width of 4 m operating in a water depth of 5 m, whereas the attenuator is two interconnected raft-type modules, where each module has a length of 4 m and floating in a water depth of 10 m.

The radiated waves from different types of WECs are obviously different; thus, the $\mathcal{A}$ and $\mathcal{B}$ changes with the geometry of the floating body. In view of this, in the design of WEC, it is important to optimise the geometry of the floating body in order to obtain $\mathcal{A}$ and $\mathcal{B}$ that can maximise the energy generation. In cases where WEC has to be tuned to its natural frequencies to maximise the energy production, the geometry of the WEC plays an important role such that the ω_N can be tuned to the dominant encountering wave frequencies of the sea state.

3.6 Boundary Element Method

While analytical or semi-analytical method can be used to solve for floating bodies of simple geometry, such as a cylindrical point absorber or the flat-type OWSC, the performance of floating bodies of complex geometry or multiple-WECs in array may be easier to be evaluated numerically. As shown in the evaluation of the velocity potential φ_D and φ_R in (3.99), the BIE only involves the evaluation over the wetted surface of the floating bodies S_w. Thus, a special kind of numerical method, known as the BEM or the constant panel method, can be used in evaluating the BIE. The BEM is a powerful tool, as it requires only the modelling of the wetted surface of the floating body, thus requires significantly smaller computation resources as compared to other CFD methods. In addition, it has the capability to predict the performance of floating bodies with relatively good accuracy for first-order and up to second-order linear motion.

Writing (3.99) for φ_I and φ_D separately in the form of

$$\varphi_I(\mathbf{x}) = \frac{1}{2\pi}\left[1 + \frac{1}{2\pi}\int_{S_w}\left(\frac{\partial G(\mathbf{x},\boldsymbol{\xi})}{\partial \mathbf{n}}\right)dS\right]^{-1} \times \int_{S_w} G(\mathbf{x},\boldsymbol{\xi})\cdot\mathbf{n}\cdot dS, \tag{3.113a}$$

$$\varphi_D(\mathbf{x}) = 2\left[1 + \frac{1}{2\pi}\int_{S_w}\left(\frac{\partial G(\mathbf{x},\boldsymbol{\xi})}{\partial \mathbf{n}}\right)dS\right]^{-1} \times \varphi_I(\mathbf{x}). \tag{3.113b}$$

The lower-order constant panel method or the higher-order panel method can be used to evaluate Green's function, as shown hereafter. In both cases, the floating body is discretised into N_e number of elements, also called panels.

3.6.1 Lower-Order Constant Panel Method

In the lower-order constant panel method (or simply the constant panel method), the evaluation of Green's function over the panel's area is executed by using the Gauss quadrature method as described in Appendix E. The constant panel method assumes that there is only one Gauss integration point in the centre of the panel, and therefore the integral is evaluated at $(\bar{\xi},\bar{\eta}) = (0,0)$ with weight $\bar{w}_k = 1$ (please refer to Appendix E).

The evaluation of the free surface Green's function can be described in matrix form. Referring to Figure 3.11, the surface S_W is discretised into N_e number of panels, where each panel is a quadrilateral rectangular element.

The evaluation of $G(\mathbf{x},\boldsymbol{\xi}) = G_{ij}$ requires the distribution of source points $\boldsymbol{\xi}$ at all the panels and evaluate their influences on the field points $\mathbf{x}$. In matrix form, this can be written as

$$\mathbf{g}_{N_e \times N_e} = G_{ij} = \begin{bmatrix} G_{11} & G_{12} & \cdots & G_{1,N_e} \\ G_{21} & G_{22} & \cdots & G_{2,N_e} \\ \vdots & \vdots & \ddots & \vdots \\ G_{N_e,1} & G_{N_e,2} & \cdots & G_{N_e,N_e} \end{bmatrix}, \tag{3.114}$$

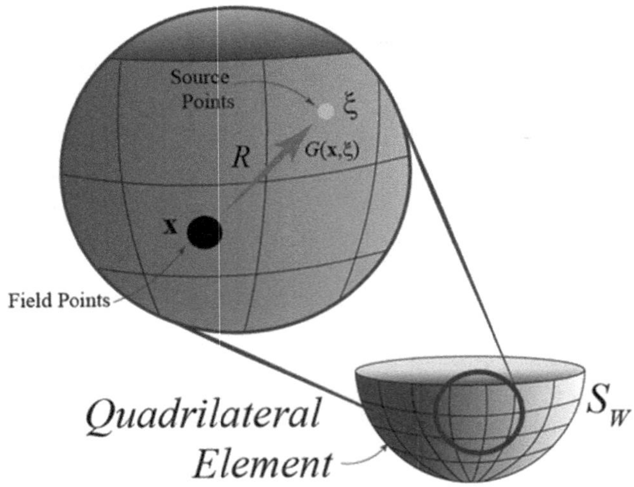

FIGURE 3.11
Discretisation of wetted surface into N_e number of panel and evaluation of $G(\mathbf{x},\boldsymbol{\xi})$ for low-order constant panel method.

where the first subscript for $G(\mathbf{x},\boldsymbol{\xi})$ in Equation (3.114) represents the source point $\mathbf{x}$, whereas the second subscript represents the field point $\boldsymbol{\xi}$. For example, G_{11} means the evaluation of source in source point 1 (or panel 1) on field point 1 (or panel point 1), G_{12} means the evaluation of source in source point 1 on field point 2, so on and so forth. The matrix $\mathbb{g}$ has a dimension of $N_e \times N_e$. From here onwards, the subscript of the matrix variable when indicated represents the size of the matrix, i.e., $\mathbb{g}_{N_e \times N_e}$ as in Equation (3.115).

The evaluation of $G(\mathbf{x},\xi)$ over the integral is simply computed by multiplying $\mathbb{g}$ with the area matrix $\mathbb{A}$, i.e.,

$$\int_{S_w} G(\mathbf{x},\boldsymbol{\xi}) \cdot \mathbf{n} \cdot dS = \sum_{i=1}^{N_e} \sum_{j=1}^{N_e} G_{ij} \Delta a_i = \mathbb{g}\mathbb{A} = \mathbb{G}, \tag{3.115}$$

where Δa_{ij} is the area for each panel (i, j) in the surface S_w.

Similarly, the evaluation of the derivative of the free surface Green's function following the steps in Appendix D is given as

$$\int_{S_w} \left(\frac{\partial G(\mathbf{x},\boldsymbol{\xi})}{\partial \mathbf{n}} \right) \cdot dS = \sum_{i=1}^{N_e} \sum_{j=1}^{N_e} \frac{\partial G_{ij}}{\partial \mathbf{n}} \Delta a_i = \frac{\partial \mathbb{G}}{\partial \mathbf{n}}. \tag{3.116}$$

3.6.2 Higher-Order Panel Method

The higher-order panel method on the other hand evaluates the free surface Green's function over each element by taking several Gauss integration points. For example, for the same quadrilateral element, if $(N_g)_s$ Gauss integration points are used for the source panels and $(N_g)_f$ Gauss integration points are used for the field panels, the evaluation of Green's function over the panels' area would produce a Green's function matrix $\mathbb{g}$ of size $(N_s \times N_f)$, where $N_s = N_e \times (N_g)_s$ and $N_f = N_e \times (N_g)_f$. For instance, if two-by-two integration points ($(N_g)_s = 4$) are taken for the source points and three-by-three ($(N_g)_f = 9$) for the field points (shown in Figure 3.12), this would produce a $\mathbb{g}$ matrix with size of $4N_e \times 9N_e$, i.e.,

$$\mathbb{g}_{N_s \times N_f} = \begin{bmatrix} \hat{\mathbb{g}}_{p1,p1} & \hat{\mathbb{g}}_{p1,p2} & \cdots & \hat{\mathbb{g}}_{p1,pN_e} \\ \hat{\mathbb{g}}_{p2,p1} & \hat{\mathbb{g}}_{p1,p2} & \cdots & \hat{\mathbb{g}}_{p2,pN_e} \\ \vdots & \vdots & \ddots & \vdots \\ \hat{\mathbb{g}}_{pN_e,p1} & \hat{\mathbb{g}}_{pN_e,p2} & \cdots & \hat{\mathbb{g}}_{pN_e,pN_e} \end{bmatrix}, \tag{3.117}$$

where $\hat{\mathbb{g}}(\xi, \mathbf{x})$ is the free surface Green's function matrix for each panel where $\boldsymbol{\xi}$ and $\mathbf{x}$ are the source and field points taken at the Gauss integration points. Therefore, each $\hat{\mathbb{g}}$ is a $(N_g)_s \times (N_g)_f$ matrix. The subscript for each $\hat{\mathbb{g}}$ denotes the influence of source panels (denoted by the first subscript) on the field panels (denoted by the second subscript). For example, $\hat{\mathbb{g}}_{p1,p2}$ implies the influence of sources at the Gauss integration points of panels p_1 has on the field points at Gauss integration point of panels p_2. In general, the $\hat{\mathbb{g}}$ for the influence of the source panel denoted as p_s and field panel denoted as p_f is given as

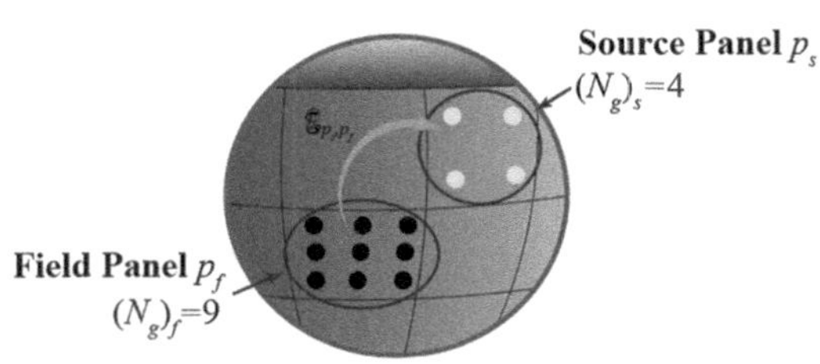

FIGURE 3.12
Evaluation of $\hat{\mathbb{g}}_{p_s,p_f}$ for HOBEM.

$$\left[\hat{\mathbb{g}}_{ps,pf}\right]_{(N_g)_S\times(N_g)_f}=\begin{bmatrix} G_{11} & G_{12} & \cdots & G_{1,(N_g)_f} \\ G_{21} & G_{22} & \cdots & G_{2,(N_g)_f} \\ \vdots & \vdots & \ddots & \vdots \\ G_{(N_g)_s,1} & G_{(N_g)_s,2} & \cdots & G_{(N_g)_S,(N_g)_f} \end{bmatrix}. \quad (3.118)$$

We can first evaluate the integration of $\hat{\mathbb{g}}$ over the area of one panel Δa by using the Gauss quadrature as presented in Appendix E. Writing Equation (3.115) for the evaluation of Green's function over the area of one panel, expressed in its natural coordinate ($\bar{\xi}$-$\bar{\eta}$-coordinate) and using Equation (E.4) in Appendix E,

$$\int_{\Delta a} G\left(\bar{\mathbf{x}},\bar{\xi}\right)\cdot\mathbf{n}\cdot d\Delta a=\int_{-1}^{1}\int_{-1}^{1} G\left(\bar{\mathbf{x}},\bar{\boldsymbol{\xi}}\right)\cdot d\bar{\xi}\,d\bar{\eta}=\sum_{j=1}^{N_{ps}}\sum_{k=1}^{N_{pf}}\bar{w}_j\cdot G\left(\bar{\mathbf{x}}_j,\bar{\boldsymbol{\xi}}_j\right)\cdot\bar{w}_k, \quad (3.119)$$

where $\bar{\mathbf{x}}=\left(\bar{x},\bar{y}\right)$ and $\bar{\boldsymbol{\xi}}=\left(\bar{\xi},\bar{\eta}\right)$ are the natural coordinates at the field and source, respectively, in a single panel.

Writing Equation (3.119) in matrix form gives us

$$\hat{\mathbb{G}}_{1\times1}=\left[\mathbb{w}_{ps}\right]_{1\times(N_g)_S}\times\left[\hat{\mathbb{g}}_{ps,pf}\right]_{(N_g)_S\times(N_g)_f}\times\left[\mathbb{w}_{pf}\right]_{(N_g)_f\times1} \quad (3.120)$$

where $\mathbb{w}_{p_s}$ is a $1\times\left(N_g\right)_S$ row vector with value equals to the Gauss quadrature weight $\bar{w}_j$ used for the source panel, whereas $\mathbb{w}_f$ is a $\left(N_g\right)_f\times1$ row vector with value equals to the weight $\bar{w}_k$ used for the field panel.

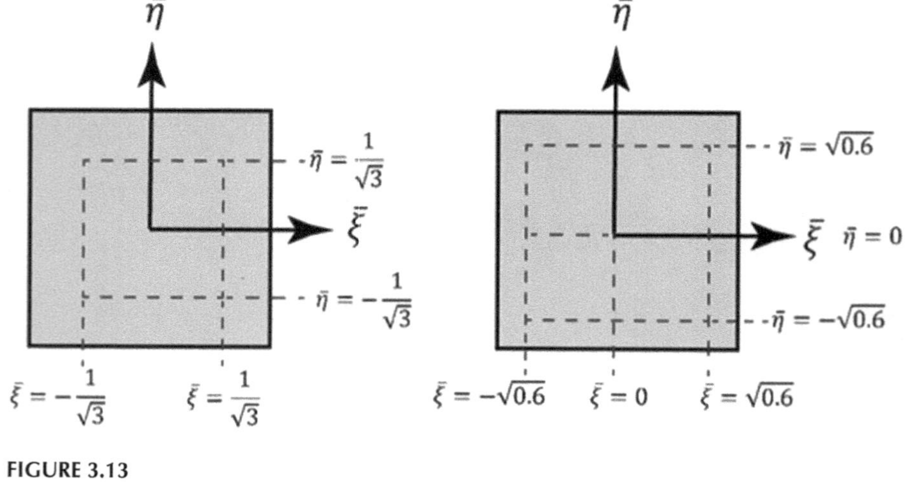

FIGURE 3.13
Higher-order panel method numerical scheme.

After obtaining $\hat{\mathbb{G}}$ for each panel, Green's function matrix is then assembled to yield

$$\mathbb{G}_{N_e \times N_e} = \begin{bmatrix} \hat{\mathbb{G}}_{11} & \hat{\mathbb{G}}_{12} & \cdots & \hat{\mathbb{G}}_{1,N_e} \\ \hat{\mathbb{G}}_{21} & \hat{\mathbb{G}}_{22} & \cdots & \hat{\mathbb{G}}_{2,N_e} \\ \vdots & \vdots & \ddots & \vdots \\ \hat{\mathbb{G}}_{N_e,1} & \hat{\mathbb{G}}_{N_e,2} & \cdots & \hat{\mathbb{G}}_{N_e,N_e} \end{bmatrix}, \tag{3.121}$$

in which $\mathbb{G}$ in Equation (3.121) has the same size as $\mathbb{G}$ in Equation (3.115) for the constant panel method.

3.6.3 Evaluation of Excitation Force and Added Mass, Radiated Damping

With the free surface Green's function evaluated, either using the constant panel method or higher-order panel method, $\mathbb{G}$ can then be used to obtain the radiated and diffracted potential, φ_I and φ_D, respectively, given in matrix form as

$$\{\varphi_I(\mathbf{x})\} = \frac{1}{2\pi}\left[\mathbb{I} + \frac{1}{2\pi}\frac{\partial \mathbb{G}}{\partial \mathbf{n}}\right]^{-1} \times \mathbb{G}, \tag{3.122a}$$

$$\{\varphi_D(\mathbf{x})\} = 2\left[\mathbb{I} + \frac{1}{2\pi}\frac{\partial \mathbb{G}}{\partial \mathbf{n}}\right]^{-1} \times \{\varphi_I\}, \tag{3.122b}$$

where the curl bracket $\{\blacksquare\}$ denotes the vector variable.

The excitation force matrix can be obtained from Equation (3.100) as

$$\mathbb{F}_E = \mathrm{i}\omega\rho\left\{2\left[\mathbb{I} + \frac{1}{2\pi}\frac{\partial \mathbb{G}}{\partial \mathbf{n}}\right]^{-1} \times \mathbb{G}\right\}. \tag{3.123}$$

The added mass and radiated damping are

$$\mathbb{M}_A = \mathrm{Re}\left\{\frac{\rho}{2\pi}\left[\mathbb{I} + \frac{1}{2\pi}\frac{\partial \mathbb{G}}{\partial \mathbf{n}}\right]^{-1} \times \mathbb{G}\right\}, \tag{3.124a}$$

$$\mathbb{B}_B = \mathrm{Im}\left\{\frac{\omega\rho}{2\pi}\left[\mathbb{I} + \frac{1}{2\pi}\frac{\partial \mathbb{G}}{\partial \mathbf{n}}\right]^{-1} \times \mathbb{G}\right\}. \tag{3.124b}$$

Similarly, the global matrices for mass and stiffness are given as

$$\mathbb{M} = \begin{bmatrix} m & m(r_g)_y & -m(r_g)_x \\ m(r_g)_y & I_{xx} & I_{xy} \\ -m(r_g)_x & I_{yx} & I_{yy} \end{bmatrix}, \mathbb{B} = \begin{bmatrix} k_z & 0 & 0 \\ 0 & k_{x'} & 0 \\ 0 & 0 & I_{y'} \end{bmatrix}, \mathbb{K} = \begin{bmatrix} k_z & 0 & 0 \\ 0 & k_{x'} & 0 \\ 0 & 0 & I_{y'} \end{bmatrix}, \tag{3.125a}$$

where $(r_g)_x$ and $(r_g)_y$ are the radius of gyration r_g in the x- and y-axes, respectively, given as

$$(r_g)_x = \sqrt{\frac{\sum_{j=1}^{n} m_j (r_x)_j^2}{\sum_{j=1}^{n} m_j}} \text{ and } (r_g)_y = \sqrt{\frac{\sum_{j=1}^{n} m_j (r_y)_j^2}{\sum_{j=1}^{n} m_j}}, \tag{3.126}$$

where m is the mass, r_x the x-distance measured from the centre of gravity and r_y the y-distance measured from the centre of gravity.

Therefore, we have the displacement solved using the equation of motion:

$$\{\mathbb{W}\} = \frac{\mathbb{F}_E}{-\omega^2(\mathbb{M} + \mathbb{M}_A) + \mathrm{i}\omega(\mathbb{B} + \mathbb{B}_{pto}) + \mathbb{K}}, \tag{3.127}$$

where $\mathbb{M}$, $\mathbb{B}$, $\mathbb{K}$, and $\mathbb{B}_{pto}$ are the global matrix for mass, damping, stiffness and PTO damping.

3.7 Case Example

To demonstrate the performance of the WEC, we consider a generic cylindrical WEC with a diameter $\varnothing = 4$ m and draft $d_{wec} = 1$ m similar to Figure 1.9(a). The WEC is subjected to regular and irregular wave analyses. A constant water depth $D = 10$ m is considered. The BEM described in Section 3.6 is used to assess the performance of the WEC. The following sections present the key assessment parameters, i.e., response amplitude operator (RAO), absorbed power P_a, capture width CW, and q-factor, used to assess the performance of the WECs. The wave elevation surrounding the WECs will also be presented.

3.7.1 Performance of Single WEC

3.7.1.1 Optimal PTO Damping

According to Equation (3.55), the optimal damping of the PTO $(B_{pto})_{opt}$ depends on the added mass $\mathcal{A}$ and radiated damping $\mathcal{B}$, which change with different wave periods (i.e., wave frequencies). The $(B_{pto})_{opt}$ for the cylindrical PA-WEC with different diameter $\varnothing$ and draft d_{wec} is presented in Figure 3.14.

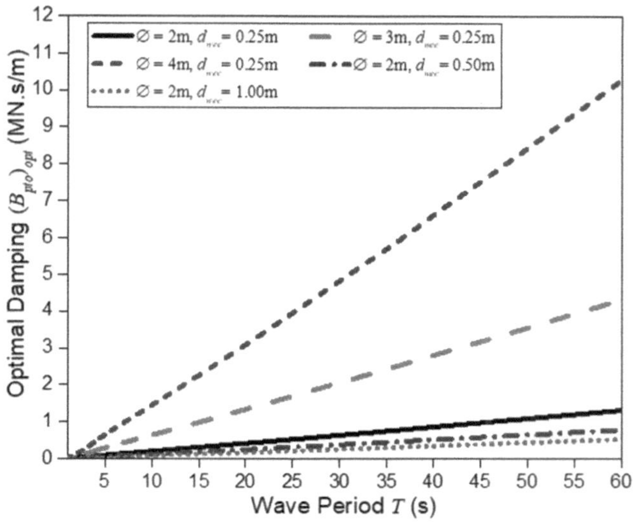

FIGURE 3.14
Optimal damping $(B_{pto})_{opt}$ for cylindrical PA-WEC with different diameter ∅ and draft d_{wec}.

3.7.1.2 RAO, Absorbed Power, and Capture Width

As using a variable $\left(B_{pto}\right)_{opt}$ is time-consuming, a constant $\left(B_{pto}\right)_{opt}$ is assumed for the WEC. The heave RAO is presented in Figure 3.15(a). Once the RAO is obtained, the corresponding P_a can be derived by using (3.57) and presented in Figure 3.15(b). In Figure 3.15, the $\left(B_{pto}\right)_{opt}$ evaluated at $T = 0$, 2s, 4s, 6s, 8s and at large T (i.e., $T = \infty$) is considered to study its effect on the heave RAO and P_a. It can be seen that the heave RAO increases when a smaller $\left(B_{pto}\right)_{opt}$ is used where the $\left(B_{pto}\right)_{opt}$ reduces with the decrease in the wave period. However, this does not translate to a corresponding larger P_a, as the absorbed power according to (3.55) depends on the $\left(B_{pto}\right)_{opt}$. For example, the heave RAO for a constant $\left(B_{pto}\right)_{opt}$ taken at T close to zero may be the largest, but this corresponds to $P_a = 0$ as $\left(B_{pto}\right)_{opt} = 0$. Similarly, the P_a with constant $\left(B_{pto}\right)_{opt}$ taken at $T = 2$s is the largest when the wave period is smaller than 12s but reduces when T exceeds 12s.

The $\left(B_{pto}\right)_{opt}$ as mentioned earlier depends on $\mathcal{A}$ and $\mathcal{B}$, which in turn are affected by the geometry of the WEC. Figure 3.16 shows the heave RAO and P_a for the cylindrical PA-WEC with various diameter ∅ and draft d_{wec} when the $\left(B_{pto}\right)_{opt}$ is taken at $T = 4$s. Figure 3.16(a) shows that the increase/decrease

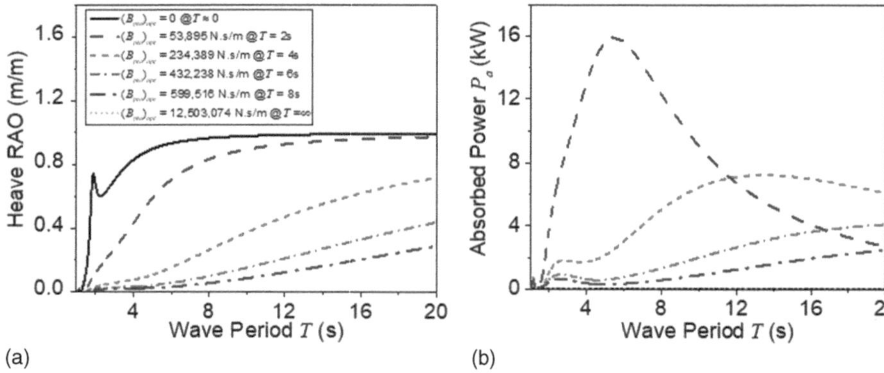

FIGURE 3.15
(a) Heave RAO and (b) Absorbed power of PA-WEC for different $(B_{pto})_{opt}$. $\varnothing = 4\text{m}$ and $d_{wec} = 0.5\text{m}$.

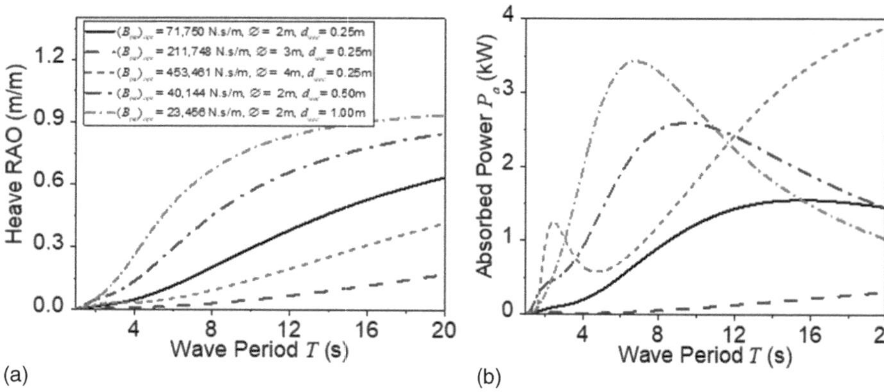

FIGURE 3.16
RAO and absorbed power for PA-WECs with different diameter $\varnothing$ and draft d_{wec}. $(B_{pto})_{opt}$ obtained at $T \approx 4\text{s}$.

of the heave RAO has a direct correlation with the $(B_{pto})_{opt}$ similar to that presented in Figure 3.15. However, there is no direct correlation between the RAO with the geometry of the WEC, as the computation of $(B_{pto})_{opt}$ depends on Equation (3.55), which varies unpredictably with $\mathcal{A}$ and $\mathcal{B}$. For example, the increase in $\varnothing$ for $d_{wec} = 0.25$ m does not result in a direct increase or decrease in $(B_{pto})_{opt}$, i.e., $(B_{pto})_{opt}$ for $\varnothing = 4$ m is higher for the counterpart for $\varnothing = 3\text{m}$ but lower than the counterpart for $\varnothing = 2$ m, as presented in Figure 3.14. Similarly, a higher heave RAO does not translate directly to a greater P_a according to Equation (3.57), as shown in Figure 3.16(b). Therefore, the choice of the constant $(B_{pto})_{opt}$ depends on the operational sea state where the WEC is operating so that the power generation can be optimised. Alternatively, a variable $(B_{pto})_{opt}$ as given in Figure 3.14, can be incorporated in the WEC by using a control system to maximise the power generation at various wave periods.

A means to measure the effectiveness of the WEC in wave energy extraction is via the capture width CW measurement. The CW expresses the wave power in terms of the width of wave fronts that is captured by the device and has a length unit (usually in metre). The CW of the WEC can be measured by using Equation (3.59), which expresses the ratio of the wave power to the wave resource. The corresponding CW of the same point absorber considered in Figure 3.16 is presented in Figure 3.17(a), which shows a similar trend as the P_a presented in Figure 3.16(b). By selecting an appropriate $(B_{pto})_{opt}$, CW of greater than 0.5 m can be achieved at certain wave periods. Note that the CW can be converted to capture width ratio CWR by dividing CW with the characteristic length of the WEC, which is the diameter ∅ of the WEC. Also, the CW can be used to quantify the wave energy extraction bandwidth of the WEC, e.g., WEC with $\varnothing = 2$ m and $d_{wec} = 1.00$ m may have a higher CW at lower wave periods but the CW decreases with the increase in wave periods. On the other hand, WEC with $\varnothing = 4$ m and $d_{wec} = 0.25$ m has a wider energy absorption bandwidth as compared to its counterpart.

The CWR can be expressed in ratio or percentage, as shown in Figure 3.17(b), which quantifies the effectiveness of the WEC in extracting wave energy, i.e., how much energy is being absorbed by the WEC from

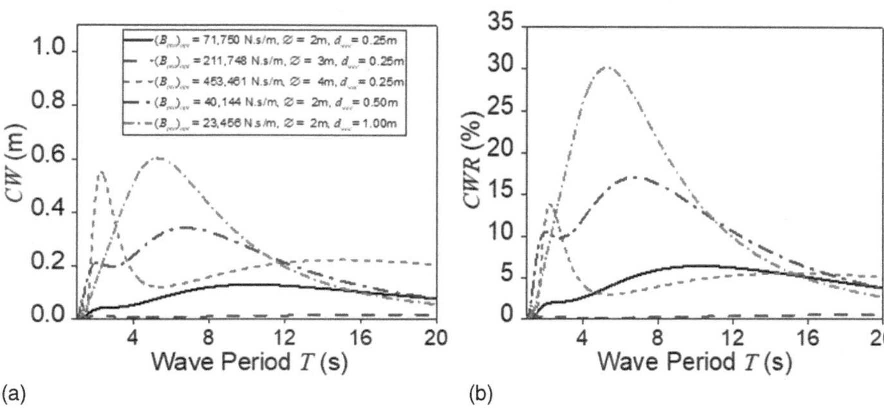

FIGURE 3.17
(a) Capture width and (b) capture width ratio for PA-WECs with different diameter ∅ and draft d_{wec}.

one wave front. By comparing Figures 3.17(a) and (b), the maximum *CW* for $\varnothing = 2$ m and $d_{wec} = 1.00$ m, as well as $\varnothing = 4$ m and $d_{wec} = 0.25$ m may be comparable, but the *CWR* shows that the former is more effective in extracting energy from the wave front.

3.7.1.3 Wave Elevation

The wave elevation surrounding the WEC can be computed using Equation (3.110), as shown in Figure 3.18. The wave climate can be used to study the impact of wave disturbances by the WEC has on the environment. Figure 3.18. shows the wave climate surrounding the three different types of WECs, i.e., point absorber, terminator, and attenuator. As can be seen from the figures, various types of WECs with different power generation mechanisms cause different wave disturbances. The wave elevation upstream of the WEC is generally higher due to the wave reflection from the WEC, whereas the wave elevation downstream is smaller due to the attenuation of wave force. The wave diffraction on the other hand varies according to the geometry and energy generation mechanism of the WEC.

3.7.1.4 Irregular Wave Analysis

By using the short-term wave analysis as presented in Sections 2.3 and 3.3.8, the performance of the WEC under irregular wave condition can be assessed. Figure 3.19(a) shows the response spectrum of the WEC computed from Equation (3.69) under the developing sea modelled using the JONSWAP wave spectrum. The developing sea corresponds to the local wind wave generated sea that is relevant to the tropical climate. The corresponding average $\overline{w}$ for different peak period T_p is summarised in the legend. The response spectrum presents a means to analyse the energy absorption

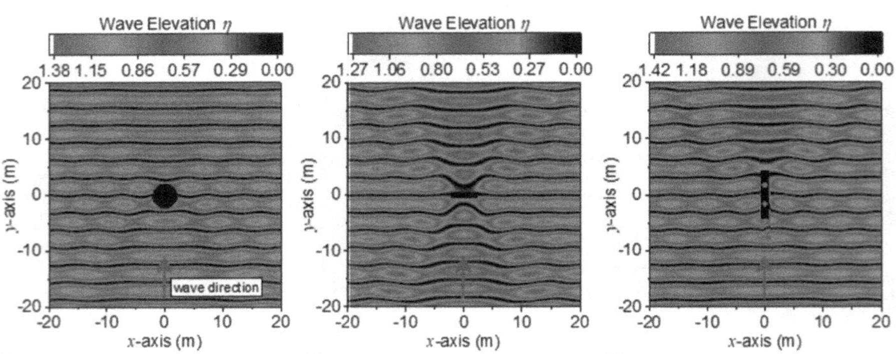

FIGURE 3.18
Wave elevation $\hat{\eta}$ surrounding (a) point absorber, (b) terminator, and (c) attenuator WECs.

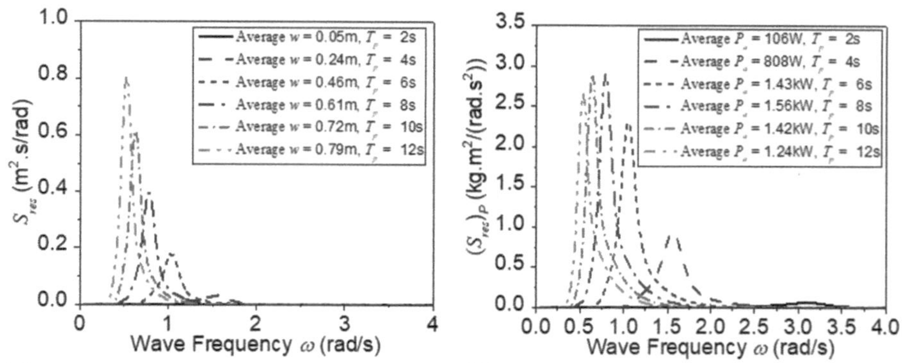

(a) (b)

FIGURE 3.19
Comparison of (a) average heave response w and (b) average absorbed power $\overline{P}_a$ for PA-WECs under JONSWAP (developing sea) wave spectrum. $\varnothing = 2\text{m}$ and $d_{wec} = 1\text{m}$. $H_S = 2\text{m}$.

bandwidth of the WEC under irregular wave conditions. For example, the peak in S_{res} represents the wave frequencies when the response of the WEC is maximum. For conventional floating structures, such as ship, oil platform, or VLFS (which will be presented in Chapter 4), the coincidence of floating bodies' natural frequencies with the dominant wave frequencies of the sea state is undesirable, as this would cause magnification of the amplitude due to resonance, thus the floating structures have to be designed with ω_N further away from these wave frequencies. The ω_N of some WECs, on the other hand, have to be tuned to the dominant wave frequencies of the sea state so that maximum wave energy can be generated. However, the structure integrity of the WEC has be taken into consideration at these sea states due to the amplification of the response during resonance. A wider bandwidth is also preferable as it indicates that the WEC has the capability to generate energy at a wider range of wave frequency.

By using Equation (3.70), the response spectrum of P_a for the same sea states considered in Figure 3.19(a) can be computed and presented in Figure 3.19(b). The corresponding average power of Equation (3.71) is shown in the legend of the figure. This allows us to quantify $\overline{P}_a$ that can be generated at a particular sea state. Figure 3.19(b) shows that although the peaks of $T_p = 8\text{s}$, 10s, and 12s are similar, the $\overline{P}_a$ for $T_p = 12\text{s}$ is the highest due to the wider power absorption bandwidth.

Figure 3.20 compares the S_{res} and $\left(S_{res}\right)_p$ for the same WEC as considered in Figure 3.19 under the fully developed sea (modelled using the BS wave spectrum). The fully developed sea usually has a longer wavelength but lower wave height, such as a swell that has been travelling thousands of kilometres before reaching the shore. The average P_a in a fully developed sea state (modelled by the BS wave spectrum) is lower as compared to its counterparts of a

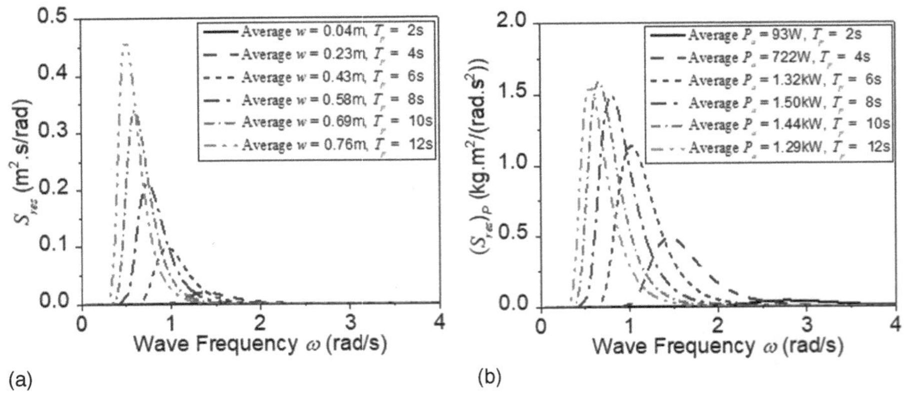

FIGURE 3.20
Comparison of (a) average heave response w and (b) average absorbed power $\overline{P_a}$ for PA-WECs under BS wave spectrum. $\varnothing = 2$ m and $d_{wec} = 1$ m. $H_s = 1$m.

developing sea (modelled by the JONSWAP wave spectrum) at lower wave peak periods. However, as the T_p increases, the average P_a under the BS spectrum is higher than that under the JONSWAP because of the wider power absorption bandwidth.

For the tropical climate, it is common that there exists the occurrence of swell (fully developed sea) and localised wind wave (developing sea) at the same time for the same coastal region. The Ochi-Hubble (OH) wave spectrum introduced in Section 2.3.6 can be used to model this sea state where there is a superposition of swell and wind wave. Figure 3.21 shows the average P_a (i.e., $\overline{P_a}$ in Equation (3.71)) and average w (i.e., $\overline{w}$ in Equation (3.69)) of the same WEC considered in the previous figure under the OH wave spectrum. It is assumed here that the swell has a wave height $H_{s1} = 4$ m and $T_{p1} = 10$ s, whereas the wind wave has a wave height $H_{s2} = 2$ m and $T_{p2} = 2$ s. The $\overline{P_a}$ and $\overline{w}$ under the OH wave spectrum are higher, as compared to the WEC under single peak wave spectrum (refer to Figures 3.19 and 3.20) as the wave energy is higher due to the superposition of swell and wind wave.

3.7.2 Performance of WEC Array

The WEC is usually deployed in array in order to increase the energy output at the allocated deployment sea space. However, it is important to ensure that the energy generation for the wave farm is optimised from the selected array configuration, as the wave reflection and diffraction from the WEC array may result in destructive or constructive interferences. The latter is preferred, as this indicates that the interaction between WEC in array results in greater energy generation, as compared to when they

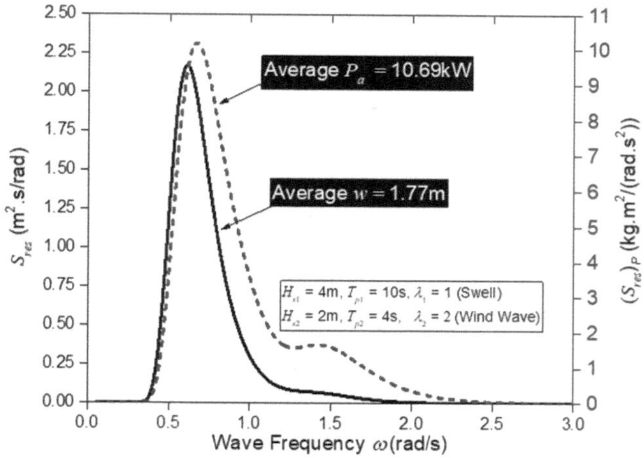

FIGURE 3.21
Comparison of absorbed power for PA-WECs under OH double peaked wave spectrum. ∅ = 2m and d_{wec} = 1m.

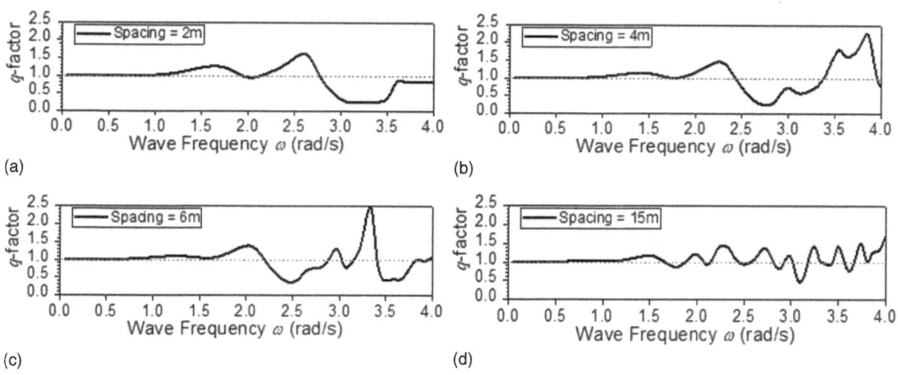

FIGURE 3.22
Comparison of q-factor for PA-WECs with different spacing under regular wave.

are deployed in isolation. The q-factor presented in Section 3.4.2 can be used to quantify the effectiveness of the array configuration in energy generation.

Figure 3.22 shows the q-factor for the PA-WEC array under different wave frequencies. Four different spacings are considered, i.e., spacing = 2, 4, 6, and 15 m in Figure 3.22(a) to (d), respectively. The q-factor greater than 1.0 is desirable as it implies that the average power generation from the array is greater than a single WEC at the same wave frequencies. Figure 3.22 shows that the q-factor of the WEC array varies significantly when the spacing is small but approaches 1.0 when the spacing is large, i.e., when spacing = 15 m, as shown in Figure 3.22(d). The q-factor at smaller wave frequencies is almost 1.0 for all case as the WECs in the array heave at the same amplitude and phase as the incoming waves, thereby causing little wave disturbances and interactions between the WECs. The investigation of the optimal spacing and array configuration is therefore important in the deployment of the wave farm in order to ensure that maximum wave energy can be extracted from the wave farm.

Similarly, the wave elevation surrounding the WEC arrays can also be computed using the same equation (3.110). The wave elevations downstream from the WEC arrays are a lot smaller as compared to their counterparts of a single WEC (see Figure 3.18) due to the shielding effect of the arrays. However, the wave elevations downstream of the attenuator WEC arrays are not as small as their counterparts for the point absorber and terminator. This is because the attenuator is always weathervane with the incoming wave direction.

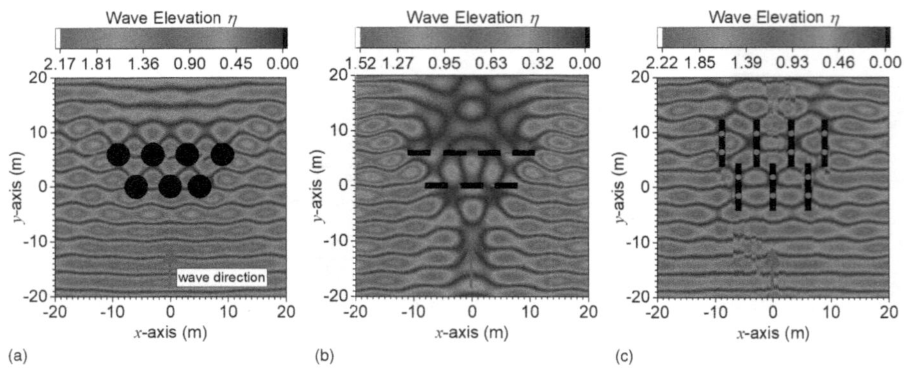

FIGURE 3.23
Wave elevation $\hat{\eta}$ surrounding arrays of point absorber, terminator, attenuator WECs.

Note: The hat in $\hat{\eta}$ is omitted in the figures.

4

Hydroelasticity of VLFS and Integration with WEC

This chapter first presents the mathematical formulation of hydroelastic response of a VLFS, followed by the numerical solution scheme using the coupled finite element boundary element (FE-BE) method. The state-of-the-art validation of the numerical code is then presented. This is followed by the evaluation of the performance of WEC when integrated with the floating structures. Two types of floating structures are considered here, i.e., floating breakwater and VLFS, where the vertical and horizontal type raft-type WECs are hinge connected to the structures. The integrated WEC and VLFS allow the conversion of wave energy into electricity while protecting the coast from erosion due to the wave impact. The effects of varying dimensions of WEC, mooring system, wave periods, and wave direction are studied to suggest the optimal design of the WEC to be most effective in generating energy and attenuating the wave force. Both regular and irregular waves of a tropical climate are considered. The q-factor and CWR of the WEC arrays are presented. In addition, the wave climate surrounding the floating structures are studied.

4.1 State of the Art

The motion of a VLFS differs from conventional floating bodies such as boats and WECs because the structural deformation of the former has to be considered in the hydrodynamic analysis.[119] The structure thus "deflects" under wave action, and such elastic deformation under hydrodynamic loading is termed "hydroelasticity." In this chapter, the mathematical modelling and numerical scheme for solving the hydroelastic response of VLFS are presented. The wave theory outlined in Chapter 2 and relevant hydrodynamics in Chapter 3 will be referred to when presenting the theory of hydroelasticity.

Similar to the hydrodynamic response of a rigid body presented in Chapter 3, the deformation of a VLFS could be assumed to be harmonic and oscillating with the wave frequency ω such that a steady-state response

DOI: 10.1201/9781003387855-4

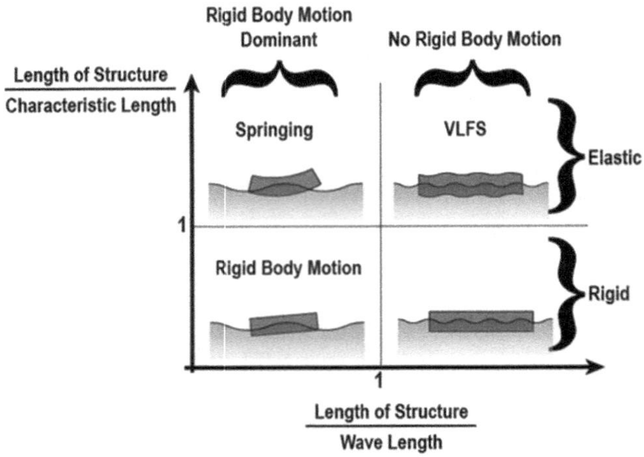

FIGURE 4.1
Definition of VLFS according to Suzuki et al.[123]

can be assumed to allow the solution of the hydroelastic analysis using the frequency domain approach. The frequency domain approach is computationally efficient and proven to predict good results particularly for long term response.[94] Usually, the frequency domain approach is used instead of the time domain when determining the hydroelastic response of the VLFS because of its simplicity and ability to capture the pertinent response parameters in a steady-state condition. Nevertheless, interested readers on transient response of VLFS such as motion due to aircraft landing or wave impact problems may refer to Kim and Webster,[120] Watanabe et al.,[121] and Kashiwagi[102,122] who have investigated the transient response using the time domain approach.

According to Suzuki et al.'s[123] definition of a VLFS, the hydroelastic response is only dominant when the following two ratios are larger than 1.0:

i. Ratio (i): Structural length/Wavelength (L/λ)
ii. Ratio (ii): Structural length/Characteristic length (L/λ_c)

The characteristic length λ_c is given as

$$\lambda_c = \sqrt[4]{4\pi^2 \frac{\mathcal{D}}{k}}, \tag{4.1}$$

where $\mathcal{D}$ is the flexural rigidity given as $\mathcal{D} = \frac{\mathcal{E} d_p^3}{12}$ and k is the spring constant of the hydrostatic restoring f $\mathcal{E}$ is the Young's modulus and d_p is the thickness of the VLFS modelled as plate.

4.2 Mathematical Formulation for Hydroelasticity

In modelling the hydroelastic response, the water and the structure are assumed to oscillate in a steady-state harmonic motion with the circular frequency ω. The water is modelled using the linear potential theory and the VLFS is assumed to be a pontoon-type (or mat-like) VLFS commonly modelled as a solid plate by using the plate theory. The detailed numerical model and mathematical formulations for the equation of motions of the fluid and plate are shown in the subsequent sections.

4.2.1 Pictorial Description of VLFS Model

Consider a pontoon-type VLFS with a waterplane area A_{wp} and thickness d_p as shown in Figure 4.2. The VLFS is commonly modelled as an equivalent solid plate with thickness d_p for the hydroelastic analysis. The VLFS is subjected to an incoming wave of wave period T and a wave height $H = 2A$ that impacts the structure at a wave angle θ with respect to the negative x-axis. The seabed is assumed to be levelled and the water depth is D.

The symbol Δ denotes the plate domain, whereas the water domain is denoted by Ω. The plates are assumed to be perfectly flat with free edges. Similar to the notations used in Chapter 3, the symbols S_F, S_B, S_W, and $S_{\pm\infty}$ are used to represent the free surface, the seabed, the wetted surface of the VLFS and the artificial boundary at infinity, respectively. Here, the wetted surface is further split into the bottom wetted surface S_{WB} and the side wetted surface S_{WS} of the VLFS, i.e., $S_W \ni S_{WB} \cup S_{WS}$. The free and undisturbed water surface is at $z = 0$.

As the body deformation has to be taken into consideration, a rigid body motion should not be assumed for the VLFS. Instead, the VLFS will deform as shown in Figure 4.3. The equivalent plate will deform by the plate displacement

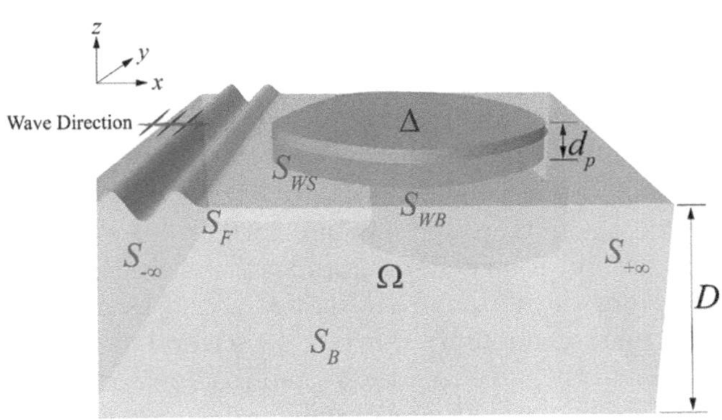

FIGURE 4.2
Figure depicting elevated view of arbitrary-shape VLFS.

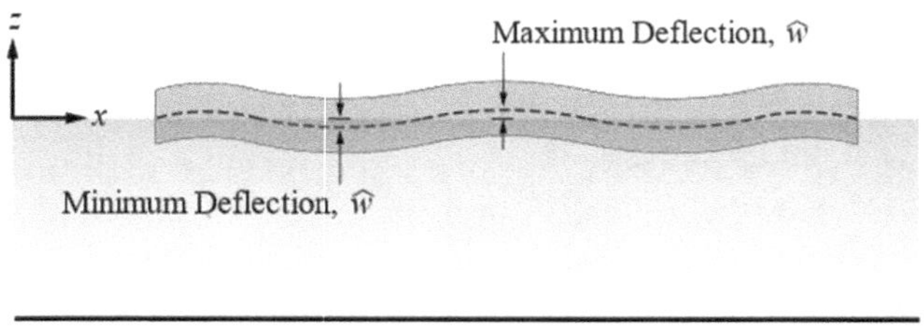

FIGURE 4.3
Deflection of VLFS.

$\hat{W}$, which includes the vertical deflection $\hat{w}(x,y)$, the rotation about the y-axis $\psi^x(x,y)$, and the rotation about the x-axis $\psi^y(x,y)$, i.e., $\hat{W}=(\hat{w},\psi^x,\psi^y)$ (more details in Section 4.2.2, please refer to Appendix A for the schematic diagram). Note that $\hat{W}$ and $\hat{w}$ are used for the hydroelastic response formulation to differentiate from the rigid body motion of a floating body, i.e., W and w.

4.2.2 Structure Motion

The box-like VLFS in Figure 4.2 is modelled as an equivalent solid plate. The equivalent plate has a Young's modulus $\mathcal{E}$ and Poisson's ratio ν, and is assumed to be perfectly flat with free edges. The plate has a thickness d_p, and its material is commonly assumed to be isotropic or orthotropic and obeys Hooke's law. In the hydroelastic analysis, the plate is assumed to be restrained in the $x-y$ plane by the station keeping system and the plate can only deform in the vertical direction (i.e., z-direction). Thus, when one uses the finite element method (FEM) for the plate analysis, the nodes of the elements are constrained from moving in the x-y direction (i.e., surge and sway) and are only allowed to move in the vertical z-direction.

For the hydroelastic analysis, it is common to model the plate according to the classical thin plate theory, where the plate motion is described solely by the vertical plate deflection $\hat{w}(x,y)$ and the plate rotation about x- and y-axes, i.e., θ_x and θ_y, are derived from $\hat{w}$. The vertical deflection $\hat{w}$ is measured from the free and undisturbed water surface, as shown in Figure 4.3.

More recently, the adoption of the Mindlin plate theory for equivalent floating plate is getting popular due to the theory's ability to provide a better prediction of the stress resultants and its allowance for the effects of transverse shear deformation and rotary inertia, unlike the classical thin plate theory where the equations of motion are described solely by the vertical deflection $\hat{w}(x,y)$, the rotation about the x-axis $\psi^y(x,y)$, and the rotation about the y-axis $\psi^x(x,y)$. The additional rotation variables allow for the effect of transverse shear deformation and also reduce the order of derivatives needed for computation of the stress resultants.

The thin plate and thick plate theory are presented in the subsequent chapters.

4.2.2.1 Thin Plate Theory

Kirchhoff-Love plate theory is a mathematical theory that describes the bending of thin plates. It is also known as the classical thin plate theory (CPT) devised by British mathematician Augustus Edward Hough Love (1863–1940) and developed based on assumption by German physicist Gustav Kirchhoff (1824–1887).

The governing equation of the CPT is given in Appendix A as

$$\mathcal{D}\left(\frac{\partial^4 \hat{w}}{\partial x^4}+2\frac{\partial^4 \hat{w}}{\partial x^2 \partial y^2}+\frac{\partial^4 \hat{w}}{\partial y^4}\right)=-\rho_p d_p \omega^2 \hat{w}+\mathrm{i}\rho_p \omega \varphi+\rho g \hat{w}, \tag{4.2}$$

where $(x,y)\in\Delta$, $\mathcal{G}=\frac{\mathcal{E}}{\left[2(1+\nu)\right]}$ is the shear modulus, $\mathcal{D}=\frac{\mathcal{E}d_p^3}{\left[12\left(1-\nu^2\right)\right]}$ the flexural rigidity, d_p the thickness of the plate, and ρ_p the mass density of the plate.

The rotation about the x- and y-axes can be derived from $\hat{w}$,

$$\theta^y = \frac{\partial \hat{w}}{\partial x}, \tag{4.3a}$$

$$\theta^x = \frac{\partial \hat{w}}{\partial y}. \tag{4.3b}$$

The stress resultants for the CPT are given as

$$M_{xx} = \mathcal{D}\frac{\partial^2 \hat{w}}{\partial x^2}, \tag{4.4a}$$

$$M_{yy} = \mathcal{D}\frac{\partial^2 \hat{w}}{\partial y^2}, \tag{4.4b}$$

$$M_{xy} = \mathcal{D}\frac{\partial^2 \hat{w}}{\partial x \partial y}, \tag{4.4c}$$

$$Q_x = \frac{\partial M_{xx}}{\partial x}+\frac{\partial M_{xy}}{\partial y}, \tag{4.4d}$$

$$Q_y = \frac{\partial M_{yy}}{\partial y}+\frac{\partial M_{xy}}{\partial x}, \tag{4.4e}$$

where M_{xx} and M_{yy} are the bending moment about the x- and y-axes, M_{xy} the twisting moment, Q_x and Q_y the shear forces in the x- and y-directions.

4.2.2.2 Thick Plate Theory

Mindlin (1951)[124] used the variational approach for deriving the governing equation of motion for the first-order shear deformable plate (or commonly known as Mindlin plate or the first-order shear deformation plate theory (FSDT)), which includes the effect of rotary inertia. The detailed mathematical formulation of the governing equation is given in Appendix B. The governing equations in time-harmonic motion for an isotropic plate are[125,126]

$$\kappa^2 \mathcal{G} d_p \left[\left(\frac{\partial^2 \hat{w}}{\partial x^2} + \frac{\partial^2 \hat{w}}{\partial y^2} \right) + \left(\frac{\partial \psi^x}{\partial x} + \frac{\partial \psi^y}{\partial y} \right) \right] = -\omega^2 \rho_p h_p \hat{w} + \mathrm{i}\omega\rho\varphi + \rho g \hat{w}, \quad (4.5a)$$

$$\begin{aligned} &\frac{\mathcal{D}(1-\nu)}{2}\left(\frac{\partial^2 \psi^x}{\partial x^2} + \frac{\partial^2 \psi^x}{\partial y^2} \right) + \frac{\mathcal{D}(1+\nu)}{2}\left(\frac{\partial^2 \psi^x}{\partial x^2} + \frac{\partial^2 \psi^y}{\partial x \partial y} \right) \\ &\qquad - \kappa^2 \mathcal{G} d_p \left(\psi^x + \frac{\partial \hat{w}}{\partial x} \right) = -\omega^2 \frac{\rho_p d_p^3}{12} \psi^x, \end{aligned} \quad (4.5b)$$

$$\begin{aligned} &\frac{\mathcal{D}(1-\nu)}{2}\left(\frac{\partial^2 \psi^y}{\partial x^2} + \frac{\partial^2 \psi^y}{\partial y^2} \right) + \frac{\mathcal{D}(1+\nu)}{2}\left(\frac{\partial^2 \psi^x}{\partial x \partial y} + \frac{\partial^2 \psi^y}{\partial y^2} \right) \\ &\qquad - \kappa^2 \mathcal{G} d_p \left(\psi^y + \frac{\partial \hat{w}}{\partial y} \right) = -\omega^2 \frac{\rho_p d_p^3}{12} \psi^y. \end{aligned} \quad (4.5c)$$

The RHS of Equation (4.5a) to Equation (4.5c) are the rotary inertia terms, whereas the shear correction factor κ^2 (usually taken as 5/6) is introduced to compensate for the error due to the assumption of a constant shear strain (and thus constant shear stress) through the plate thickness that violates the zero shear stress condition at the free surface.[125.]

Equation (4.5) may be expressed in the following compact matrix equation:

$$\left[\mathcal{D}(\mathbb{B}_1 + \mathbb{B}_2) + \omega^2 \rho_p \mathbb{B}_3 \right] \cdot \{\hat{\mathrm{w}}\} = \{\hat{\mathrm{f}}\}, \quad (4.6)$$

where $\{\hat{\mathrm{w}}\} = \{\hat{w}\ \psi^x\ \psi^y\}^{\mathrm{T}}$ are the displacement vectors and $\{\hat{\mathrm{f}}\} = \{\mathrm{i}\omega\rho_f\varphi + \rho_f g\hat{w} \quad 0 \quad 0\}$ the forcing vectors; where ρ_f is the fluid mass density. The differential operators $\mathbb{B}_1$ and $\mathbb{B}_2$ are given by

$$\mathbb{B}_1 = \frac{6\kappa^2(1-\nu)}{d_p^2} \begin{bmatrix} \frac{\partial^2}{\partial x^2} + \frac{\partial^2}{\partial y^2} & \frac{\partial}{\partial x} & \frac{\partial}{\partial y} \\ -\frac{\partial}{\partial x} & -1 & 0 \\ -\frac{\partial}{\partial y} & 0 & -1 \end{bmatrix}, \quad (4.7a)$$

$$\mathbb{B}_2 = \begin{bmatrix} 0 & 0 & 0 \\ 0 & \frac{1-\nu}{2}\left(\frac{\partial^2}{\partial x^2}+\frac{\partial^2}{\partial y^2}\right)+\frac{1+\nu}{2}\left(\frac{\partial^2}{\partial x^2}\right) & \frac{1+\nu}{2}\left(\frac{\partial^2}{\partial x \partial y}\right) \\ 0 & \frac{1+\nu}{2}\left(\frac{\partial^2}{\partial x \partial y}\right) & \frac{1-\nu}{2}\left(\frac{\partial^2}{\partial x^2}+\frac{\partial^2}{\partial y^2}\right)+\frac{1+\nu}{2}\left(\frac{\partial^2}{\partial y^2}\right) \end{bmatrix}, \tag{4.7b}$$

and finally, $\mathbb{B}_3$ is the matrix of constant multiplier

$$\mathbb{B}_3 = \begin{bmatrix} d_p & 0 & 0 \\ 0 & \frac{d_p^3}{12} & 0 \\ 0 & 0 & \frac{d_p^3}{12} \end{bmatrix}. \tag{4.8}$$

Based on the strain-displacement relation and assuming a plane-stress distribution in accordance with Hooke's law, the boundary conditions at the free edges are given by[125]

$$M_{xx} = \mathcal{D}\left(\frac{\partial \psi^x}{\partial x}+\nu\frac{\partial \psi^y}{\partial y}\right), \tag{4.9a}$$

$$M_{yy} = \mathcal{D}\left(\frac{\partial \psi^y}{\partial y}+\nu\frac{\partial \psi^x}{\partial x}\right), \tag{4.9b}$$

$$M_{xy} = \frac{\mathcal{D}(1-\nu)}{2}\left(\frac{\partial \psi^x}{\partial y}+\frac{\partial \psi^y}{\partial x}\right), \tag{4.9c}$$

$$Q_x = \kappa^2 \mathcal{G} d_p\left(\psi^x+\frac{\partial \hat{w}}{\partial x}\right), \tag{4.9d}$$

$$Q_y = \kappa^2 \mathcal{G} d_p\left(\psi^y+\frac{\partial \hat{w}}{\partial y}\right). \tag{4.9e}$$

4.2.3 Water Motion

4.2.3.1 Governing Equation and Boundary Conditions

Similar to the hydrodynamic analysis in Chapter 3, the water is assumed to be a perfect fluid with no viscosity and incompressible, and the fluid motion to be irrotational. Based on these assumptions, the fluid motion may be represented by a velocity potential $\varphi(x,y,z)$. Therefore φ must satisfy the Laplace equation[116,89]

$$\nabla^2\varphi = 0 \text{ in } \Omega. \tag{4.10}$$

The boundary conditions on the seabed and free surface for the VLFS are the same as that of a floating body in rigid body motion, as presented in Equations (3.33b) and (3.33c). For ease of reference, they are reproduced as follows.

$$\text{BC on Seabed: } \frac{\partial\varphi}{\partial z} = 0, \text{ at } z = -D \tag{4.11a}$$

$$\text{Free Surface BC: } \omega^2\varphi + g\frac{\partial\varphi}{\partial z} = 0, \text{ at } z = 0 \tag{4.11b}$$

However, there is a slight difference in the boundary conditions on the wetted surface. Following the boundary condition (3.33d), the boundary condition on the wetted surface S_W is

$$\text{BC on Wetted Surface: } \frac{\partial\varphi}{\partial \mathbf{n}} = \begin{cases} \mathrm{i}\omega\hat{W} & \text{on } S_{WB} \\ 0 & \text{on } S_{WS} \end{cases}, \tag{4.12}$$

where $\hat{W} = (\hat{w}, \theta^y, \theta^x)$ for CPT and $\hat{W} = (\hat{w}, \psi^y, \psi^x)$ for FSDT.

The boundary condition Equation (4.12) implies that the velocity of the wave particles at the bottom wetted surface of the floating body must be equivalent to the vertical velocity of the structure motion. This also means that the VLFS must always be in contact with the water, i.e., no air gap exists between the VLFS and the water.

Similar to the case of the rigid body motion, the velocity potential must also satisfy the radiation condition, which in our case for the three-dimensional domain uses the Sommerfeld radiation condition[127] at the artificial fluid boundary at infinity $S_{\pm\infty}$:

$$\lim_{|\mathbf{x}|\to\infty} \sqrt{|\mathbf{x}|}\left(\frac{\partial}{\partial|\mathbf{x}|} - \mathrm{i}k\right)\left[\varphi(x,y,z) - \varphi_I(x,y,z)\right] = 0, \tag{4.13}$$

where $\mathbf{x} = (x,y)$ and $\mathrm{i} = \sqrt{-1}$. The wave number k satisfies the dispersion relationship Equation (2.28) and φ_I is the incident velocity potential given in Equation (3.44).

4.2.3.2 Hydrostatic and Hydrodynamic Pressure

The hydrostatic and hydrodynamic pressure $P(x,y,z)$ acting on the bottom of the structure (i.e., $z = 0$) are given by the linearised Bernoulli equation, i.e.,

$$P(x,y,-D) = \mathrm{i}\omega\rho_f\varphi(x,y,-D) + \rho_f g\hat{w}(x,y). \tag{4.14}$$

4.2.4 Decoupling of Governing Equations Using Modal Expansion Method

The plate governing Equations (4.2) and (4.5) indicate that the response of the plate $\hat{w}$ is coupled with the fluid motion (or velocity potential $\varphi(x,y,z)$). On the other hand, the fluid motion can only be obtained when the plate deflection $\hat{w}(x,y)$ is specified in the boundary condition Equation (4.12). In order to decouple this interaction problem into a hydrodynamic problem in terms of the velocity potential and a plate vibration problem in terms of the generalised displacement, we shall adopt the modal expansion method as proposed by Newman,[92] which has been introduced in Section 3.5.1. In this method, the deflection of the plate $\hat{W}(x,y)$ is expanded by a series of the products of the modal functions $\mathcal{Z}_l$ and their corresponding complex amplitudes $\zeta_l(x,y)$:

$$\hat{W}(x,y) = \sum_{l=1}^{N_m} \mathcal{Z}_l(x,y) \cdot \zeta_l(x,y), \tag{4.15}$$

where N_m denotes the total number of modes taken in the plate analysis. $\mathcal{Z}_l = \left(\mathcal{Z}_l^w, \mathcal{Z}_l^{\theta^y}, \mathcal{Z}_l^{\theta^x}\right)$ and $\zeta_l = \left(\zeta_l^w, \zeta_l^{\theta^y}, \zeta_l^{\theta^x}\right)$ for CPT and $\mathcal{Z}_l = \left(\mathcal{Z}_l^w, \mathcal{Z}_l^{\psi^y}, \mathcal{Z}_l^{\psi^x}\right)$ and $\zeta_l = \left(\zeta_l^w, \zeta_l^{\psi^y}, \zeta_l^{\psi^x}\right)$ for FSDT.

As the problem is linear, the total velocity potential can be represented by a linear superposition of the diffracted part $\varphi_D(x,y,z)$ and the radiated part $\varphi_R(x,y,z)$, where $\varphi_D(x,y,z)$ is computed from the sum of the incident wave $\varphi_I(x,y,z)$ and scattered wave $\varphi_S(x,y,z)$ shown in Section 3.3.4. By using the modal expansion method, the total velocity potential $\varphi(x,y,z)$ in Equation (3.42) may be expressed as[128]

$$\varphi(x,y,z) = \varphi_D(x,y,z) + \varphi_R(x,y,z) = \varphi_D(x,y,z) + \mathrm{i}\omega \sum_{l=1}^{N_m} \mathcal{Z}_l(x,y) \cdot \varphi_l(x,y). \tag{4.16}$$

Here, $\varphi_{l=1,2,\ldots N_m}$ is the radiated potential corresponding to the unit-amplitude motion of the l-th modal function. Note that the complex amplitudes $\mathcal{Z}_l$ in Equation (4.16) are assumed to be the same values as $\mathcal{Z}_l$ in (4.15).[92]

By substituting Equations (4.15) and (4.16) into the Laplace Equation (4.10) and the fluid boundary conditions Equations (4.11) to (4.13), we arrive at the following decoupled governing equation and boundary conditions for each of the unit-amplitude radiation potentials (i.e., for $l = 1,2,\ldots,N_m$) and the diffraction potential (i.e., for $l = D$).

$$\nabla^2 \varphi_l = 0, \text{ in } \Omega \tag{4.17a}$$

$$\frac{\partial \varphi_l}{\partial z} = 0, \text{ on } S_B \tag{4.17b}$$

$$\frac{\partial \varphi_l}{\partial z} = \begin{cases} \mathrm{i}\omega \mathcal{Z}_l & \text{for } l = 1,2,\ldots,N_m \\ 0 & \text{for } l = D \end{cases}, \text{ on } S_{WB} \tag{4.17c}$$

$$\frac{\partial \varphi_l}{\partial z} = 0, \text{ on } S_{WS} \tag{4.17d}$$

$$g\frac{\partial \varphi_l}{\partial z} = -\omega^2 \varphi_l, \text{ on } S_{WB} \tag{4.17e}$$

$$\lim_{|\mathbf{x}|\to\infty} \sqrt{|\mathbf{x}|}\left(\frac{\partial}{\partial |\mathbf{x}|} - \mathrm{i}k\right)\left[\varphi_l(x,y,z) - \varphi_l(x,y,z)\right] = 0, \text{ on } S_{\pm\infty} \tag{4.17f}$$

The BVP for the diffracted potential and each of the unit-amplitude radiation potential are defined by Equations (4.17b) to (4.17f) in an uncoupled form. This BVP could be solved by using the BEM as presented in Section 3.6 in order to compute the velocity potential. The water-plate Equation (4.2) for CPT and Equation (4.5) for FSDT can then be solved by using the FEM once the velocity potential is obtained.

4.3 Numerical Scheme: BE-FE Method

4.3.1 Overview

Recall in the previous chapter that we first described the equation of motion of the water using the linear wave theory while the equation of motion for the plate by using either the CPT or the FSDT in a coupled form. We then employed the modal expansion method to decouple the plate deflections and water motions so that the BVP for the diffracted potentials and each of the unit-amplitude radiation can be represented in an uncoupled form.

The problem at hand is to devise numerical techniques to solve these decoupled equations of motion for the plate deflections and water motions (i.e., the velocity potential). There are several methods for solving the equations of motions. To name a few, Mamidupudi and Webster[129] have used the finite difference method for solving the governing equations of the plate (modelling of a pontoon-type floating airport) and Green's function for solving the fluid's Laplace equation. The eigenfunctions method (analytical method) has also been employed by researchers such as Wu et al.,[130] Nagata et al.,[131] and Ohmatsu[104] for determining the hydroelastic response of a VLFS. However, such analytical methods are only applicable for VLFSs with simple geometries such as circular and rectangular shapes.

In this chapter, the hybrid FE-BE method for solving the equations of motion for the water and plate is presented. The details on the FEM could be found in books by Petyt[132] and Cook et al.,[133] whereas details on BEM appear in books by Brebbia[134] and Becker.[135] The flow chart describing the algorithm of this numerical method is given in Figure 4.4. The FEM is used to solve the governing equations of motion for the plate because of its versatility in handling complicated plate geometries.

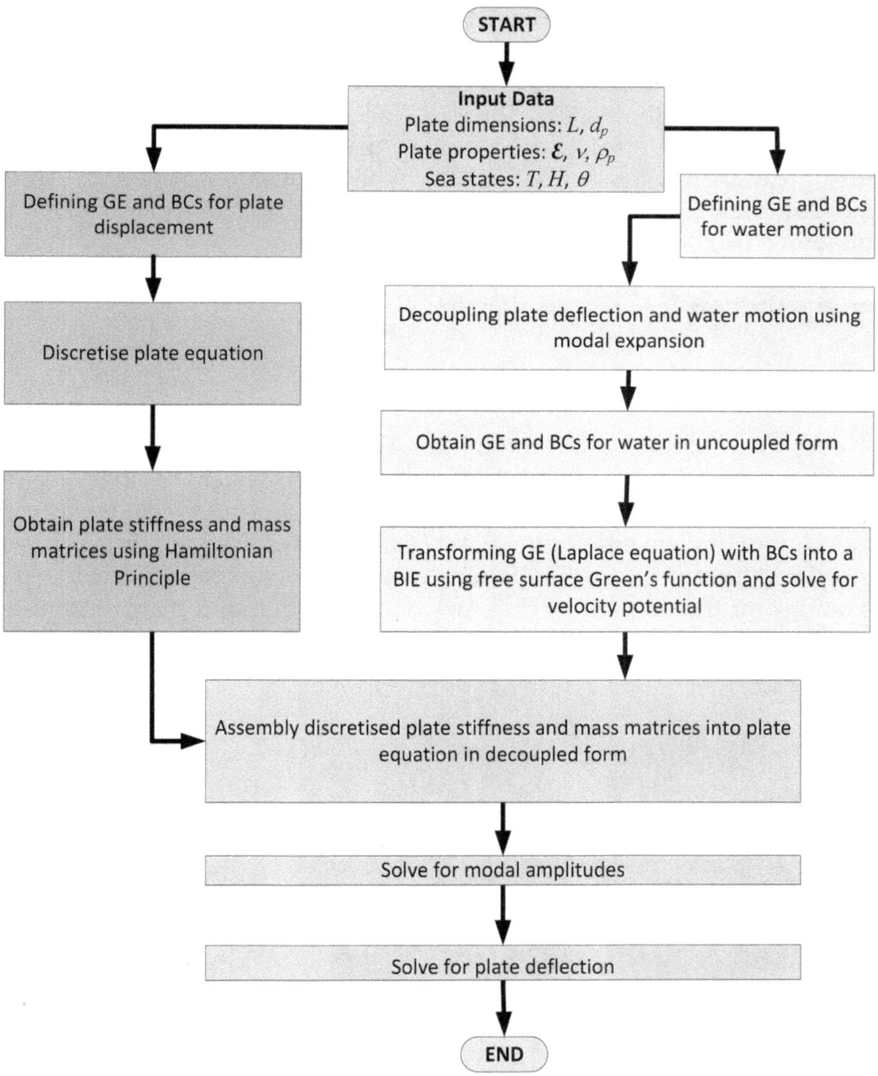

FIGURE 4.4
Flow chart describing algorithm for hybrid FE-BE method.

On the other hand, for determining the fluid motion, the BEM is adopted to transform the Laplace Equation (4.17a) together with the boundary conditions Equations (4.17b) to (4.17f) into a BIE. This means that only the boundaries of the computational fluid domain need to be discretised. The free surface Green's function derived by Linton[136] that automatically satisfies the boundary conditions of the fluid domain at the seabed, free surface and the Sommerfeld condition is used. So, the remaining unknown parameters to be determined for the fluid part are only those associated with the wetted surface of the floating body. The BEM thus significantly reduces the computational time needed for the hydroelastic analysis. The velocity potential of the water is obtained by solving the BIE. The computed velocity potentials are used to derive the exciting force as well as the added mass and added damping for the water-plate equations. By using the FEM, the modal amplitudes ζ_l can then be obtained. Together with the modal functions $\mathcal{Z}_l$, these modal amplitudes can be used to derive the plate deflections $\hat{W}$ as given in Equation (4.15).

In the sequel, we will present the method of solutions for the plate deflection and the flow field surrounding the VLFSs.

4.3.2 Discretisation of Plate

In order to use the FE-BE method, the solid plate has to be first discretised into N_e number of quadrilateral elements, with each element having N_d number of nodes thereby producing a total of N_D number of nodes in the model. The number of elements for the plate domain Δ and wetted surface S_W are being kept the same but with different integration points to solve the integral equations for the FEM and BEM. An example of the solid plate discretised into N_e number of elements is shown in Figure 4.5.

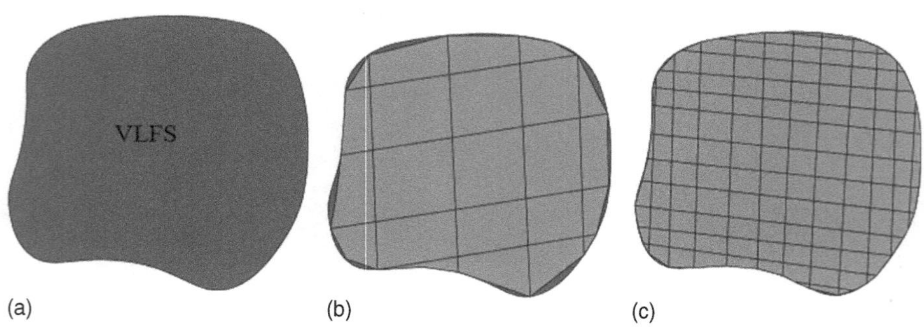

FIGURE 4.5
Discretisation of solid plate (VLFS) into a different number of quadrilateral elements of different sizes.

4.3.3 BEM for Solving Boundary Integral Equation

The solution to the velocity potential $\varphi(x,y,z)$ follows the same BEM scheme as shown in Section 3.5.5 and Section 3.6 where the Laplace equation is transformed into a BIE by imposing the Green's theorem. By substituting the boundary conditions Equation (4.17c) into Equation (3.98), we obtain

$$\varphi_k(\mathbf{x}) = \left[2\pi + \int_{S_w}\left(\frac{\partial G(\mathbf{x},\xi)}{\partial \mathbf{n}}\right)\cdot dS_W\right]^{-1} \times \begin{cases} \mathrm{i}\omega\int_{S_w} G(\mathbf{x},\xi)\cdot \mathcal{Z}_l\cdot\zeta_l\cdot dS_W & \text{for } k=R \\ 4\pi\varphi_I(\mathbf{x}) & \text{for } k=D \end{cases}. \tag{4.18}$$

Here, it is assumed that the floating body is discretised into N_e number of elements, also called as panels (as in the constant panel method). Writing Equation (3.99) for φ_l and φ_D separately in the form

$$\varphi_l(\mathbf{x}) = \frac{1}{2\pi}\left[1 + \frac{1}{2\pi}\int_{S_w}\left(\frac{\partial G(\mathbf{x},\xi)}{\partial \mathbf{n}}\right)\cdot dS_W\right]^{-1} \times \int_{S_w} G(\mathbf{x},\xi)\cdot \mathcal{Z}_l\cdot dS_W, \tag{4.19a}$$

$$\varphi_D(\mathbf{x}) = 2\left[1 + \frac{1}{2\pi}\int_{S_w}\left(\frac{\partial G(\mathbf{x},\xi)}{\partial \mathbf{n}}\right)\cdot dS_W\right]^{-1} \times \varphi_I(\mathbf{x}). \tag{4.19b}$$

Note here that φ_l is the unit-amplitude radiated potential having a unit of metre where the radiated potential φ_R has a unit of metre-square-per-second. Therefore, the complex amplitude ζ_l is omitted in the integral for φ_l in Equation (3.113a). It is to note that the BEM can easily handle the Sommerfeld radiation condition by using the free surface Green's function G Equation (3.82). For details on the transformation of the Laplace equation into a BIE using the free surface Green's function, please refer Section 3.5.4.

Similarly, by using Equations (3.113a) and (3.112b) in Section 3.6, the radiated and diffracted potentials, i.e., φ_l and φ_D, can then be re-written in the matrix form as

$$\{\varphi_l(\mathbf{x})\} = \frac{1}{2\pi}\left[\mathbb{I} + \frac{1}{2\pi}\frac{\partial \mathbb{G}}{\partial \mathbf{n}}\right]^{-1} \times \mathbb{G}\cdot\boldsymbol{\mathcal{Z}}_l, \text{ for } l = 1,2,\ldots,N_m, \tag{4.20a}$$

$$\{\varphi_D(\mathbf{x})\} = 2\left[\mathbb{I} + \frac{1}{2\pi}\frac{\partial \mathbb{G}}{\partial \mathbf{n}}\right]^{-1} \times \{\varphi_I\}, \tag{4.20b}$$

where $\mathbb{G}$ is the global matrix for the free surface Green's function given in Equation (3.120), $\mathbb{I}$ is the identity matrix, and $\mathcal{Z}_l$ the mode shapes (eigenvectors) obtained by performing a free vibration analysis on the plate.

4.3.4 FEM for Solving Plate Equation

The FEM is a powerful numerical method for solving differential equations in various fields of engineering and mathematical modelling. It involves breaking down a structure into smaller elements and then reconnecting them at nodes to simulate physical phenomena. The discretised model shown in Section 4.3.2 will be used to demonstrate the methodology. The generic displacement in an element could be represented by a summation of nodal displacement as

$$\{\mathbb{u}_e\}_{3N_d\times 3} = \mathbb{n}(\xi,\eta)\cdot\{\mathbb{u}_d\} = \sum_{j=1}^{N_d} N_j(\xi,\eta)\cdot(u_d)_j, \tag{4.21}$$

where $\mathbb{u}_e$ is the displacement for each element and $\mathbb{u}_d$ the displacement vector at each node. $\mathbb{n}$ is the basis function given in Appendix D which could be of a four-node or eight-node quadrilateral elements for our case study.[137] Note here that the vector $\{\mathbb{u}_d\}$ is given as

$$\{\mathbb{u}_d\}^{\mathrm{T}} = \left\{(u_d)_1 \quad (u_d)_2 \quad \cdots \quad (u_d)_{N_d}\right\}. \tag{4.22}$$

For the plate theory, each u_d comprises the plate deflection, rotation about x-axis, and rotation about y-axis, i.e.,

$$u_d = \hat{W} = \left(\hat{w},\theta^y,\theta^x\right)\text{for CPT}, \tag{4.23a}$$

$$u_d = \hat{W} = \left(\hat{w},\psi^y,\psi^x\right)\text{for FSDT}. \tag{4.23b}$$

Therefore, the elemental matrix size will be 12×12 and 24×24 for a four-node quadrilateral and an eight-node quadrilateral element, respectively. The $\mathbb{n}$ for element with N_d number of nodes is thus

$$\mathbb{n} = \mathrm{diag}\left[N_1 \quad N_1 \quad N_1 \quad N_2 \quad N_2 \quad N_2 \quad \cdots \quad N_{N_d} \quad N_{N_d} \quad N_{N_d}\right]_{3N_d\times 3N_d}, \tag{4.24}$$

where $N_1, N_2,\ldots,N_{N_d}$ are the basis functions for each node.

The displacement vectors are given as

$$\{\mathbb{u}_d\}^{\mathrm{T}} = \left\{\hat{w}_1 \quad \theta_1^y \quad \theta_1^x \quad \hat{w}_2 \quad \theta_2^y \quad \theta_2^x \quad \cdots \quad \hat{w}_{N_d} \quad \theta_{N_d}^y \quad \theta_{N_d}^x\right\}_{1\times 3N_d}\text{for CPT}, \tag{4.25a}$$

$$\{\mathbb{u}_d\}^{\mathrm{T}} = \left\{\hat{w}_1 \quad \psi_1^y \quad \psi_1^x \quad \hat{w}_2 \quad \psi_2^y \quad \psi_2^x \quad \cdots \quad \hat{w}_{N_d} \quad \psi_{N_d}^y \quad \psi_{N_d}^x\right\}_{1\times 3N_d}\text{for FSDT}. \tag{4.25b}$$

The global displacement $\mathbb{u}$ is then given as

$$[\mathbb{u}]_{3N_D\times 3} = [\mathbb{N}]_{3N_D\times 3N_D}\cdot\{\mathbb{u}_d\}_{3N_D\times 3}, \tag{4.26}$$

where $\mathbb{N}$ is the basis function arranged in global matrix.

The following chapters demonstrate the used of these basis functions in expanding the plate motion to obtain the elemental stiffness and mass matrices.

4.3.4.1 Classical Thin Plate Theory

4.3.4.1.1 Energy Functional in Matrices

The FEM requires the transformation of the energy functional involved in the plate theory (i.e., the potential energy functional U_E, kinetic energy functional T_E and work done by the forces on an element W_{nc}) into the matrix form. The evaluation of U_E and T_E will produce the structural and mass stiffnesses, respectively. Here we will demonstrate the transformation of energy functional into the matrices for the CPT and FSDT plate theory.

The generic formulation for U_E, T_E and W_{nc} are given as

$$U_E = \frac{1}{2}\mathcal{D}\int_{\Delta}\left[\left(\frac{\partial^2\hat{w}}{\partial x^2}\right)^2 + 2\left(\frac{\partial^2\hat{w}}{\partial x\partial y}\right)^2 + \left(\frac{\partial^2\hat{w}}{\partial y^2}\right)^2\right]\cdot d\Delta, \tag{4.27a}$$

$$T_E = -\frac{\omega^2}{2}\int_{\Delta}\rho_p d_p\left(\hat{w}\right)^2\cdot d\Delta, \tag{4.27b}$$

$$W_{nc} = \int_{\Delta}\left(\mathring{\imath}\rho_p\omega\varphi + \rho_p g\hat{w}\right)\cdot\hat{w}\cdot d\Delta, \tag{4.27c}$$

where the definition of the variables is given in Appendix A. The derivative of the energy functional can also be found in Appendix A.

4.3.4.1.1.1 Potential Energy Functional The potential energy functional U_E can be written in the matrix form as

$$U_E = \frac{1}{2}\int_{\Delta}\mathcal{D}\left[\left(\frac{\partial^2\hat{w}}{\partial x^2}\right)^T\left(\frac{\partial^2\hat{w}}{\partial x^2}\right) + 2\left(\frac{\partial^2\hat{w}}{\partial x\partial y}\right)^T\left(\frac{\partial^2\hat{w}}{\partial x\partial y}\right) + \left(\frac{\partial^2\hat{w}}{\partial y^2}\right)^T\left(\frac{\partial^2\hat{w}}{\partial y^2}\right)\right]\cdot d\Delta. \tag{4.28}$$

4.3.4.1.1.2 Kinetic Energy Functional Similarly, the kinetic energy functional can be written in matrix form as

$$\begin{aligned} T_E &= -\frac{\omega^2}{2}\int_{\Delta}\rho_p\cdot d_p\cdot\left(\hat{w}\right)^2\cdot d\Delta, \\ T_E &= -\frac{\omega^2\rho_p}{2}\int_{\Delta}\hat{w}^T\cdot d_p\cdot\hat{w}\cdot d\Delta. \end{aligned} \tag{4.29}$$

4.3.4.1.1.3 Non-conservative Work Done Lastly, the work done in matrix form:

$$W_{nc} = \mathrm{i}\rho_p\omega\int_{\Delta}\hat{w}^T\cdot\varphi\cdot d\Delta + \rho g\int_{\Delta}\hat{w}^T\cdot\hat{w}\cdot d\Delta, \tag{4.30}$$

where the first integral contributes to the excitation force F_E, added mass $\mathcal{A}$ and radiated damping $\mathcal{B}$ whereas the second integral contributes to the hydrostatic stiffness K.

4.3.4.1.2 Evaluation of Integral

We first evaluate the integration over the elemental area for U_E in Equation (4.28) where the approximate integration method as mentioned for the CPT in Appendix A will be used. However, instead of using the full integration method, the selective integration method is employed, where the full integration method (using the 3×3 Gauss quadrature method) is used to evaluate the first integral for U_E and the reduced integration method (using the 2×2 Gauss quadrature) for the second integral for U_E. The evaluation of integral using the Gauss quadrature is given in Appendix E where the evaluation of the integral for a function $f(\bar{\xi},\bar{\eta})$ in its natural coordinate $(\bar{\xi},\bar{\eta})$ is given as

$$\int_{\Gamma} f(x,y)d\Gamma = \int_{-1}^{1}\int_{-1}^{1} f(\bar{\xi},\bar{\eta})\cdot\left|J(\bar{\xi}_k,\bar{\eta}_k)\right|\cdot d\bar{\xi}\,d\bar{\eta} = \sum_{k=1}^{N_g} f(\bar{\xi}_k,\bar{\eta}_k)\cdot\left|J(\bar{\xi}_k,\bar{\eta}_k)\right|\cdot\varpi_k, \tag{4.31}$$

where $\left|J(\bar{\xi},\bar{\eta})\right|$ is the Jacobian given as

$$\det J = \begin{vmatrix} \dfrac{\partial x}{\partial\bar{\xi}} & \dfrac{\partial x}{\partial\bar{\eta}} \\ \dfrac{\partial y}{\partial\bar{\xi}} & \dfrac{\partial y}{\partial\bar{\eta}} \end{vmatrix}. \tag{4.32}$$

The Jacobian is used to transform the coordinate of the element from the cartesian coordinate into the natural coordinate. Cook et al.[133] have demonstrated that the use of the selective integration method is able to display the correct bending behaviour of the element (no shear locking phenomena) while at the same time having the contribution of normal strain in the bending modes to resist nodal loads (no spurious modes).

On the other hand, the integral for T_E will be evaluated using the full integration method. The evaluation of integral of the energy function will produce the elemental matrices for the finite element analysis. In order to obtain the elemental matrix for each functional, we need to represent the displacement involved in Equations (4.28)–(4.30) by the summation of nodal displacement using (4.21) as shown in the subsequent sub-sections.

4.3.4.1.2.1 Stiffness Matrices The stiffness matrix of the CPT can be derived from U_E in Equation (4.28). By using the shape function as described in Equation (4.21), U_E in Equation (4.28) expressed in the matrix from in its natural coordinates yields

$$U_E = \hat{w}^T \cdot \left\{ \frac{D}{2} \int_{-1}^{1}\int_{-1}^{1} \left[\left(\frac{\partial^2 \mathbb{n}}{\partial x^2}\right)^T \left(\frac{\partial^2 \mathbb{n}}{\partial x^2}\right) + 2\left(\frac{\partial^2 \mathbb{n}}{\partial x \partial y}\right)^T \left(\frac{\partial^2 \mathbb{n}}{\partial x \partial y}\right) + \left(\frac{\partial^2 \mathbb{n}}{\partial y^2}\right)^T \left(\frac{\partial^2 \mathbb{n}}{\partial y^2}\right) \right] \cdot \left|J\left(\xi_k, \eta_k\right)\right| \cdot d\xi \, d\eta \right\} \cdot \hat{w}^T. \tag{4.33}$$

Please refer to Appendix E on the transformation from cartesian coordinates to natural coordinates. The integral for U_E can then be evaluated using the Gauss quadrature method given in the Appendix E as,

$$U_E = \hat{w}^T . \left\{ \sum_{k=1}^{N_g} \frac{D}{2} \left[\begin{array}{l} \left(\dfrac{\partial^2 \mathbb{n}\left(\bar{\xi}_k, \bar{\eta}_k\right)}{\partial \bar{\xi}^2}\right)^T \left(\dfrac{\partial^2 \mathbb{n}\left(\bar{\xi}_k, \bar{\eta}_k\right)}{\partial \bar{\xi}^2}\right) \\ + 2\left(\dfrac{\partial^2 \mathbb{n}\left(\bar{\xi}_k, \bar{\eta}_k\right)}{\partial \bar{\xi} \partial \bar{\eta}}\right)^T \left(\dfrac{\partial^2 \mathbb{n}\left(\bar{\xi}_k, \bar{\eta}_k\right)}{\partial \bar{\xi} \partial \bar{\eta}}\right) \\ + \left(\dfrac{\partial^2 \mathbb{n}\left(\bar{\xi}_k, \bar{\eta}_k\right)}{\partial \bar{\eta}^2}\right)^T \left(\dfrac{\partial^2 \mathbb{n}\left(\bar{\xi}_k, \bar{\eta}_k\right)}{\partial \bar{\eta}^2}\right) \end{array} \right] . \left|J\left(\bar{\xi}_k, \bar{\eta}_k\right)\right| . \bar{w}_k \right\} . \hat{w}, \tag{4.34}$$

where the terms in the curl bracket give the elemental bending stiffness matrices,

$$\mathbb{k}^b = \sum_{k=1}^{N_g} \frac{D}{2} \left[\begin{array}{l} \left(\dfrac{\partial^2 \mathbb{n}\left(\bar{\xi}_k, \bar{\eta}_k\right)}{\partial \bar{\xi}^2}\right)^T \left(\dfrac{\partial^2 \mathbb{n}\left(\bar{\xi}_k, \bar{\eta}_k\right)}{\partial \bar{\xi}^2}\right) \\ + 2\left(\dfrac{\partial^2 \mathbb{n}\left(\bar{\xi}_k, \bar{\eta}_k\right)}{\partial \bar{\xi} \partial \bar{\eta}}\right)^T \left(\dfrac{\partial^2 \mathbb{n}\left(\bar{\xi}_k, \bar{\eta}_k\right)}{\partial \bar{\xi} \partial \bar{\eta}}\right) \\ + \left(\dfrac{\partial^2 \mathbb{n}\left(\bar{\xi}_k, \bar{\eta}_k\right)}{\partial \bar{\eta}^2}\right)^T \left(\dfrac{\partial^2 \mathbb{n}\left(\bar{\xi}_k, \bar{\eta}_k\right)}{\partial \bar{\eta}^2}\right) \end{array} \right] \cdot \left|J\left(\bar{\xi}_k, \bar{\eta}_k\right)\right| . \bar{w}_k, \tag{4.35}$$

where N_g is the number of Gauss interpolation points.

4.3.4.1.2.2 Mass Matrix The mass matrix can be derived from the kinetic energy functional T_E Equation (4.29). By representing the displacement using Equation (4.21), T_E can be written in the natural coordinates as

$$T_E = -\hat{w}^T \cdot \left\{ \frac{\omega^2 \rho_p d_p}{2} \int_{-1}^{1}\int_{-1}^{1} \mathrm{n}^T \cdot \mathrm{n} \cdot \left| J\left(\bar{\xi}_k, \bar{\eta}_k\right) \right| \cdot d\bar{\xi}\, d\bar{\eta} \right\} \cdot \hat{w}. \tag{4.36}$$

The integral can then be solved using the Gauss quadrature method which yields

$$T_E = -\omega^2 \cdot \hat{w}^T \cdot \left\{ \frac{\rho_p d_p}{2} \sum_{k=1}^{N_g} \mathrm{n}\left(\bar{\xi}_k, \bar{\eta}_k\right)^T \cdot \mathrm{n}\left(\bar{\xi}_k, \bar{\eta}_k\right) \cdot \left| J\left(\bar{\xi}_k, \bar{\eta}_k\right) \right| \cdot \bar{w}_k \right\} \cdot \hat{w}. \tag{4.37}$$

The elemental mass matrix is given by the terms in the curl bracket, i.e.,

$$\mathbb{m} = \frac{\omega^2 \rho_p h_p}{2} \sum_{k=1}^{N_g} \mathrm{n}\left(\bar{\xi}_k, \bar{\eta}_k\right)^T \cdot \mathrm{n}\left(\bar{\xi}_k, \bar{\eta}_k\right) \cdot \left| J\left(\bar{\xi}_k, \bar{\eta}_k\right) \right| \cdot \bar{w}_k. \tag{4.38}$$

4.3.4.2 First-Order Shear Deformation Plate Theory

4.3.4.2.1 Energy Functional in Matrices

Following similar procedures, the potential energy functional U_E, kinetic energy function T_E and non-conservative work done W_{nc} for FSDT are given as

$$U_E = \frac{1}{2} \int_{\Delta} \left\{ \begin{array}{l} \mathcal{D} \left[\left(\dfrac{\partial \psi^x}{\partial x} + \nu \dfrac{\partial \psi^y}{\partial y} \right)^2 + \left(\dfrac{\partial \psi^y}{\partial y} + \nu \dfrac{\partial \psi^x}{\partial x} \right)^2 \right] \\ + \dfrac{\mathcal{D}(1-\nu)}{2} \left(\dfrac{\partial \psi^x}{\partial y} + \dfrac{\partial \psi^y}{\partial x} \right)^2 \\ + \kappa^2 \mathcal{G} d_p \left[\left(\psi^x + \dfrac{\partial \hat{w}}{\partial x} \right)^2 + \left(\psi^y + \dfrac{\partial \hat{w}}{\partial y} \right)^2 \right] \end{array} \right\} \cdot d\Delta, \tag{4.39a}$$

$$T_E = -\frac{\omega^2 \rho_p}{2} \int_{\Delta} \left\{ d_p \left(\hat{w}\right)^2 + \frac{d_p^3}{12} \left[\left(\psi^x\right)^2 + \left(\psi^y\right)^2 \right] \right\} \cdot d\Delta, \tag{4.39b}$$

$$W_{nc} = \int_{\Delta} \left(\mathrm{i} \rho_p \omega \cdot \varphi + \rho_p g \cdot \hat{w} \right) \cdot \hat{w} \cdot d\Delta, \tag{4.39c}$$

where the definition of the variables is given in Appendix B. The derivative of the energy functional can also be found in Appendix B.

4.3.4.2.2 Potential Energy Functional

The potential energy function U_E Equation (4.39a) can be written in the matrix form as

$$U_E = \frac{1}{2}\int_{\Delta} \frac{d_p^3}{12}\cdot\left(\chi^f\right)^T\cdot\mathbb{D}^f\cdot\left(\chi^f\right)\cdot d\Delta + \frac{1}{2}\int_{\Delta}\kappa^2 h_p\cdot\left(\chi^s\right)^T\cdot\mathbb{D}^s\cdot\left(\chi^s\right)\cdot d\Delta, \tag{4.40}$$

where

$$\chi^f = \begin{bmatrix} \frac{\partial\psi_1^x}{\partial x} & 0 & 0 & & \frac{\partial\psi_{N_d}^x}{\partial x} & 0 & 0 \\ 0 & \frac{\partial\psi_1^y}{\partial y} & 0 & \cdots & 0 & \frac{\partial\psi_{N_d}^y}{\partial y} & 0 \\ 0 & 0 & \frac{\partial\psi_1^x}{\partial y}+\frac{\partial\psi_1^y}{\partial x} & & 0 & 0 & \frac{\partial\psi_{N_d}^x}{\partial y}+\frac{\partial\psi_1^y}{\partial x} \end{bmatrix}_{3\times 3N_d}, \tag{4.41}$$

$$\chi^s = \begin{bmatrix} \psi_1^x+\frac{\partial\hat{w}_1}{\partial x} & 0 & & \psi_{N_d}^x+\frac{\partial\hat{w}_{N_d}}{\partial x} & 0 \\ 0 & \psi_1^y+\frac{\partial\hat{w}_1}{\partial y} & \cdots & 0 & \psi_{N_d}^y+\frac{\partial\hat{w}_{N_d}}{\partial y} \end{bmatrix}_{2\times 2N_d}, \tag{4.42}$$

$$\mathbb{D}^f = \frac{\mathcal{E}}{\left(1-\nu^2\right)}\begin{bmatrix} 1 & \nu & 0 \\ \nu & 1 & 0 \\ 0 & 0 & \frac{1-\nu}{2} \end{bmatrix}_{3\times 3}, \quad \mathbb{D}^s = \frac{\mathcal{E}}{2(1+\nu)}\begin{bmatrix} 1 & 0 \\ 0 & 1 \end{bmatrix}_{2\times 2}. \tag{4.43}$$

4.3.4.2.3 Kinetic Energy Functional

Similarly, the kinetic energy functional Equation (4.39b) can be written in matrix form as

$$T_E = -\frac{\omega^2\rho_p}{2}\int_{\Delta}\left\{\mathbb{u}_e\right\}^T\cdot\mathbb{D}^m\cdot\left\{\mathbb{u}_e\right\}\cdot d\Delta, \tag{4.44}$$

where $\mathbb{u}_e$ is given in Equation (4.21), and

$$\mathbb{D}^m = \begin{bmatrix} d_p & 0 & 0 \\ 0 & \frac{d_p^3}{12} & 0 \\ 0 & 0 & \frac{d_p^3}{12} \end{bmatrix}_{3\times3}. \tag{4.45}$$

4.3.4.2.4 Non-conservative Work Done

Lastly, the work done in Equation (4.39c) in matrix form is

$$W_{nc} = \mathbb{i}\rho_p\omega\int_\Delta \{\mathbb{u}_e\}^T \cdot \varphi \cdot d\Delta + \rho_p g\int_\Delta \{\mathbb{u}_e\}^T \cdot \{\mathbb{u}_e\} \cdot d\Delta, \tag{4.46}$$

where the first integral contributes to the excitation force F_E, added mass $\mathcal{A}$, and radiated damping $\mathcal{B}$, whereas the second integral contributes to the hydrostatic stiffness K.

4.3.4.2.5 Evaluation of Integral

4.3.4.2.5.1 Stiffness Matrices The stiffness matrix of the FSDT can be derived from U_E in Equation (4.40). The vector $\{\chi^f\}$ and $\{\chi^s\}$ in Equations (4.41) and (4.42), respectively, can be represented by the summation of nodal displacement as

$$\chi^f = \mathbb{B}^f \cdot \{\mathbb{u}_d\} = \begin{bmatrix} 0 & \frac{\partial N_1}{\partial \bar{\xi}} & 0 & & 0 & \frac{\partial N_{N_d}}{\partial \bar{\xi}} & 0 \\ 0 & 0 & \frac{\partial N_1}{\partial \bar{\eta}} & \dots & 0 & 0 & \frac{\partial N_{N_d}}{\partial \bar{\eta}} \\ 0 & \frac{\partial N_1}{\partial \bar{\eta}} & \frac{\partial N_1}{\partial \bar{\xi}} & & 0 & \frac{\partial N_{N_d}}{\partial \bar{\eta}} & \frac{\partial N_{N_d}}{\partial \bar{\xi}} \end{bmatrix}_{3\times3N_d} \times \mathrm{diag}\begin{Bmatrix} \hat{w}_1 \\ \psi_1^x \\ \psi_1^y \\ \vdots \\ \hat{w}_{N_d} \\ \psi_{N_d}^x \\ \psi_{N_d}^y \end{Bmatrix}^T_{3N_d\times3N_d}, \tag{4.47}$$

$$\chi^s = \mathbb{B}^s \cdot \{\mathbb{u}_d\} = \begin{bmatrix} \frac{\partial N_1}{\partial \bar{\xi}} & N_1 & 0 & & \frac{\partial N_1}{\partial \bar{\xi}} & N_1 & 0 \\ \frac{\partial N_1}{\partial \bar{\eta}} & 0 & N_1 & \dots & \frac{\partial N_1}{\partial \bar{\eta}} & 0 & N_1 \end{bmatrix}_{2\times3N_d} \times \mathrm{diag}\begin{Bmatrix} \hat{w}_1 \\ \psi_1^x \\ \psi_1^y \\ \vdots \\ \hat{w}_{N_d} \\ \psi_{N_d}^x \\ \psi_{N_d}^y \end{Bmatrix}^T_{3N_d\times3N_d}. \tag{4.48}$$

Substituting Equations (4.47) and (4.48) into Equation (4.40), U_E can be represented in natural coordinate as

$$\begin{aligned} U_E = & \frac{1}{2}\int_{-1}^{1}\int_{-1}^{1}\frac{d_p^3}{12}\cdot\left\{\chi^f\left(\bar{\xi}_k,\bar{\eta}_k\right)\right\}^T\cdot\mathbb{D}^f\cdot\left\{\chi^f\left(\bar{\xi}_k,\bar{\eta}_k\right)\right\}\cdot\left|J\left(\bar{\xi}_k,\bar{\eta}_k\right)\right|d\bar{\xi}\,d\bar{\eta} \\ & +\frac{1}{2}\int_{-1}^{1}\int_{-1}^{1}\kappa^2 d_p\cdot\left\{\chi^T\left(\bar{\xi}_k,\bar{\eta}_k\right)\right\}\cdot\mathbb{D}^s\cdot\left\{\chi^s\left(\bar{\xi}_k,\bar{\eta}_k\right)\right\}\cdot\left|J\left(\bar{\xi}_k,\bar{\eta}_k\right)\right|d\bar{\xi}\,d\bar{\eta}. \end{aligned} \tag{4.49}$$

By evaluating the integral using the Gauss quadrature method (4.31),

$$\begin{aligned} U_E = & \frac{1}{2}\cdot\sum_{k=1}^{N_g}\frac{d_p^3}{12}\cdot\left[\chi^f\left(\bar{\xi}_k,\bar{\eta}_k\right)\right]^T\cdot\mathbb{D}^f\cdot\left[\chi^f\left(\bar{\xi}_k,\bar{\eta}_k\right)\right]\cdot\left|J\left(\bar{\xi}_k,\bar{\eta}_k\right)\right|\cdot\varpi_k \\ & +\frac{1}{2}\sum_{k=1}^{N_g}\kappa^2\cdot d_p\cdot\left[\chi^s\left(\bar{\xi}_k,\bar{\eta}_k\right)\right]^T\cdot\mathbb{D}^s\cdot\left[\chi^s\left(\bar{\xi}_k,\bar{\eta}_k\right)\right]\cdot\left|J\left(\bar{\xi}_k,\bar{\eta}_k\right)\right|\cdot\varpi_k. \end{aligned} \tag{4.50}$$

Equation (**4.50**) can be further written as

$$\begin{aligned} U_E = & \frac{1}{2}\cdot\{\mathbb{u}_d\}^T\cdot\left\{\sum_{k=1}^{N_g}\frac{d_p^3}{12}\cdot\left[\mathbb{B}^f\left(\bar{\xi}_k,\bar{\eta}_k\right)\right]^T\cdot\mathbb{D}^f\cdot\mathbb{B}^f\left(\bar{\xi}_k,\bar{\eta}_k\right)\cdot\left|J\left(\xi_k,\eta_k\right)\right|\cdot\varpi_k\right\}\cdot\{\mathbb{u}_d\} \\ & +\frac{1}{2}\{\mathbb{u}_d\}^T\cdot\left\{\sum_{k=1}^{N_p}\kappa^2\cdot d_p\cdot\left[\mathbb{B}^s\left(\bar{\xi}_k,\bar{\eta}_k\right)\right]^T\cdot\mathbb{D}^s\cdot\mathbb{B}^s\left(\bar{\xi}_k,\bar{\eta}_k\right)\cdot\left|J\left(\bar{\xi}_k,\bar{\eta}_k\right)\right|\cdot\varpi_k\right\}\cdot\{\mathbb{u}_d\} \end{aligned} \tag{4.51}$$

where the terms in the curl bracket are the flexural stiffness $\mathbb{k}^f$ and shear stiffness $\mathbb{k}^s$ matrices:

$$\mathbb{k}^f = \sum_{k=1}^{N_g}\frac{d_p^3}{12}\left[\mathbb{B}^f\left(\bar{\xi}_k,\bar{\eta}_k\right)\right]^T\cdot\mathbb{D}^f\cdot\mathbb{B}^f\left(\bar{\xi}_k,\bar{\eta}_k\right)\cdot\left|J\left(\bar{\xi}_k,\bar{\eta}_k\right)\right|\cdot\varpi_k,$$

$$\mathbb{k}^s = \sum_{k=1}^{N_g}\kappa^2\cdot d_p\left[\mathbb{B}^s\left(\bar{\xi}_k,\bar{\eta}_k\right)\right]^T\cdot\mathbb{D}^s\cdot\mathbb{B}^s\left(\bar{\xi}_k,\bar{\eta}_k\right)\cdot\left|J\left(\bar{\xi}_k,\bar{\eta}_k\right)\right|\cdot\varpi_k, \tag{4.52}$$

and $\mathbb{B}^f$ and $\mathbb{B}^s$ in full matrices are given as

$$\mathbb{B}^f = \begin{bmatrix} 0 & \frac{\partial N_1}{\partial\bar{\xi}} & 0 & \cdots & \cdots & \cdots & 0 & \frac{\partial N_{N_d}}{\partial\bar{\xi}} & 0 \\ 0 & 0 & \frac{\partial N_1}{\partial\bar{\eta}} & \cdots & \cdots & \cdots & 0 & 0 & \frac{\partial N_{N_d}}{\partial\bar{\eta}} \\ 0 & \frac{\partial N_1}{\partial\bar{\eta}} & \frac{\partial N_1}{\partial\bar{\xi}} & \cdots & \cdots & \cdots & 0 & \frac{\partial N_{N_d}}{\partial\bar{\eta}} & \frac{\partial N_{N_d}}{\partial\bar{\xi}} \end{bmatrix}_{3\times 3N_d} \tag{4.53a}$$

$$\mathbb{B}^s = \begin{bmatrix} \dfrac{\partial N_1}{\partial \bar{\xi}} & N_1 & 0 & \cdots & \cdots & \cdots & \dfrac{\partial N_{N_d}}{\partial \bar{\xi}} & N_{N_d} & 0 \\ \dfrac{\partial N_1}{\partial \bar{\eta}} & 0 & N_1 & \cdots & \cdots & \cdots & \dfrac{\partial N_{N_d}}{\partial \bar{\eta}} & 0 & N_{N_d} \end{bmatrix}_{2\times 3N_d} \quad (4.53b)$$

The total stiffness matrix $\mathbb{k}^t$ is thus given as

$$\mathbb{k}^t = \mathbb{k}^f + \mathbb{k}^s. \quad (4.54)$$

4.3.4.2.5.2 Mass Matrix The mass matrix can be derived from the kinetic energy functional T_E Equation (4.44). By representing $\mathbb{u}_e$ as summation of nodal displacement Equation (4.21) in Equation (4.44), T_E can be written in the natural coordinates as

$$T_E = -\{\mathbb{u}_d\}^T \cdot \left\{ \frac{\omega^2 \rho_p}{2} \int_{-1}^{1}\int_{-1}^{1} \mathbb{n}^T \cdot \mathbb{D}^m \cdot \mathbb{n} \cdot \left| J\left(\bar{\xi}_k, \bar{\eta}_k\right) \right| \cdot d\bar{\xi}\, d\bar{\eta} \right\} \cdot \{\mathbb{u}_d\}. \quad (4.55)$$

The integral can then be solved using the Gauss quadrature method, which yields

$$T_E = -\omega^2 \cdot \{\mathbb{u}_d\}^T \cdot \left\{ \frac{\rho_p}{2} \sum_{k=1}^{N_g} \mathbb{n}\left(\bar{\xi}_k, \bar{\eta}_k\right)^T \cdot \mathbb{D}^m \cdot \mathbb{n}\left(\bar{\xi}_k, \bar{\eta}_k\right) \cdot \left| J\left(\bar{\xi}_k, \bar{\eta}_k\right) \right| \cdot \bar{w}_k \right\} \cdot \{\mathbb{u}_d\}. \quad (4.56)$$

The mass matrix is given by the terms in the curl bracket, i.e.,

$$\mathbb{m} = \frac{\rho_p}{2} \sum_{k=1}^{N_g} \mathbb{n}\left(\bar{\xi}_k, \bar{\eta}_k\right)^T \cdot \mathbb{D}^m \cdot \mathbb{n}\left(\bar{\xi}_k, \bar{\eta}_k\right) \cdot \left| J\left(\bar{\xi}_k, \bar{\eta}_k\right) \right| \cdot \bar{w}_k. \quad (4.57)$$

4.3.4.2.5.3 Excitation Force, Added Mass, Radiated Damping and Hydrostatic Stiffness Matrices By substituting $\{\mathbb{u}_e\}$ in Equation (4.21) into the Work Done W_{nc} Equation (4.46), we have the W_{nc} expressed in its natural coordinate as

$$\begin{aligned} W_{nc} = &\{\mathbb{u}_d\}^T \cdot \left\{ \mathbb{i}\rho\omega \int_{-1}^{1}\int_{-1}^{1} \mathbb{n}\left(\bar{\xi}_k, \bar{\eta}_k\right)^T \cdot \varphi \cdot \left| J\left(\bar{\xi}_k, \bar{\eta}_k\right) \right| \cdot d\bar{\xi}\, d\bar{\eta} \right\} \\ &+ \{\mathbb{u}_d\}^T \cdot \left\{ \rho g \int_{-1}^{1}\int_{-1}^{1} \mathbb{n}\left(\bar{\xi}_k, \bar{\eta}_k\right)^T \cdot \mathbb{n}\left(\bar{\xi}_k, \bar{\eta}_k\right) \cdot \left| J\left(\bar{\xi}_k, \bar{\eta}_k\right) \right| \cdot d\bar{\xi}\, d\bar{\eta} \right\} \{\mathbb{u}_d\}. \end{aligned} \quad (4.58)$$

Upon evaluation using the Gauss quadrature method, Equation (4.58) becomes

$$
\begin{aligned}
W_{nc} &= \{\mathbb{u}_d\}^T \cdot \left\{ \mathbb{i}\omega\rho_p \sum_{k=1}^{N_g} \mathbb{n}\left(\bar{\xi}_k,\bar{\eta}_k\right)^T \cdot \varphi\left(\bar{\xi}_k,\bar{\eta}_k\right) \cdot \left| J\left(\bar{\xi}_k,\bar{\eta}_k\right) \right| \cdot \bar{w}_k \right\} \\
&\quad + \{\mathbb{u}_d\}^T \cdot \left\{ \rho_p g \sum_{k=1}^{N_g} \mathbb{n}\left(\bar{\xi}_k,\bar{\eta}_k\right)^T \cdot \mathbb{n}\left(\bar{\xi}_k,\bar{\eta}_k\right) \cdot \left| J\left(\bar{\xi}_k,\bar{\eta}_k\right) \right| \cdot \bar{w}_k \right\}.
\end{aligned} \tag{4.59}
$$

The second term in the RHS of Equation (4.59) is the hydrostatic stiffness $\mathbb{k}$:

$$
\mathbb{k} = \rho_p g \sum_{k=1}^{N_g} \mathbb{n}\left(\bar{\xi}_k,\bar{\eta}_k\right)^T \cdot \mathbb{n}\left(\bar{\xi}_k,\bar{\eta}_k\right) \cdot \left| J\left(\bar{\xi}_k,\bar{\eta}_k\right) \right| \cdot \bar{w}_k = \rho_p g \cdot \Delta a. \tag{4.60}
$$

Note that the summation terms yield the area of one element Δa.

On the other hand, the velocity potential $\varphi\left(\bar{\xi},\bar{\eta}\right)$ in (4.59) is made up of the diffracted and radiated part (see Equation (4.16)). The diffracted velocity potential contributes to the excitation force, i.e.,

$$
\tilde{\mathbb{F}}_E = \{\mathbb{u}_d\}^T \cdot \left\{ \mathbb{i}\omega\rho_f \sum_{k=1}^{N_g} \mathbb{n}\left(\bar{\xi}_k,\bar{\eta}_k\right)^T \cdot \varphi_D\left(\bar{\xi}_k,\bar{\eta}_k\right) \cdot \left| J\left(\bar{\xi}_k,\bar{\eta}_k\right) \right| \cdot \bar{w}_k \right\} = \{\mathbb{u}_d\}^T \cdot \mathbb{F}_E, \tag{4.61}
$$

where the evaluation of φ_D in Equation (4.61) is the same as presented in Section 3.6, i.e.,

$$
\{\varphi_D(\mathbf{x})\} = 2\left[\mathbb{I} + \frac{1}{2\pi}\frac{\partial \mathbb{G}}{\partial \mathbf{n}}\right]^{-1} \times \{\varphi_{In}\}. \tag{4.62}
$$

The velocity potential $\varphi(\xi,\eta)$ in Equation (4.59), when replaced with the radiated potential, will contribute to the added mass $\mathbb{M}_A^c$ and radiated damping $\mathbb{B}_B^c$ given as

$$
\mathbb{M}_A^c = \mathrm{Re}\left\{ \frac{\rho_f}{2\pi}\left[\mathbb{I} + \frac{1}{2\pi}\frac{\partial \mathbb{G}}{\partial \mathbf{n}}\right]^{-1} \times \mathbb{G} \right\}, \tag{4.63a}
$$

$$
\mathbb{B}_B^c = \mathrm{Im}\left\{ \frac{\omega\rho_f}{2\pi}\left[\mathbb{I} + \frac{1}{2\pi}\frac{\partial \mathbb{G}}{\partial \mathbf{n}}\right]^{-1} \times \mathbb{G} \right\}. \tag{4.63b}
$$

The superscript c in $\mathbb{M}_A^c$ and $\mathbb{B}_B^c$ denotes that the global matrix for these added mass and radiated damping is evaluated at the centroid of each panel, resulting in $N_e \times N_e$ matrices, instead of $3N_D \times 3N_D$ matrices. $\mathbb{M}_A^c$ and $\mathbb{B}_B^c$ can be transformed into $3N_e \times 3N_e$ matrices by using a transformation matrix shown later in Equation (4.69).

The evaluation of $\mathbb{G}$ and $\frac{\partial \mathbb{G}}{\partial \mathbf{n}}$ in Equation (3.124) are given in Section 3.6. It is however noted here that, according to Equation (3.81b), $\frac{\partial \mathbb{G}}{\partial \mathbf{n}}$ can be further simplified as Equation (4.64) for a VLFS due to the small thickness to length ratio.

$$\frac{\partial \mathbb{G}}{\partial \mathbf{n}} = \frac{\omega^2}{g} \mathbb{G} \tag{4.64}$$

4.3.4.3 Haskind-Hanoaka Relations

According to Haskind's relation in Section 3.5.6, the excitation force $\mathbb{F}_E$ can be expressed in terms of unit-amplitude radiated potential φ_l and the incident potential φ_I, thereby eliminating the need to solve for the scattered potential φ_S. Hanoaka extends the Haskind's relation and applied it to compute the excitation force for hydroelasticity.

By using the Haskind's relation to express the excitation force Equation (4.61) in integral from as in Equation (3.102), the excitation force can be written as

$$F_E = \mathrm{i}\omega\rho_f \int_{S_w} \left(\varphi_I \frac{\partial \varphi_l}{\partial \mathbf{n}} - \varphi_l \frac{\partial \varphi_I}{\partial \mathbf{n}} \right) \cdot dS. \tag{4.65}$$

By substituting the bottom wetted surface boundary condition for the unit-amplitude radiated potential Equation (4.17c) into Equation (3.102), the excitation force can be written as

$$\begin{aligned} F_E &= \mathrm{i}\omega\rho_f \cdot \int_{S_w} \left(\mathrm{i}\omega\varphi_I \mathcal{Z}_l - \varphi_l \frac{\partial \varphi_I}{\partial \mathbf{n}} \right) \cdot dS \\ &= -\omega^2 \rho_f \cdot \int_{S_w} (\varphi_I \cdot \mathcal{Z}_l) \cdot dS + \mathrm{i}\omega\rho_f \int_{S_w} \left(\varphi_l \frac{\partial \varphi_I}{\partial \mathbf{n}} \right) \cdot dS. \end{aligned} \tag{4.66}$$

The exciting force expression Equation (4.66), in terms of the unit-amplitude radiated and incident potential, is referred to as the Haskind-Hanaoka relation.[118]

4.3.4.4 Global Matrices for Mass and Stiffness

The FSDT shall be used to demonstrate the assembly of elemental matrix into global matrix. The elemental matrix involves in the FSDT are the flexural stiffness, shear stiffness, mass stiffness and hydrostatic stiffness, denoted by $\mathbb{k}^f$, $\mathbb{k}^s$, $\mathbb{m}$ and $\mathbb{k}$, respectively. Each of these matrices has a dimension of $N_d \times N_d$, where N_d is the size of the quadrilateral element.

Consider, for example, an irregular rectangular plate discretised into four elements of four-node quadrilateral plate elements ($N_e = 4, N_d = 4$) with a total of nine nodes ($N_D = 9$), as shown in Figure 4.6. It is noted here that the numbers outside the nodes (labelled in red) represent the node number for each quadrilateral panel, whereas the numbers inside the nodes denote the node number when the quadrilateral panels are merged/assembled. Each node contains three dofs. Elements 1, 2, 3 and 4 are represented by e1, e2, e3, and e4.

For each element, the elemental matrix has a size of $3N_d \times 3N_d = 12 \times 12$. Taking mass as an example, the elemental mass matrix is thus given as

$$\mathbb{m} = \begin{bmatrix} \hat{\mathbb{m}}_1 & 0 & 0 & 0 \\ 0 & \hat{\mathbb{m}}_2 & 0 & 0 \\ 0 & 0 & \hat{\mathbb{m}}_3 & 0 \\ 0 & 0 & 0 & \hat{\mathbb{m}}_4 \end{bmatrix}_{12\times 12}, \tag{4.67}$$

where $\hat{\mathbb{m}}$ denotes the matrix associated with each node where the subscript denotes the node number. Therefore, the elemental matrices can be assembled according to Figure 4.6, which yields the global mass matrix given as

$$[\mathrm{M}]_{27\times 27} = \begin{bmatrix} \hat{\mathbb{m}}_1^{e1} & 0 & 0 & 0 & 0 & 0 & 0 & 0 & 0 \\ 0 & \hat{\mathbb{m}}_2^{e1}+\hat{\mathbb{m}}_1^{e2} & 0 & 0 & 0 & 0 & 0 & 0 & 0 \\ 0 & 0 & \hat{\mathbb{m}}_3^{e2} & 0 & 0 & 0 & 0 & 0 & 0 \\ 0 & 0 & 0 & \hat{\mathbb{m}}_3^{e1}+\hat{\mathbb{m}}_1^{e3} & 0 & 0 & 0 & 0 & 0 \\ 0 & 0 & 0 & 0 & \begin{matrix}\hat{\mathbb{m}}_4^{e1}+\hat{\mathbb{m}}_3^{e2} \\ +\hat{\mathbb{m}}_2^{e3}+\hat{\mathbb{m}}_1^{e4}\end{matrix} & 0 & 0 & 0 & 0 \\ 0 & 0 & 0 & 0 & 0 & \hat{\mathbb{m}}_4^{e2}+\hat{\mathbb{m}}_2^{e4} & 0 & 0 & 0 \\ 0 & 0 & 0 & 0 & 0 & 0 & \hat{\mathbb{m}}_3^{e3} & 0 & 0 \\ 0 & 0 & 0 & 0 & 0 & 0 & 0 & \hat{\mathbb{m}}_4^{e3}+\hat{\mathbb{m}}_3^{e4} & 0 \\ 0 & 0 & 0 & 0 & 0 & 0 & 0 & 0 & \hat{\mathbb{m}}_4^{e4} \end{bmatrix} \tag{4.68}$$

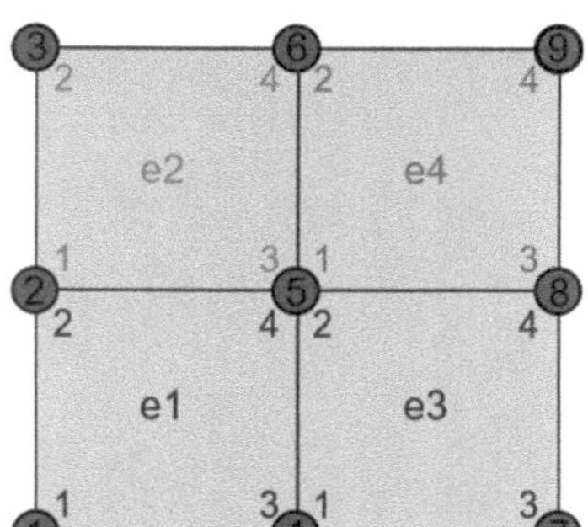

FIGURE 4.6
Rectangular plate discretised into four quadrilateral elements.

where the superscript denotes the element number, i.e., e1 for element 1, e2 for element 2, so on and so forth. Similarly, the global matrices for flexural stiffness $\mathbb{K}^f$, shear stiffness $\mathbb{K}^s$, and hydrostatic stiffnes $\mathbb{K}$ can be assembled using the same method.

However, the added mass $\mathbb{M}_{\mathcal{A}}$ and radiated damping $\mathbb{B}_{\mathcal{B}}$ are $N_e \times N_e$ matrices as they are derived from $\mathbb{G}$ which is an $N_e \times N_e$ matrix with its field point located at the centroid of each element. Therefore, the $\mathbb{M}_{\mathcal{A}}$ and $\mathbb{B}_{\mathcal{B}}$ have to be distributed to the node of the element using a transformation matrix $[\mathbb{o}]$:

$$\mathbb{M}_{\mathcal{A}} = [\mathbb{o}]^T_{N_e \times 3N_D} \left[\mathbb{M}^c_{\mathcal{A}}\right]_{N_e \times N_e} [\mathbb{o}]_{N_e \times 3N_D}, \tag{4.69a}$$

$$\mathbb{B}_{\mathcal{B}} = [\mathbb{o}]^T_{N_e \times 3N_D} \left[\mathbb{B}^c_{\mathcal{B}}\right]_{N_e \times N_e} [\mathbb{o}]_{N_e \times 3N_D}, \tag{4.69b}$$

where $[\mathbb{o}]$ is a relation matrix that maps the centroid of the element to its node. E.g., by referring to Figure 4.6, the following relations could be established to map the $\mathbb{M}_{\mathcal{A}}$ from the centroid of each element to the nodes:

(i) Centroid of e1 is related to node 1, node 2, node 4, and node 5.
(ii) Centroid of e2 is related to node 2, node 3, node 5, and node 6.
(iii) Centroid of e3 is related to node 4, node 5, node 7, and node 8.
(iv) Centroid of e4 is related to node 5, node 6, node 8, and node 9.

As each node has three dofs, $[\mathbb{o}]$ has to map the centroid of the element denoted by the row of the matrix to the nodes denoted by matrix $\mathbb{q}$ where $\mathbb{q} = \begin{bmatrix} 1 & 1 & 1 \end{bmatrix}$. $[\mathbb{o}]$ is thus given as

$$[\mathbb{o}] = \begin{bmatrix} \mathbb{q}_1 & \mathbb{q}_2 & 0 & \mathbb{q}_4 & \mathbb{q}_5 & 0 & 0 & 0 & 0 \\ 0 & \mathbb{q}_2 & \mathbb{q}_3 & 0 & \mathbb{q}_5 & \mathbb{q}_6 & 0 & 0 & 0 \\ 0 & 0 & 0 & \mathbb{q}_4 & \mathbb{q}_5 & 0 & \mathbb{q}_7 & \mathbb{q}_8 & 0 \\ 0 & 0 & 0 & 0 & \mathbb{q}_5 & \mathbb{q}_6 & 0 & \mathbb{q}_8 & \mathbb{q}_9 \end{bmatrix}_{N_e \times 3N_D} \tag{4.70}$$

4.3.4.5 Solving Equation of Motion

The equation of motion for the plate can be written in the global form as

$$\{\mathbb{u}_d\}^T \cdot \left\{-\omega^2 [\mathbb{M} + \mathbb{M}_a] + \mathrm{i}\omega(\mathbb{B}_{\mathcal{B}} + \mathbb{B}_{pto}) + \mathbb{K}^f + \mathbb{K}^s + \mathbb{K}\right\} \cdot \{\mathbb{u}_d\} = \{\mathbb{u}_d\}^T \cdot \mathbb{F}_{E}, \tag{4.71}$$

which can be further simplified as

$$\left\{-\omega^2 [\mathbb{M} + \mathbb{M}_a] + \mathrm{i}\omega(\mathbb{B}_{\mathcal{B}} + \mathbb{B}_{pto}) + \mathbb{K}^f + \mathbb{K}^s + \mathbb{K}\right\} \cdot \{\mathbb{u}_d\} = \mathbb{F}_E. \tag{4.72}$$

By using the modal expansion method, the displacement vector $\{\mathbb{u}_d\}$ may be expanded in an appropriate set of modes following Equation (4.15), i.e.,

$$\{\mathbb{u}_d\} = \mathbb{Z} \cdot \{\mathbb{c}\}, \tag{4.73}$$

where $\mathbb{Z}$ is the matrix containing the eigenvectors of the plate

$$\mathbb{Z} = \begin{bmatrix} (\mathcal{Z}_1)_1 & (\mathcal{Z}_1)_2 & \cdots & (\mathcal{Z}_1)_{N_m-1} & (\mathcal{Z}_1)_{N_m} \\ (\mathcal{Z}_2)_1 & (\mathcal{Z}_2)_2 & \cdots & (\mathcal{Z}_2)_{N_m-1} & (\mathcal{Z}_2)_{N_m} \\ \vdots & \vdots & \ddots & \vdots & \vdots \\ (\mathcal{Z}_{N_D})_1 & (\mathcal{Z}_{N_D})_2 & \cdots & (\mathcal{Z}_{N_D})_{N_m-1} & (\mathcal{Z}_{N_D})_{N_m} \end{bmatrix}_{3N_d \times N_m} \tag{4.74}$$

where $(\mathcal{Z}_i)_j = \begin{pmatrix} \mathcal{Z}_i^W & \mathcal{Z}_i^{\psi^y} & \mathcal{Z}_i^{\psi^x} \end{pmatrix}^T$ is jth mode at node i. $\mathbb{Z}$ has a dimension of $3N_d \times N_m$, and $\mathbb{c}$ is a $3N_d \times 3$ matrix, whereas N_m is the number of modes taken in the free vibration analysis. The eigenvalue and eigenvector can be solved from

$$[\mathbb{K} - \omega_n \mathbb{M}] \cdot \mathbb{Z} = 0. \tag{4.75}$$

By substituting Equation (4.73) into Equation (4.72) and upon multiplying both side by $\mathbb{Z}^T$, we have

$$\mathbb{Z}^T \cdot \left\{ -\omega^2 [\mathbb{M} + \mathbb{M}_a] + \mathbb{i}\omega(\mathbb{B}_B + \mathbb{B}_{pto}) + \mathbb{K}^f + \mathbb{K}^s + \mathbb{K} \right\} \cdot \{\mathbb{Z} \cdot \mathbb{c}\} = \mathbb{Z}^T \cdot \mathbb{F}_E. \tag{4.76}$$

The multiplication of $\mathbb{Z}^T \cdot [\blacksquare] \cdot \mathbb{Z}$, where $[\blacksquare]$ is any $3N_d \times 3N_d$ global matrices, will produce the generalised matrix with dimension $N_m \times N_m$. This reduces the computation time required to solve the equation of motion significantly, as N_m is smaller than N_d.

Equation (4.76) can now be solved algebraically for the complex amplitude $\mathbb{c}$. This complex amplitude matrix is then back substituted into Equation (4.73) to obtain the plate deflection. The stress resultants may be readily obtained from taking appropriate derivatives of the deflection and bending rotations, as given in Equation (4.9).

4.4 Case Study I: Generic VLFS

A generic rectangular pontoon-type VLFS with a length 300 m, width 60 m, and draft 2 m as shown in Figure 4.7. The VLFS is subjected to an incoming wave of wave period T and a wave height $H = 2A$ that impacts the structure at a wave angle θ with respect to the negative x-axis. The seabed is assumed to be levelled, and the water depth is $D = 120\,\text{m}$.

Figure 4.7(b) shows the VLFS modelled as an equivalent solid plate. The Young's module of the VLFS is $\mathcal{E} = 11.9$ GPa, and $\rho_p = 256.25\ \text{kg/m}^3$.

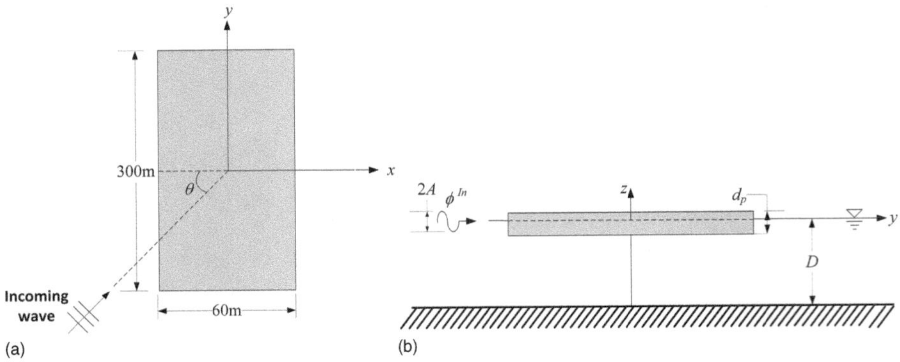

FIGURE 4.7
Schematic diagram of VLFS (a) plan view and (b) elevation view.

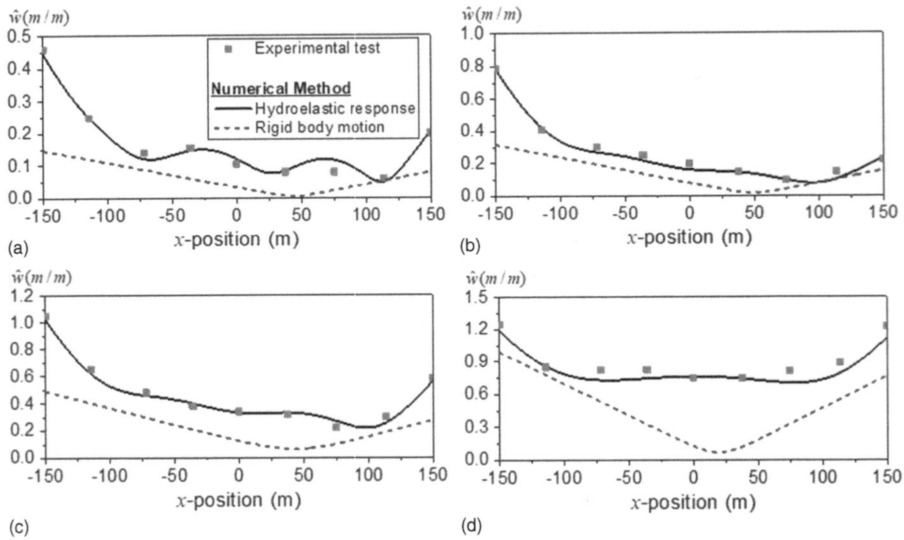

FIGURE 4.8
Hydroelastic response of VLFS under (a) $\lambda / L = 0.2$, (b) $\lambda / L = 0.4$, and (c) $\lambda / L = 0.6$ (d) $\lambda / L = 1.0$.

4.4.1 Hydroelastic Response

The hydroelastic response of the VLFS is shown in Figure 4.8. The results show that the response predicted by the FE-BE method is in good agreement with their counterparts obtained from experimental test.[100] It is also important to include the structural deformation evidenced from Figure 4.8, where the results of the rigid body motion assumption differ significantly from the experimental test. The hydroelastic response of the VLFS is profound when the λ / L is small and approaching an average value of 1.0 when the λ / L is large.

4.4.2 Stress Resultants

The stress resultants can be obtained for the CPT and FSDT given in Equations (4.4) and (4.9). Comparing the governing equation for CPT and FSDT, the former involves the fourth derivative of the vertical plate deflection Equation (A.21) in Appendix A, whereas the latter only involves the second derivative of the plate deflection Equation (B.14) in Appendix B. This means that in solving the governing equation, the assumed plate displacement must have a functional that is a fourth derivative for the CPT and second derivative for the FSDT. The requirement for a higher-order derivative will increase the computational time in the finite element computational analysis.

The CPT also assumes that there is no transverse shear deformation in the plate and considers only in-plane displacements and rotations. This assumption implies that the transverse shear stresses are zero, and the thickness of the plate remains constant before and after deformation. The rotation of the plate is derived from the plate deflection, i.e., Equation (A.1a). Also, according to the equilibrium condition, a pair of shear forces has to exist, and these shear forces Q_x, Q_y are derived from the twisting moment M_{xy}, i.e., Equation (A.6). This results in the stress resultants involving the second-order derivative of the plate deflection Equation (A.5).

On the other hand, FSDT considers the transverse shear deformation by relaxing the assumption of normality between the transverse normal and the plate's mid-surface. Thus, the transverse shear strain and shear stresses are included in the theory. As a result, the governing equation and stress resultants only involve up to first-order derivative, and this improves the stress resultants significantly, especially for the shear forces.

4.5 Case Study II: Hinged-Connected VLFS

The VLFS, due to its large size, is usually constructed in multiple modular units and connected during the assembly stage. The assembled VLFS is considered to be a continuous plate during the modelling for hydroelasticity analysis. In some cases, hinge connections are used to connect two or more modular units.[234] Such scenarios can happen during the installation stage of the VLFS, where the modular units are held together temporarily by a hinge or semi-rigid connector before being welded together to form a continuous unit.

At the node where two modules are connected, only the vertical deflection of the plate $\hat{w}$ has to be merged so that there are five dofs at this node, i.e., vertical deflection, rotation about x- and y-axes for Module 1 (see Figure 4.9), and rotation about x- and y-axes for Module 2.

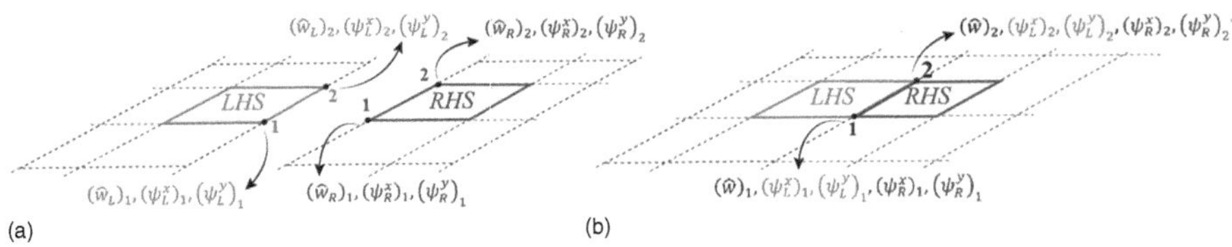

FIGURE 4.9
Schematic diagram of module 1 connected to module 2.

4.5.1 Anti-Motion Device

Figure 4.10 shows the hydroelastic response of the VLFS when two modular units are hinge connected at $x = 0$. The wavelength is under $\lambda / L = 0.2$. Note that the connected modular unit has a total length of 300 m, width of 60 m, and thickness of 2 m with $\mathcal{E} = 11.9$ GPa, and $\rho_p = 256.25$ kg/m^3, similar to the generic VLFS considered in Case Study I.

The hydroelastic response of a VLFS with hinge connections is compared to those with rigid and semi-rigid connectors. The results indicate that using hinge or semi-rigid connectors can help reduce the hydroelastic response

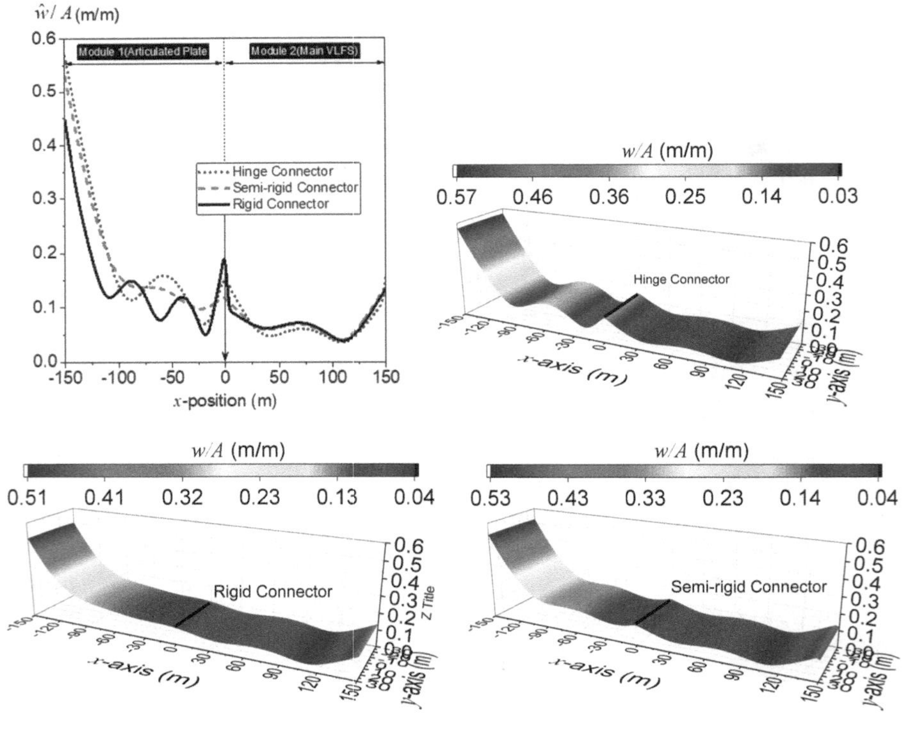

FIGURE 4.10
Hydroelastic response of VLFS with hinge, semi-rigid, and rigid connectors. $\lambda / L = 0.2$.

of the VLFS. Specifically, Module 2 experiences a smaller response compared to a continuous VLFS. Thus, if Module 1 functions as an articulated plate for an anti-motion device, it can act as a sacrificial plate, absorbing most of the energy from incoming waves to reduce the forces impacting the main structure (Module 2). Similarly, the comparison of the VLFS with rigid, semi-rigid, and hinge connectors under $\lambda / L = 0.4$ is presented in Figure 4.11. The hinged- and semi-rigid connected articulated plate are shown to mitigate the motion of the main VLFS. A comparison between Figure 4.10 and Figure 4.11 shows that the articulated plate is more effective as an anti-motion device in reducing the hydroelastic response when subjected to a smaller wavelength λ / L.

The connector can also be placed at different locations in order to effectively mitigate the hydroelastic response. Figures 4.12(a) and 4.12(b) show the hydroelastic response of the VLFS when subjected to $\lambda / L = 0.2$ and 0.4. The location of the connector is $\alpha_a = 15$ m, 30 m, 45 m, and 60 m. As the energy is absorbed by the articulated plate, there is therefore a reduction

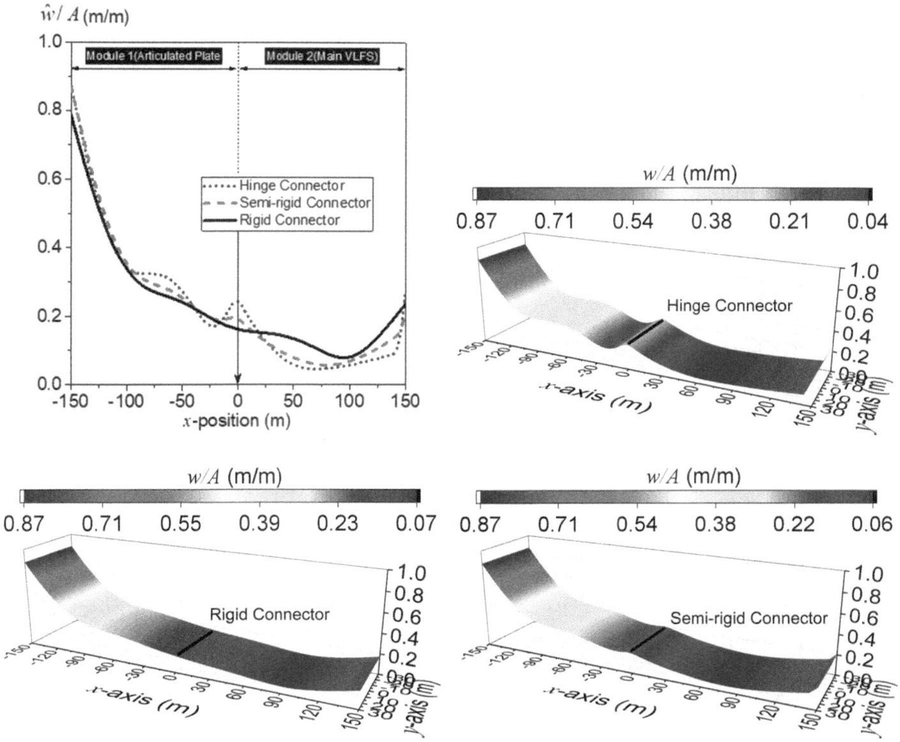

FIGURE 4.11
Hydroelastic response of VLFS with hinge, semi-rigid, and rigid connectors. $\lambda / L = 0.4$.

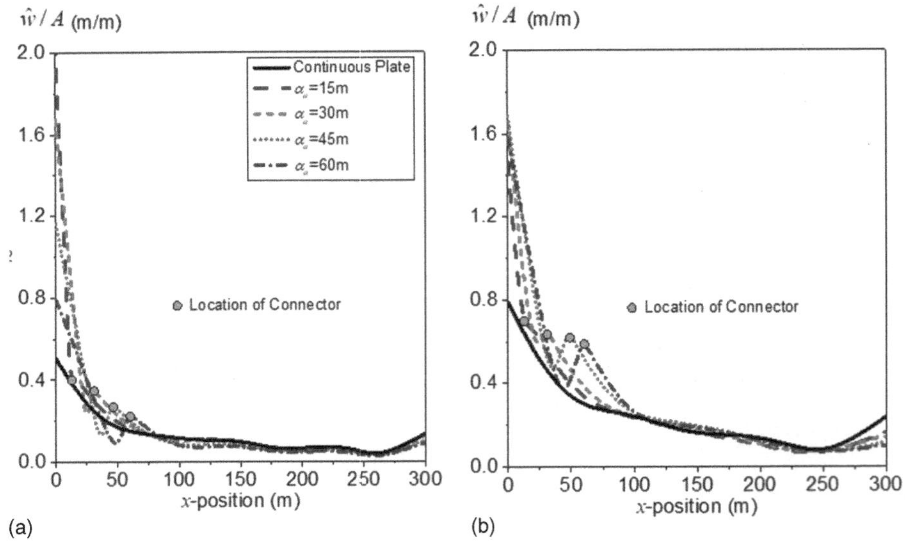

FIGURE 4.12
Hydroelastic response of VLFS with different articulated plate lengths for (a) $\lambda / L = 0.2$ and (b) $\lambda / L = 0.4$.

in the hydroelastic response of the main VLFS. The articulated plate with $L_a = 60$ m is found to be most effective in mitigating the hydroelastic response of the VLFS in this case.

The dimensions of the articulated plate could also vary, and this could have a profound effect on its effectiveness in mitigating the hydroelastic response of the main VLFS (Module 2). Figure 4.13 shows the VLFS with length $L = 300$ m connected with articulated plate with length $L_a = 15$ m, 30 m, 45 m, and 60 m. The results indicate that the articulated plate exhibits a very high response due to the incoming wave impacting the structure. The deflection in the main VLFS is significantly diminished by the presence of the articulated plate, with the longest plate proving to be the most effective in reducing the hydroelastic response.

4.5.2 Integrated Anti-Motion Device and WEC

The results from Figures 4.10 to 4.12 show that when an articulated plate is connected to a VLFS via a rigid or semi-rigid connector, it is possible to reduce the hydroelastic response of the VLFS. The articulated plate behaves like a sacrificial plate that absorbs the energy from the wave before hitting on the VLFS. If a PTO system is connected to the connector of the articulated plate, it is then possible to convert the kinetic energy arising from the motion of the articulated plate into electricity. Here, the articulated plate is analogous to a raft-type WEC.

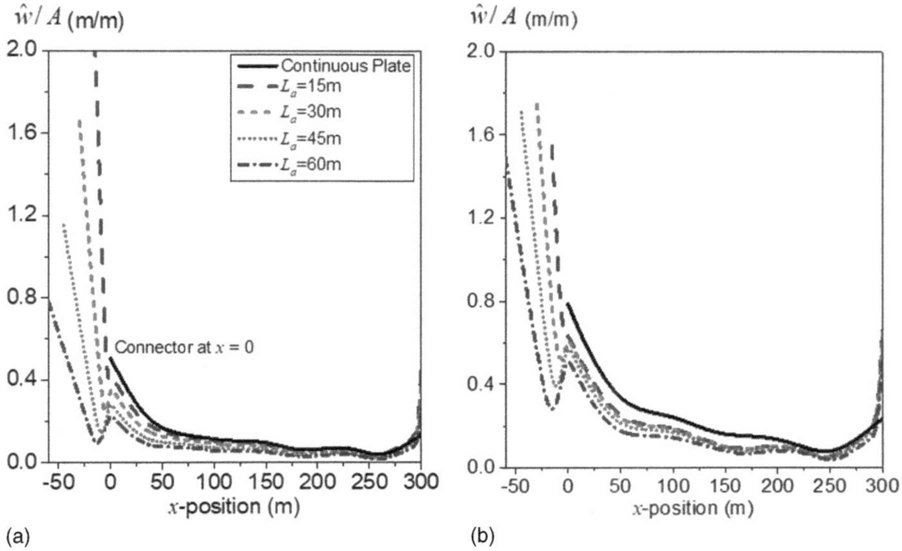

FIGURE 4.13
Hydroelastic response of VLFS with hinges at different articulate plate lengths L_a for (a) $\lambda / L = 0.2$ and (b) $\lambda / L = 0.4$.

The typical B_{pto} matrix of an interconnected node in the hinge connector is presented as follow:

$$B_{pto} = \begin{matrix} \hat{w} \\ (\psi^x)^- \\ (\psi^y)^- \\ (\psi^x)^+ \\ (\psi^y)^+ \end{matrix} \begin{bmatrix} 0 & 0 & 0 & 0 & 0 \\ 0 & 0 & 0 & 0 & 0 \\ 0 & 0 & 0 & 0 & 0 \\ 0 & 0 & 0 & +B_{pto} & -B_{pto} \\ 0 & 0 & 0 & -B_{pto} & +B_{pto} \end{bmatrix}. \tag{4.77}$$

It is noted here that for each node along the line connector, there will be five dofs – namely, $\hat{w}, (\psi^x)^-, (\psi^y)^-, (\psi^x)^+, (\psi^y)^+$. The positive (+) and negative (−) signs denote the RHS and LHS of the node in the line connector.

In order to quantify the effectiveness of the flexible line connector in mitigating the hydroelastic response, the non-dimensional compliance parameter χ is introduced, which is defined as

$$\chi = \int_{\Delta} \hat{w} \cdot d\Delta. \tag{4.78}$$

The efficiency of the VLFS compliance with and without the articulated plate, i.e., $\chi_{with\ articulated\ plate}$ and $\chi_{w/o\ articulated\ plate}$, respectively, can be computed as,

$$\eta_\chi = \frac{\chi_{w/o\,\text{articulated plate}} - \chi_{\text{with articulated plate}}}{\chi_{w/o\,\text{articulated plate}}}\%. \tag{4.79}$$

It is observed that a positive η_χ value signifies that the articulated plate anti-motion device is effective in reducing the hydroelastic response, and the opposite holds true for a negative value. Four different articulated plate lengths, L_a of 15 m, 30 m, 45 m, and 60 m, along with B_{pto} values ranging from 0 to 0.2 MN in increment of 75 kN.m.s for wavelengths $\lambda / L = 0.4, 0.6, 0.8$, and 1.0, are considered. Hydroelastic analyses are conducted using these parameters, and the results for the different λ / L ratios are shown in Figure 4.14. Each sub-figure in Figure 4.14 displays four lines representing the hydroelastic response of the VLFS under four different extreme conditions (EC):

(i) Solid Line – VLFS without articulated plate
(ii) Dotted Line – VLFS with hinge-connected articulated plate but without B_{pto}
(iii) Dash-Dot Line – VLFS with hinge-connected articulated plate with B_{pto} that generates the most energy

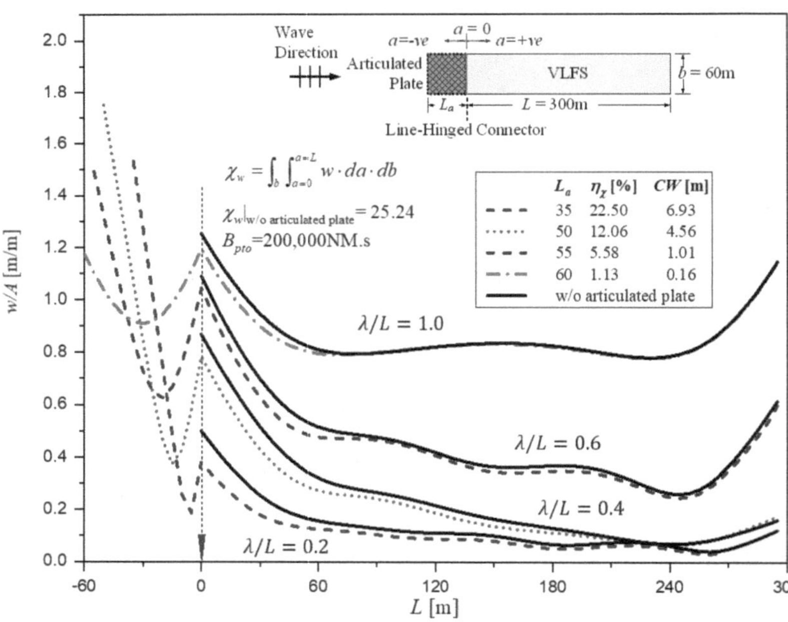

FIGURE 4.14
Performance of raft-type WEC with VLFS under different wavelength.

(iv) Dash line – VLFS with hinge-connected articulated plate with B_{pto} that is most effective in mitigating the hydroelastic response

4.5.2.1 Hydroelastic Response

Figure 4.14 shows that the hydroelastic response of the VLFS without the articulated plate, i.e., EC (i), is the largest, while the response with the articulated plate, i.e., EC (ii), is the smallest, indicating that the articulated plate anti-motion device is highly effective at reducing wave forces. However, its effectiveness diminishes as the wavelength increases. Furthermore, the figure demonstrates that in EC (iii), where the settings of L_a and B_{pto} maximise energy generation from the hinge connector, the articulated plate is generally ineffective in mitigating the hydroelastic response due to the large B_{pto} value (i.e., 0.2 MN). A large B_{pto} at the connector is similar to having a rigid or welded joint, which fails to adequately reduce wave forces. In fact, for $\lambda / L = 0.4$ and 0.6, the η_x value is negative, meaning that the hydroelastic response of the VLFS increases due to the anti-motion device. However, under EC (iv), where L_a and B_{pto} are optimised for reducing the hydroelastic response, a significant amount of wave energy can still be extracted from the hinge's rotational motion when the wavelength is small, i.e., $\lambda / L = 0.4$ and 0.6.

4.5.2.1.1 Capture Width and Compliance Efficiency

The capture width (CW) and plate compliance efficiency (η_χ) of the VLFS for the given L_a and B_{pto} are shown side by side in Figure 4.14. Generally, it can be observed that CW (i.e., the power generated) decreases as λ / L increases. This is because the VLFS, when exposed to longer wavelengths, tends to oscillate in phase with the wave orbital motion, resulting in reduced rotational movement at the hinge connector. Similarly, η_χ decreases with increasing λ / L, as the hydroelastic response can only be effectively mitigated when energy is dissipated through radiated wave damping produced by the rotational motion of the anti-motion device.

4.6 Case Study III: Power Enhancement via Elastic Deformation

The energy generation from the attenuator WEC attached to a VLFS can be enhanced by using the elastic properties of the material, e.g., the Wave Carpet,[138] Anaconda WEC,[139] SBM S3 WEC,[140] Bombora WEC,[141] Water Level Carpet (WLC) WEC,[142] and flexible runway supported by an array of circular buoy.[143]

A pontoon-type WEC, comprising a grid of N_{fm} interconnected floating modules, is examined here. Each module is connected to the others by $(N_{fm}-1)$ line hinge connectors (refer to Figure 4.15). For instance, Figure 4.15(a) shows Type-A with $N_{fm}=12$ (11 hinges), Figure 4.15(b) illustrates Type-B with $N_{fm}=6$ (5 hinges), Figure 4.15(c) depicts Type-C with $N_{fm}=4$ (3 hinges), and Figure 4.15(d) displays Type-D with $N_{fm}=1$ (0 hinges). When assembled, the total length of the pontoon-type WEC is $L=N_{fm}\times l_{fm}$ represents the length of each module. The PTO system is positioned at equal intervals of $L/12$ along the horizontal axis (x-axis) and $B/8$ along the transverse axis (y-axis), as illustrated in Figure 4.15, resulting in a total of $N_{pto}=99$ PTO systems, as shown in Figure 4.15(a) to (d).

Three groups of pontoon-type WECs with varying L values are examined: Group I with $L=100$ m, Group II with $L=200$ m, and Group III with $L=300$ m. For each group, four types of WECs (Type-A, -B, -C, and -D), each featuring N_{fm}, different numbers of modules, are analysed, as shown in Figure 4.15(a) to 4.15(d). It should be highlighted that Type-A has the most line hinge connectors, while Type-B and Type-C have fewer. The Type-D pontoon-type WEC is a single module system, resembling a continuous mat-like VLFS. These WECs exhibit different elastic deformation behaviours due to variations in module length (l_{fm}) and Young's modulus ($\mathcal{E}$), facilitating an

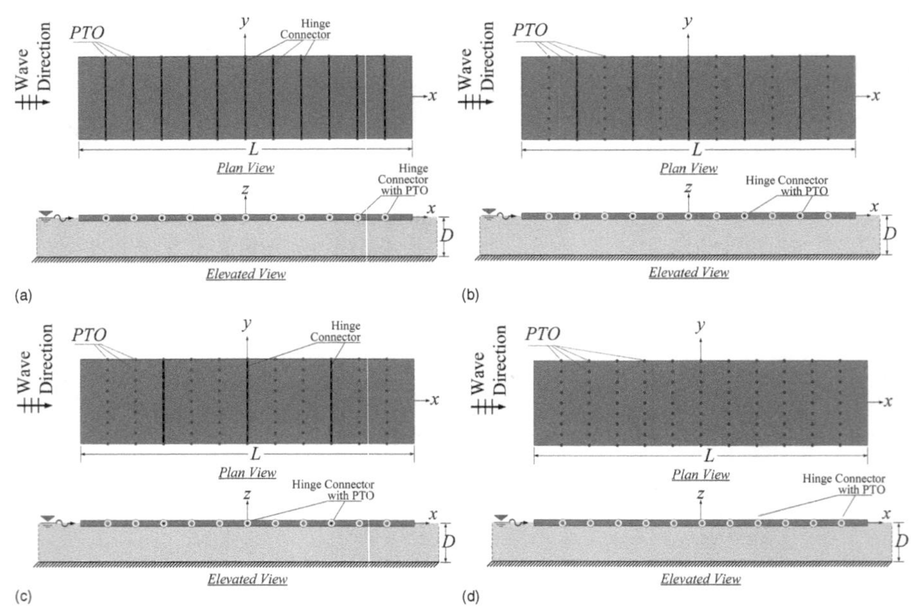

FIGURE 4.15
Pontoon-type WEC with (a) Type-A: $N_{fb}=12$ modules, (b) Type-B: $N_{fb}=6$ modules, (c) Type-C: $N_{fm}=4$ modules, and (d) Type-D: $N_{fb}=1$ module.

investigation of how structural rigidity affects power generation. To explore this, the WECs are modelled using six different $\mathcal{E}$ values: $\mathcal{E} = 0.2$ GPa, 2 GPa, 20 GPa, 200 GPa, 2,000 GPa, and 20,000 GPa. It is also important to note that the total number of PTO systems ($N_{pto} = 99$) is maintained consistently across all WECs to ensure a fair performance comparison.

4.6.1 Capture Width

Figure 4.16 illustrates the CW of the pontoon-type WEC for Groups I, II, and III, all with $N_{fm} = 1$. The figure demonstrates that the pontoon-type WEC is highly efficient at harnessing wave energy when the structure is flexible, i.e., the $\mathcal{E}$ is low ($\mathcal{E} = 0.2$ GPa). As $\mathcal{E}$ increases, both the CW and P_a decrease and become less responsive to changes in structural stiffness, signifying that the VLFS becomes more rigid. Additionally, when the pontoon-type WEC lengthens, with L equal to 200 m and 300 m, the energy generation increases, with the highest energy yield occurring when the structure is most flexible (i.e., $\mathcal{E} = 0.2$ GPa).

4.6.2 Hydroelastic Response

Figure 4.17 shows the comparison of the hydroelastic response of the Group I pontoon-type WEC for varying numbers of hinges: Type-A with 11 hinges, Type-B with 5 hinges, Type-C with 3 hinges, and Type-D with no hinges. The deflection is recorded along the WEC's centreline under head

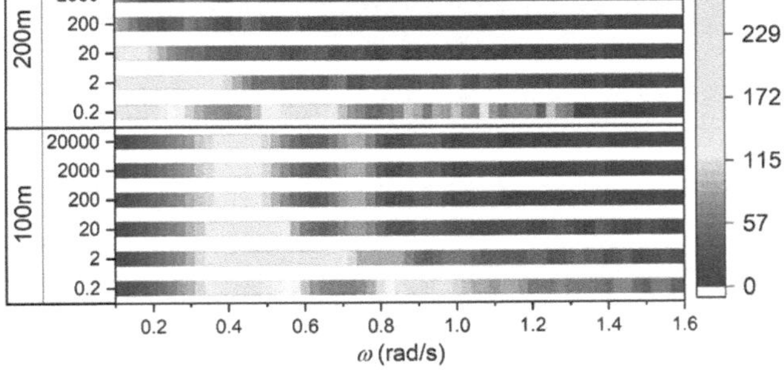

FIGURE 4.16
CW for Groups I, II, and III Pontoon-type WECs under different Young's modulus $\mathcal{E}$. $N_{fm} = 1$.

sea conditions. The results indicate that the effect of Young's modulus is minimal for WECs with a greater number of hinges, such as Type-A and Type-B, but its impact becomes more significant when fewer hinges are used, as in Type-C and Type-D. Additionally, it is evident that a continuous pontoon-type WEC (Type-D) can generate more energy than one with connectors.

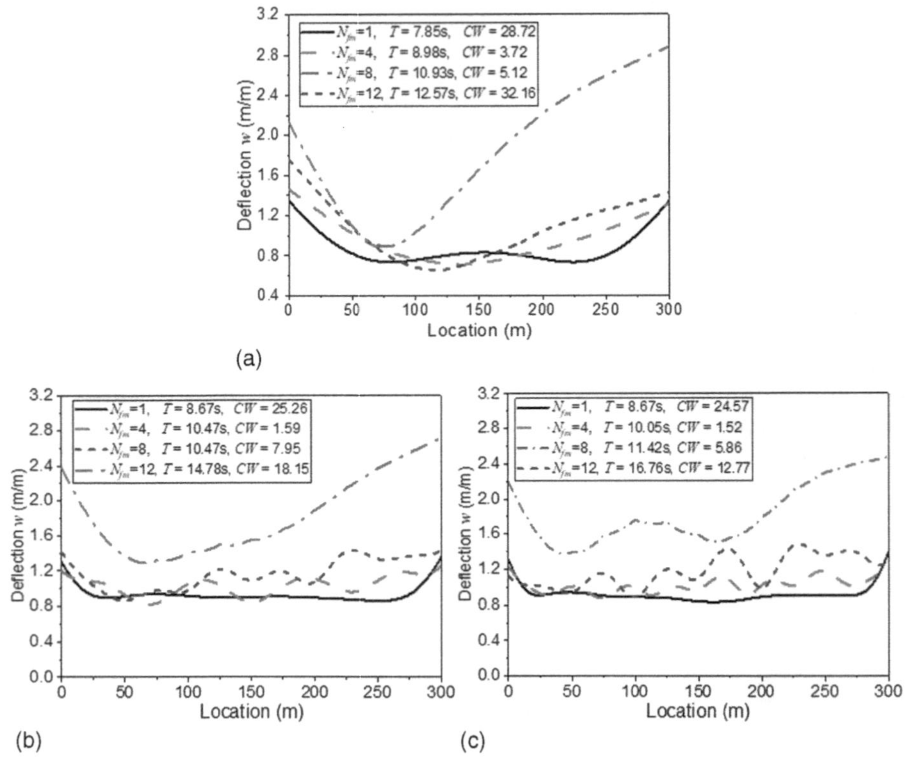

FIGURE 4.17
Hydroelastic response for (a) Group I ($L = 100$ m), (b) Group II ($L = 200$ m), (c) Group III ($L = 100$ m) pontoon-type WEC with $\mathcal{E} = G$.

5

Integration of WECS with Floating Breakwater

This chapter explores the integration of WEC with floating breakwater (FB) as a means to extract energy from the wave and attenuate the wave force. The integrated horizontal-raft WECs with a floating breakwater (iHR-WEC-FB) and the integrated vertical-flap WECs with a floating breakwater (iVP-WEC-FB) are considered. The iHR-WEC-FB and iVR-WEC-FB allow the conversion of wave energy into electricity while protecting the coast from erosion due to wave impact. A numerical model based on the FE-BE method presented in Chapter 4 is used to investigate the performance of the iHR-WEC-FB and iVP-WEC-FB. The case study examines how variations in WEC dimensions, mooring stiffness, wave periods, and wave directions affect the optimal design of the iHR-WEC-FB and iVP-WEC-FB, with the goal of maximizing energy generation and enhancing coastal protection.

5.1 Introduction

In Chapter 4, the feasibility of integrating WEC with a floating platform has been described and demonstrated via a few case studies. In this chapter, focus will be on the integration of WECs with FB for the tropical climate. Tropical countries like Singapore, the Maldives, and Indonesia, even though experiencing relatively mild sea conditions compared to their sub-tropical counterparts, are still vulnerable to coastal erosion. Over time, waves gradually remove sediments and rocks from their shorelines. Data shows that Indonesia, the world's largest archipelagic nation, has lost 29,261 hectares of coastal land—an area equivalent to the size of its capital, Jakarta—over the past 15 years due to erosion.[144] The effects of coastal erosion, exacerbated by rising sea levels, have underscored the importance of developing technologies to mitigate this damage. Conventionally, the strategies to manage coastal erosion are by retreating or relocating to regions that are higher, such as the highland, or by preserving the shoreline with hard structures, such as dikes and shore protection materials.[145–147] Although shore protection materials such as geotextile[146,147] provide a considerable advantage due to their relatively easy deployment, they may be more expensive compared to the use of

DOI: 10.1201/9781003387855-5

hard construction structures such as breakwater.[150] While bottom-mounted breakwater is commonly used, floating breakwater[151–154] has been proposed and built due to its cheaper construction costs, environmental friendliness, and greater adaptability to the rise of seawater level.[155] Sun et al.[156] provide a comprehensive review of the prospect of integrating WEC with FB. Floating breakwater made of polyethylene is also proposed – the Wavebrake and WhiprWave. For e.g., the Wavebrake (www.wavebrake.com) is a novel FB with multiple grooves to attenuate wave force and at the same time equipped with PTO to produce electricity. Another design, the WhiprWave,[156] is a high-density polyethylene (HDPE) float that exhibits low wave transmittance with a unique porous groove design to dissipate wave energy.

To maximise the use of sea space, WECs like the OWC WEC, point absorber WEC, and raft-type (attenuator) WEC are suggested for integration into breakwaters, serving the dual purpose of generating energy and reducing wave forces.[157–160] The idea of combining WECs with structures[157,161,162] like piers, breakwaters, or floating platforms offers a way to optimise ocean space and makes wave energy a more attractive and cost-effective renewable energy source. Several approaches have been developed to keep WECs competitive in energy generation, such as improving power efficiency with a trilinear-damper system[163] and integrating oscillating buoy-type WECs with vertical pile-restrained floating breakwaters.[160] There are also numerous efforts to merge WECs with floating breakwaters by placing the WECs on the exposed side of the breakwater.[161,164] A comprehensive review of the integration of WECs with FBs was provided by Mustapa et al.[165] Michailides and Angelides[166] explored the feasibility of using a flexible floating breakwater for both shoreline protection and wave energy conversion. They found that it is possible to meet the functional requirements of both the FB and WEC. Attenuator-type WECs, also called raft-type or flap WECs, which generate energy through pitching motions, have also been proposed. For example, the raft-type WEC attached to a VLFS,[167,168] as discussed in Chapter 4, showed that power generation was significantly influenced by the length of the plate-like WEC and PTO damping. The performance of the raft-type WEC could be further improved by using a modular system, as suggested by Wilkinson et al.[169] This modular approach allows for greater energy extraction from waves approaching from multiple directions. Noad and Porter[170] proposed a new model for the interaction of waves with an articulated raft, demonstrating that the capture factor increases in proportion to the number of pontoons, with shorter rafts generating more energy. A summary of the different types of WECs integrated into infrastructure is shown in the schematic diagram in Figure 5.1.

For the iHR-WEC-FB, the FB could be very large in size in the order of hundred metres, thereby making its elastic deformation significant in the assessment of the performance of WEC. The raft-type WEC could be of horizontal-raft type or vertical-raft type (known as flap)[30,168,171,172] and generates energy via its pitch motion as waves hit on it. A PTO system is used

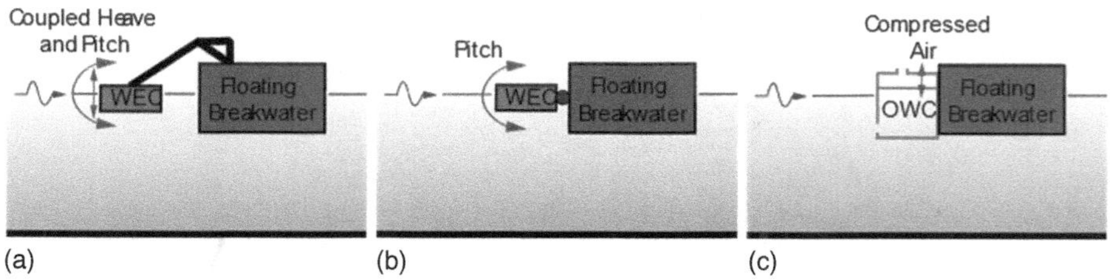

FIGURE 5.1
Schematic diagram for (a) raft-type (attenuator-type) WEC (b) point absorber WEC (c) OWC WEC, integrated to FB.

to convert mechanical motion into electricity. Two case studies will be presented in Sections 5.3 and 5.4, based on the findings found by the authors in the previous studies.[173,174] Case Study I considers horizontal-raft type attenuator WECs, whereas Case Study II uses vertical-raft type attenuator WECs, attached to FB.

5.2 Existing Prototype

There are numerous prototypes of WECs integrated into existing infrastructure available around the globe. Most of the prototypes available as of date are installed on harbour to demonstrate the feasibility of the wave energy as a sustainable source of energy for the port. As mentioned in Section 1.4, the SE@Ports and the EcoWave Power Projects were executed in the Spanish Mediterranean Port and Israel's Jaffar Port to study the feasibility of integrating WECs with port infrastructure. The Seahorse system,[175] depicted in Figure 5.2(a), was installed at the Port of Pecem, Brazil, in 2012, with a production capacity of 50 kW. It consists of two large mechanical arms mounted on the port's pier, with their kinetic movement powered by waves acting on circular buoys. In 2015, a new type of overtopping WEC, the Overtopping BReakwater for Energy Conversion (OBREC),[176] was built at the Port of Naples by incorporating it into a traditional rubble mound breakwater. Illustrated in Figure 5.3(b), the OBREC is expected to generate 3,500 MWh of energy when installed on a 500-metre breakwater. SINN Power[177] technology has installed five-point absorber WECs along the port wall of Heraklion in Greece, as shown in Figure 5.2(c). These WECs are designed to be lightweight for easy transport and installation on a frame attached to the port wall. Another notable OWC is the PICO[178] integrated into the coastline at the Azores (Portugal).

Figure 5.3(a) shows the Mutriku power plant[175] in Spain, known as the REWEC, integrate the OWC system, which comprises 16 capture chambers

FIGURE 5.2
Example of WEC integrated with (a) port (Seahorse system[175]), (b) sea wall (Overtopping BReakwater for Energy Conversion, OBREC[176]), and (c) port wall (PINN Power[177]).

FIGURE 5.3
Example of WEC integrated with (a) breakwater at Mutriku power plant (REWEC)[175] and (b) breakwater in Reggio Calabria, Italy.[179]

with the FB and has the capability to produce a total of 296 kW of energy. An OWC is also installed to a multi-caisson breakwater in Reggio Calabria, Italy,[179] as shown in Figure 5.3(b).

5.3 Case Study I: Horizontal-Raft Type WEC[173]

The first case study considers the iHR-WEC-FB, which comprises a series of an equally spaced horizontal-raft type WECs hinged to the seafront of a FB.

5.3.1 Problem Definition

Figure 5.4 illustrates the schematic of the iHR-WEC-FB. The FB is designed as a rectangular box-like structure, with its four corners anchored to the seabed via mooring lines modelled as springs with stiffness k_{mooring}. There is a total of N_{wec} WEC units evenly spaced at intervals s_p, which equals the width of each WEC ($s_p = W_{wec}$). The WECs are connected to the FB with hinge connectors, allowing them to pitch in response to incoming waves. The PTO system is located at these hinge connections to convert mechanical motion into electrical energy. The floating breakwater has a length of L_{fb} and a constant width of W_{fb}, while the raft-type WEC is modelled as a rectangular solid plate with dimensions L_{wec} and W_{wec}. The thicknesses of the FB and WEC are d_{fb} and d_{wec}, respectively.

The wave and structural characteristics of the iHR-WEC-FB are detailed in Table 5.1. The iHR-WEC-FB is positioned on water at a consistent depth of

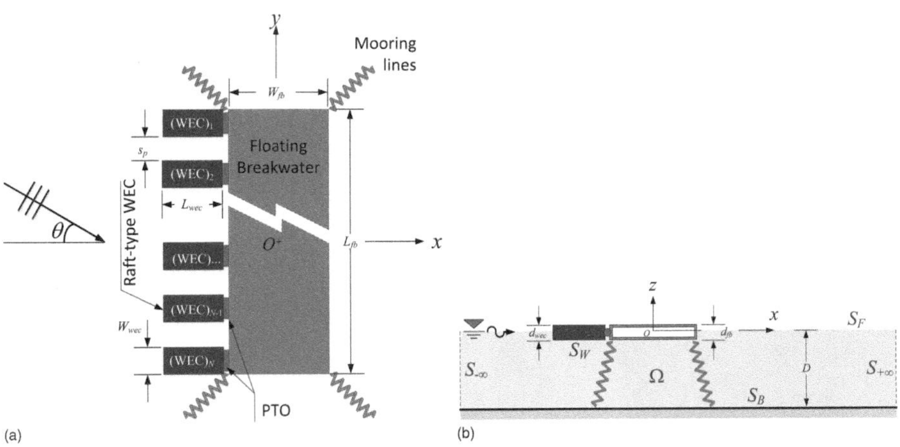

FIGURE 5.4
Schematic diagram showing iHR-WEC-FB in (a) plan view and (b) elevated view.

TABLE 5.1

Principal Dimensions and Properties for iHR-WEC-FB

		Property Symbols	Unit	Type			
				A	B**	C	D
Structure Properties	FB	L_{fb}	m	99	98	99	95
		$\overline{k}_{mooring} = k_{mooring} / \left(\rho g L^{*2}\right)$	–		0, 2 (Default), 10		
		W_{fb}	m		2		
		d_{fb}	m		2		
		$\mathcal{E}_{fb}$	GPa		11.9		
		ρ_{fb}	kg/m³		256.25		
	Raft-type WEC	N_{wec}	–	50	25	10	
		s_p	m	1	2	5	
		W_{wec}	m	1	2	5	
		L_{wec}	m		1, 2 (Default), 4		
		d_{wec}	m		2		
		B_{pto}	N.m.s		0 to 25,000 (100 interval)		
		$\mathcal{E}_{wec}$	GPa		11.9		
		ρ_{wec}	kg/m³		256.25		
Regular Wave Properties		ω	rad/s		0.52 to 2.09		
		θ	degree		0 (head sea), 60		
		A	m		1.0		
		d	m		10		

Note: $L^* = L_{wec} + W_{fb}$, **Default

$D = 10$ m. To examine its fundamental hydrodynamic properties, the impact of regular waves is analysed, assuming a wave period T varying from 3 to 12 s and a wave amplitude $A = 1$ m. The waves strike the iHR-WEC-FB at angles θ of 0°, 30°, 45°, and 60° relative to the negative x-axis, as illustrated in Figure 5.4.

Conversely, both the WEC and FB are modelled as solid, thick plates using Mindlin plate theory, taking into account their structural deformation under wave action. The dimensions of the rectangular iHR-WEC-FB are detailed in Table 5.1, where both the FB and WEC have a uniform thickness of 2 m, a Young's modulus of 11.9 GPa, and a mass density of 256.25 kg/m^3. The thicknesses of the FB and WEC are kept constant because the FB's effectiveness in attenuating wave forces is more significantly influenced by its width than its thickness.[180] The length of the FB is fixed at approximately 100 m for all four types of WECs: Type A, B, C, and D. Parametric studies are conducted by examining the four types of raft-type WECs attached to the FB to explore the impact of varying W_{wec} and N_{wec}. It is crucial to note that the total waterplane area for all four WEC types is kept constant at $A_{wp} = 100$ m^2. This ensures a fair comparison of the performance of the different iHR-WEC-FBs, as the exciting and restoring forces are influenced by A_{wp}. The performance of the WEC and FB will be evaluated using the BEM method, as described in Chapters 3 and 4, where the WEC is considered a rigid body, and the FB's elastic deformation is taken into account. Interest readers may refer to these chapters for details on the coupled FE-BE method approach. For simplicity, wave elevation $\hat{\eta}$ will be referred to as η in this chapter.

5.3.2 Performance Under Regular Waves

The performance of the iHR-WEC-FB under regular waves is first assessed. To do so, the parametric investigation of the optimal B_{pto} that produces the maximum energy is investigated. The optimal B_{pto} obtained from the parametric studies is then used to investigate the effects that various parameters such as W_{wec}, $k_{mooring}$, W_{fb}, T, and s_p have on the energy generation of the WEC and the effectiveness of the breakwater in attenuating the wave forces.

5.3.2.1 Optimal B_{PTO}

The iHR-WEC-FB produces energy through a hydraulic PTO system, where mechanical energy from the pitch motion is transformed into electricity. The force generated by the PTO is directly proportional to the B_{pto} value, which varies with different ω (i.e., T). Using the box-like Type-B iHR-WEC-FB as the reference model—comprising 25 units of WECs positioned 2 m apart, each with dimensions of $L_{fb} = W_{fb} = d_{fb} = 2$ m—a parametric study was conducted with B_{pto} values ranging from 0 to 25 kN.m.s. This study aims to identify the optimal B_{pto} value that yields the maximum wave energy (i.e., maximum CW). Tay[173] found that the optimal damping for the cases discussed in this section is 9 kN.m.s. The findings are summarised as follows:

(a) Generally, the iHR-WEC-FB generates significantly more energy when waves approach at an oblique angle. This is due to the fact that each raft-type WEC pitches with varying magnitudes and phases under oblique waves, unlike under head sea conditions, allowing it to produce more energy across a broader range of wave periods.

(b) Under oblique wave conditions, the wave power output reaches its peak when $\theta = 45^\circ$, but begins to decrease as θ increases (e.g., $\theta = 60^\circ$). This decrease occurs because the WECs exhibit smaller pitching motions when waves come from directions closer to the beam sea ($\theta = 90^\circ$).

(c) The energy output from the iHR-WEC-FB increases as T decreases, which corresponds to a smaller λ. This is because raft-type WECs pitch more significantly with smaller λ, whereas with larger λ, they oscillate in phase with the wave orbital motion, leading to lower energy generation at longer λ.

5.3.2.2 Effect of Varying WEC Widths

The impact of varying W_{wec} on the performance (energy generation) of the WEC is illustrated in Figure 5.5. The P_a of the iHR-WEC-FB for three W_{wec} values—1, 2 (default), and 5 m—is plotted for two different directions: $\theta = 0^\circ$ and 60°. Generally, both plots in Figure 5.5 indicate that P_a increases or decreases with the increase in ω. Since the performance of the iHR-WEC-FB is influenced by the relative pitching angle between the WECs, it is evident that the iHR-WEC-FB can produce more power under oblique waves compared to head sea conditions. However, if the iHR-WEC-FB operates in head sea conditions at a sea state with ω close to its natural periods ω_N, it can generate more power due to resonance.

A detailed analysis of the impact of W_{wec} reveals that the iHR-WEC-FB generates more power with a smaller W_{wec}. This is because a smaller W_{wec} has lower mass inertia, allowing it to pitch in larger magnitude when waves hit the structure. However, for the iHR-WEC-FB under head sea conditions, power production increases as W_{wec} increases. This can be attributed to the relative pitching angles between the WECs. Under head sea conditions, all the WECs pitch with nearly the same magnitude and phase, resulting in zero relative pitching angles between them. In this scenario, increasing the W_{wec} allows the WECs to better capture incoming waves, thus generating more power for the iHR-WEC-FB.

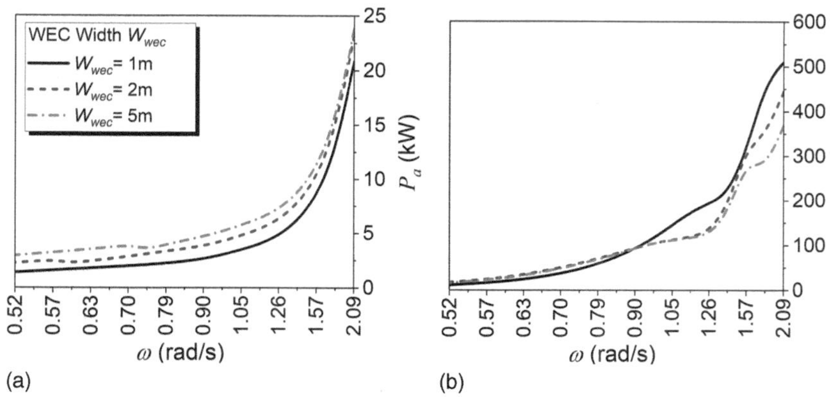

FIGURE 5.5
Power absorption of iHR-WEC-FB for varying W_{wec} under (a) $\theta = 0^{\circ}$ and (b) $\theta = 60^{\circ}$. Regular wave with A = 1.0 m.

5.3.2.3 Effect of Varying Mooring Stiffnesses

Figure 5.4(a) shows that the FB is moored at its four corners by mooring lines with the non-dimensional mooring stiffness, defined as $\bar{k}_{mooring} = \dfrac{k_{mooring}}{\rho g \left(L^*\right)^2}$, where L^* is the characteristic length taken as $L^* = L_{wec} + W_{fb}$. The effect of varying mooring stiffnesses has on the performance of the iHR-WEC-FB is investigated in Figure 5.6. Three mooring stiffnesses $\bar{k}_{mooring}$ are investigated, i.e., $\bar{k}_{mooring} = 0, 2, 10$. It can be seen from Figure 5.6 that more energy could be generated when the breakwater is moored by stiffer mooring lines. This is because the FB is more stable and has a smaller motion when the mooring stiffness is large, i.e., $\bar{k}_{mooring} = 10$, thus resulting in the larger relative average pitch motion between the WECs and FB. It is noted that $\bar{k}_{mooring} = 0$ indicates that the FB is not moored. Under such a condition, the relative average pitch motion is found to be small due to the cancelling effect between the motion of the WECs and FB.

5.3.2.4 Effect of Varying WEC Lengths

The *CW* derived from Equation (3.59) has a length unit, making it significantly influenced by the L_{wec}. Figure 5.7 illustrates how varying L_{wec} impacts the power absorption of the Type-B iHR-WEC-FB under an oblique wave at $\theta = 30°$. The P_a peaks when $L_{wec} = 2$ m but decreases when L_{wec} is either too small (i.e., $L_{wec} = 1$ m) or too large (i.e., $L_{wec} = 4$ m). Notably, a WEC can generate more energy through elastic deformation when the characteristic length λ_c of the structure (see Equation (4.1)) is larger. However, a larger

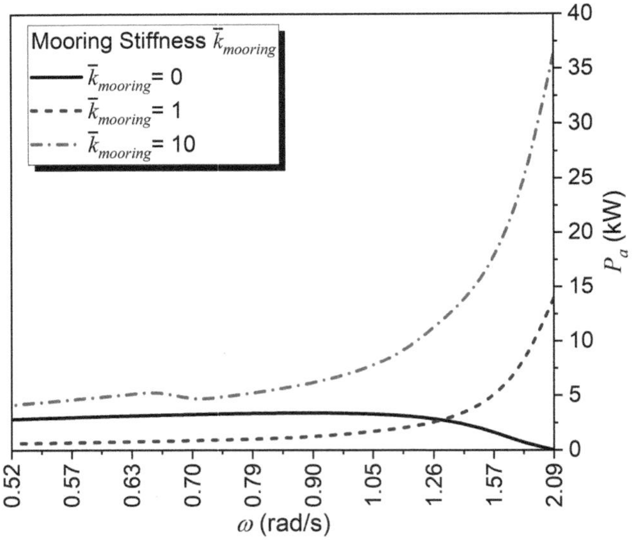

FIGURE 5.6
Power absorption of iHR-WEC-FB for varying mooring stiffness under $\theta = 0^{\circ}$. W_{wec} = 2 m and N_{wec} = 25. Regular wave with A = 10 m.

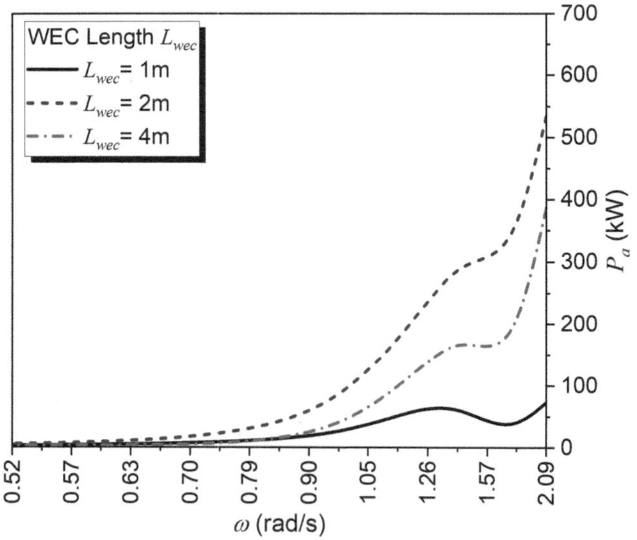

FIGURE 5.7
Power absorption of iHR-WEC-FB for varying breakwater widths under $\theta = 30^{\circ}$. $W_{wec} = 2\,\text{m}$, $N_{wec} = 25$, $\bar{k}_{mooring}$ = 2.0. Regular wave with A = 1.0 m.

length also means a heavier structure, which dampens the WEC's motion. Consequently, the optimal length for the iHR-WEC-FB to maximise energy generation is between 2 m and 3 m, as depicted in Figure 5.7.

5.3.3 Effectiveness of iHR-WEC-FB as a Breakwater

The effectiveness of the iHR-WEC-FB as a breakwater in attenuating the wave forces is investigated. The influences of three parameters, i.e., T, s_p, and W_{wec}, are being investigated in the subsequent sections. Note that for all cases, the Type-B iHR-WEC-FB is considered, i.e. $N_{wec} = 25$, $W_{wec} = 2\,\text{m}$, $L_{wec} = 2\,\text{m}$, $W_{fb} = 2\,\text{m}$, and $L_{fb} = 98\,\text{m}$.

5.3.3.1 Effect of Varying Wave Periods

According to the dispersion Equation (2.28), the wavelength λ increases with an increase in the wave period T, and vice versa. Figures 5.8(a)–5.8(c) show the wave elevations η around the iHR-WEC-FB for regular waves with periods of $T = 3.5$, 4.0, and 5.0 s, respectively.

It is evident from Figure 5.8 that wave diffraction is particularly pronounced at the corners of the iHR-WEC-FB. Additionally, the iHR-WEC-FB reflects the waves, causing higher wave elevations upstream of the structure. The wave elevation is notably high near the iHR-WEC-FB due to the interference between the incident and reflected waves, leading to standing waves in that area. Conversely, wave elevations decrease downstream of the iHR-WEC-FB. Figure 5.8 also indicates that the reflected wave elevation is greater when the iHR-WEC-FB is exposed to waves with shorter λ, i.e., smaller T, compared to waves with longer λ. Moreover, the FB proves to be more effective in reducing wave force when λ is shorter. These observations align with wave diffraction theory,[181] which explains that shorter λ waves lead to increased reflection and diffraction, causing greater wave interference. It should be noted that the FB is less effective with longer λ because the structure moves in phase with the incoming waves.

5.3.3.2 Effect of Varying Spacings Between WECs

Figure 5.9 illustrates how different spacings between the WECs impact the wave climate around the iHR-WEC-FB when it is exposed to regular waves with a period of $T = 3.5\,\text{s}$. The iHR-WEC-FB is minimally influenced by the spacing of the WECs when the incoming wave approaches from the head sea. This indicates that the wave disturbances around the iHR-WEC-FB are not significantly affected by the number of WECs, provided that the total A_{wp} and W_{fb} remain unchanged. This is because the difference in power absorption for head sea is relatively minor compared to the oblique sea (refer to Figure 5.5),

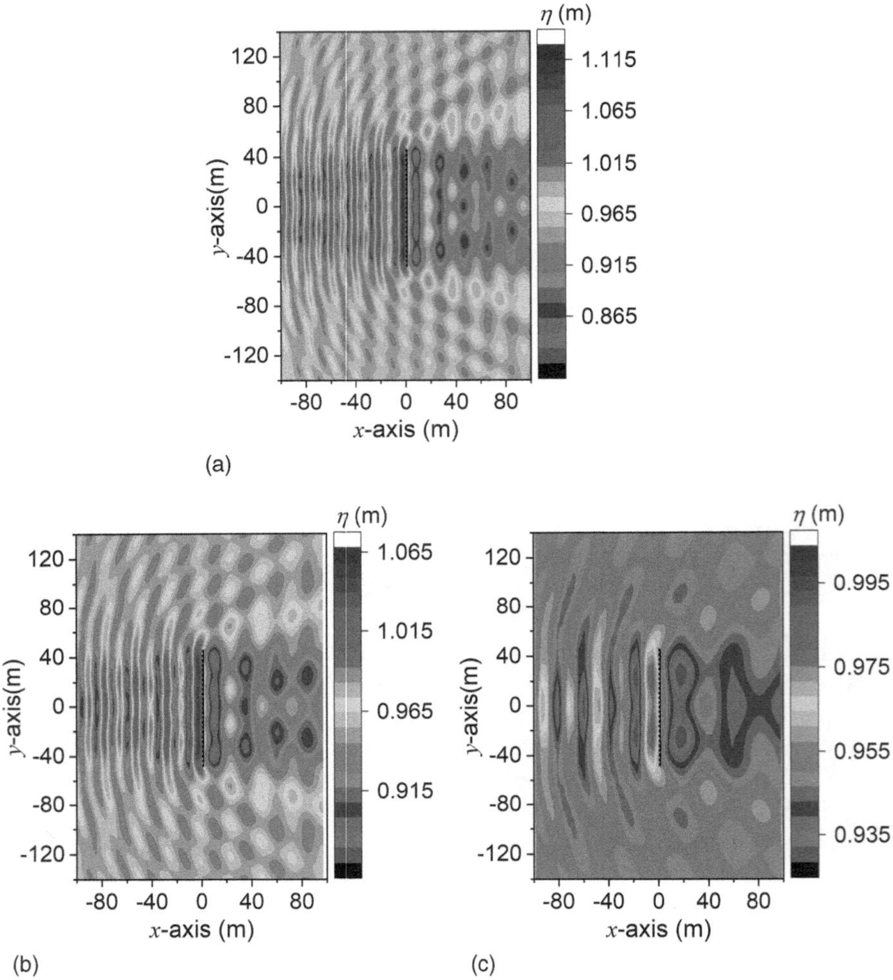

FIGURE 5.8
Wave amplitude surrounding iHR-WEC-FB with $W_{wec} = 2\,\text{m}$ and $N_{wec} = 25$ under regular wave. (a) $T = 3.5\,\text{s}$, (b) $T = 4.0\,\text{s}$, and (c) $T = 5.0\,\text{s}$. $\theta = 0^{\circ}$. $L_{wec} = 2$ m.

leading to smaller wave disturbances around the iHR-WEC-FB with varying W_{wec}. Furthermore, the transmitted wave downstream of the iHR-WEC-FB is strongly influenced by the W_{fb}, while the reflected wave upstream is primarily affected by the WEC's motion. Since the W_{fb}, which impacts wave force attenuation, and the A_{wp} of the WEC, which affects power absorption, are kept constant in this case study, the variations in wave climate are minimal across the different scenarios shown in Figure 5.9.

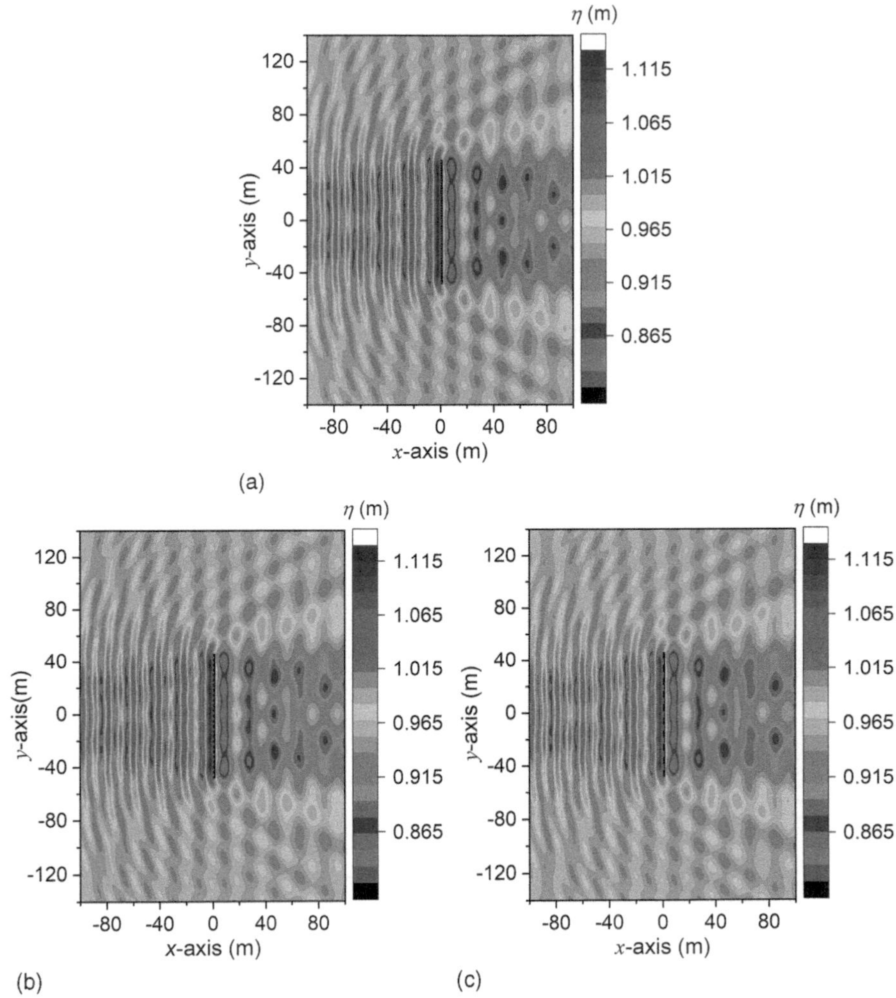

FIGURE 5.9
Wave amplitude surrounding iHR-WEC-FB with (a) $W_{wec}=1\,\text{m}$ and $N_{wec}=50$ (b) $W_{wec}=2\,\text{m}$ and $N_{wec}=25$ (c) $W_{wec}=5\,\text{m}$ and $N_{wec}=10$ under regular wave. $T=3.5\,\text{s}$, $\theta=0^\circ$, $L_{wec}=2\,\text{m}$.

5.3.3.3 Effect of Varying WEC Lengths

The performance of the iHR-WEC-FB is greatly influenced by its length in the x-axis direction, specifically the W_{fb} and L_{wec}.[182–184] Figure 5.11 illustrates the wave elevation around the iHR-WEC-FB under regular waves with a period of $T = 3.5$ s. This analysis focuses on oblique waves to evaluate the impact of varying L_{wec}. Compared to the previous wave climate

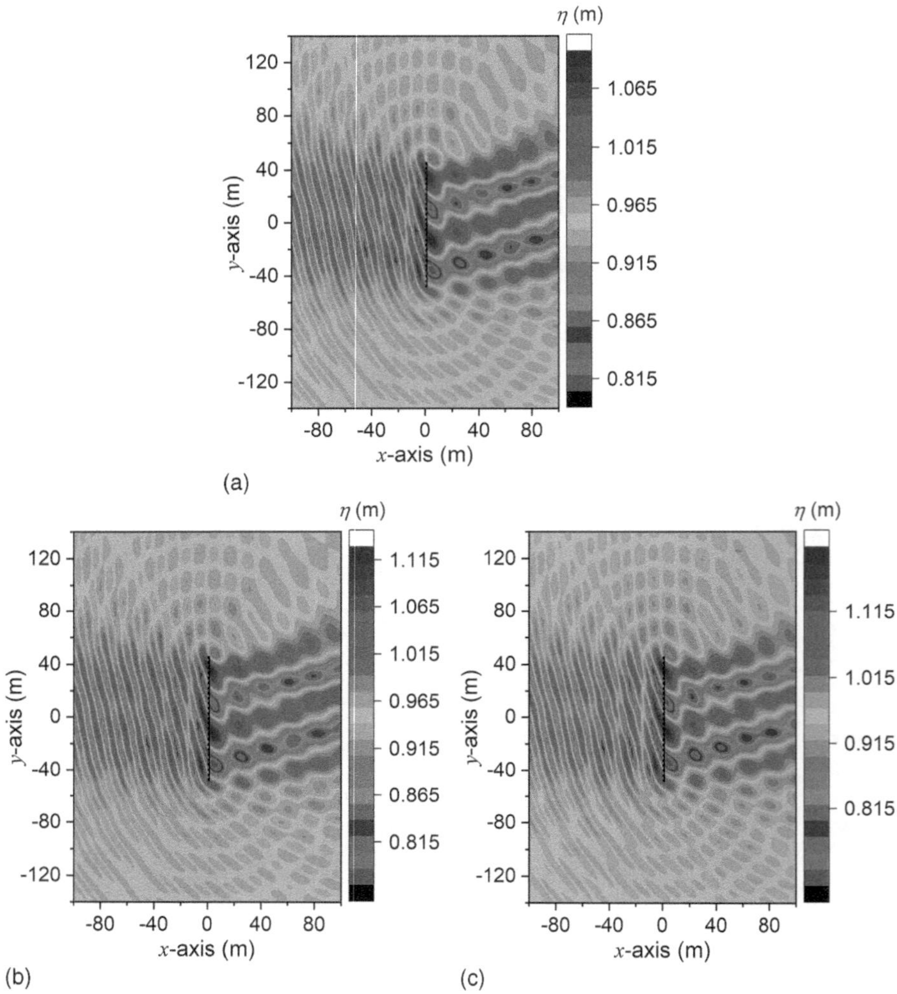

FIGURE 5.10
Wave amplitude surrounding iHR-WEC-FB with (a) $L_{wec} = 1$ m and (b) $L_{wec} = 2$ m $L_{wec} = 4$ m under regular wave. $T = 3.5$ s, $\theta = 30^{\circ}$, $W_{wec} = 2$ m, $N_{wec} = 25$.

results, there is a noticeable difference in the wave disturbance around the iHR-WEC-FB with different WEC lengths. From the minimum wave elevation η shown in Figure 5.10, it can be inferred that a longer WEC length is more effective in reducing wave forces, as it dampens more wave energy. Consequently, the iHR-WEC-FB with a larger L_{wec} reflects

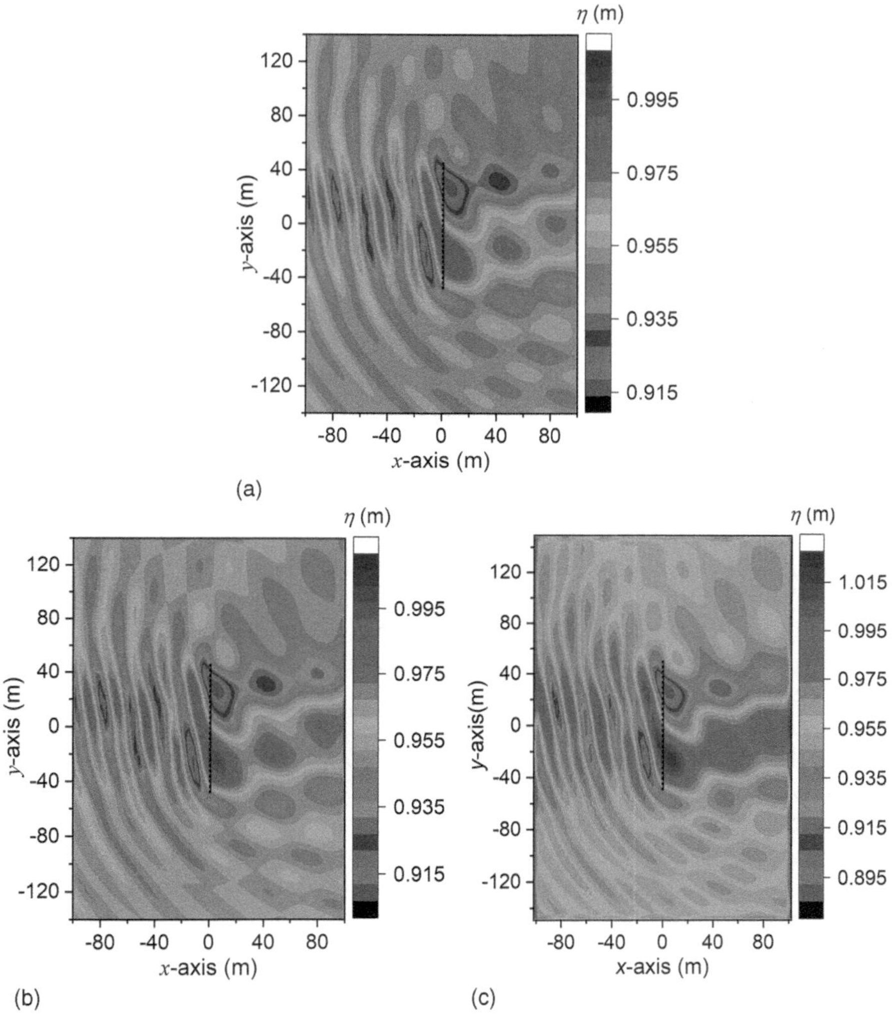

FIGURE 5.11
Wave amplitude surrounding iHR-WEC-FB with (a) $L_{wec} = 1\,\text{m}$ and (b) $L_{wec} = 2\,\text{m}$ (c) $L_{wec} = 4\,\text{m}$ under regular wave. $T = 5.0\,\text{s}$, $\theta = 30^{\circ}$, $W_{wec} = 2$ m, $N_{wec} = 25$.

more waves, as evidenced by the increased wave elevation upstream of the floating structure in Figure 5.10. Conversely, Figure 5.11 reveals that the effect of L_{wec} is minimal under $T = 5$ s, as the iHR-WEC-FB proves ineffective against longer incoming waves with higher energy, which pass the floating structure with minimal diffraction and reflection.

5.4 Case Study II: Vertical-Flap Type WEC

The second case study considers a series of vertical-flap type WECs integrated into a FB, termed the integrated vertical-flap WEC-floating breakwater (iVF-WEC-FB). Similar to iHR-WEC-FB, the iVF-WEC-FB is also made of a series of equally spaced vertical WECs hinged at the seafront of a FB.

5.4.1 Problem Definition

The iVF-WEC-FB comprises an array of equally spaced vertical flaps (VFs) hinged on supporting frames attached to the box-like FB as shown in Figure 5.12. The VF operates following the same mechanism as the OWSC, except that it is an inverted OWSC. As waves hit on the VF, the VF rotates at an angle Θ_y about the hinge (rotation about the local y-axis, see Figure 5.12) where the PTO system is used to convert the mechanical motion into electricity. Twenty-five VFs are considered in the 50-m long iVF-WEC-FB, where each VF has a length L_{wec}, width W_{wec}, and thickness d_{wec}. The dimension and properties of the VF summarised in Table 5.2 are the optimal value obtained following the approach in Section 3.3.6. The FB is allowed to move vertically (heave motion) but with its horizontal motions (sway and surge) constrained.

The iVF-WEC-FB is subjected to both regular and irregular waves. For both the regular wave and irregular wave analyses, a constant water depth of $D = 10$ m is considered. A parametric study is first performed where a single VF attached to a short FB under a regular wave is considered. The objective of the parametric study is to determine the suitable mooring system of the FB, as well as the optimal length and width of the VF. The regular wave and unidirectional irregular wave, modelled using the BS wave spectrum, are used to study the performance of the iVF-WEC-FB and wave climate surrounding the structure. The regular wave frequency ω ranges from 1.0 to 6.0 rad/s, whereas the peak wave period T_p ranges from 3 to 6 s.

Both FB and VF are designed as hollow boxes to allow for sufficient buoyancy, and their outer hulls are made of thin-shell structures. The buoyancy force in the VF must be sufficient to ensure sufficient restoring moment for continuous pitching motion under surging force. It is to be noted that the L_{fb} is adjusted according to the optimal width of the VF with equal spacing s_p obtained from the parametric study. Also, the length of the supporting frame is determined according to the optimal length of the VF to allow for the VF to pitch without clashing with the FB. The inputs for the numerical study of the full-scale iVF-WEC-FB are summarised in Table 5.2.

The performance of the iVF-WEC-FB is assessed using the ANSYS AQWA software, which also adopts the BEM. Here, the FB and WEC are modelled as rigid body motion where only the wetted surfaces of the structure need to

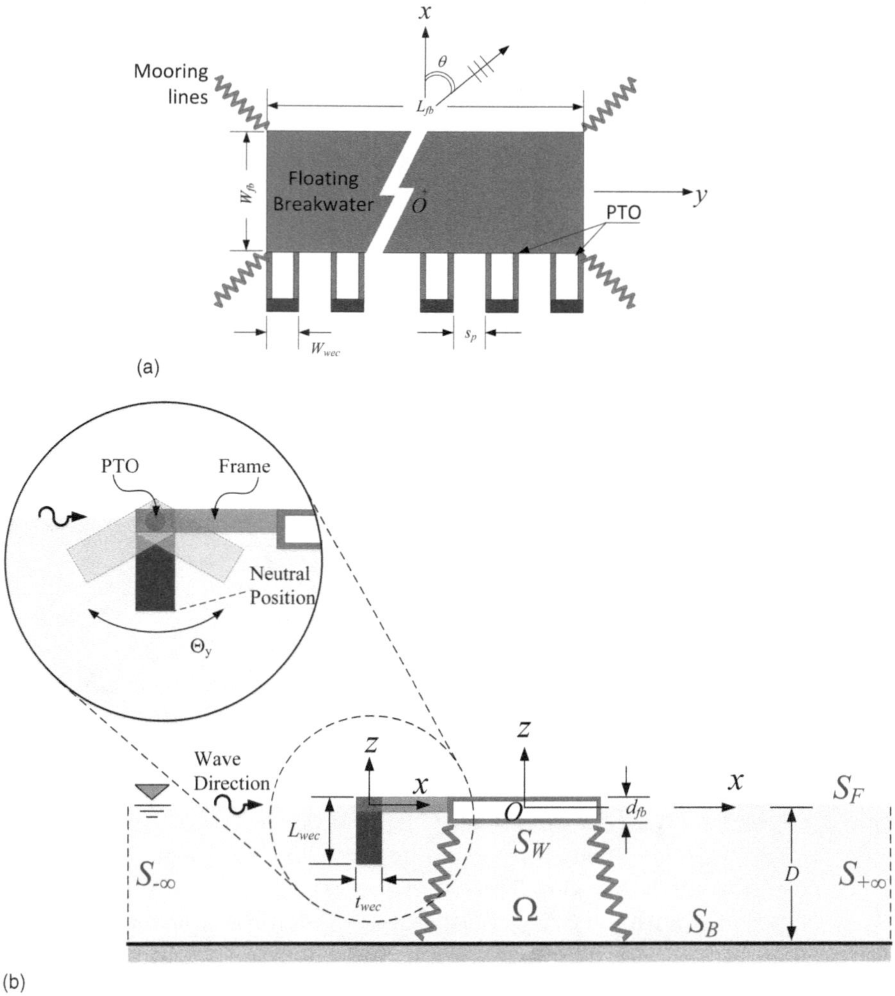

FIGURE 5.12
Schematic diagram showing iVF-WEC-FB in (a) plan view and (b) elevated view.

be modelled. The VF is hinged connected to the FB, and the PTO systems are modelled as damping in the software.

5.4.2 Performance of Single Vertical Flap

Parametric studies are conducted on a single VF to obtain the optimal VF's dimensions in maximising the performance of the WEC. The single VF attached to a FB is considered under regular waves. The effect the mooring system of the FB, width and length of the WECs, has on the performance of the WEC is presented hereafter.

TABLE 5.2

Dimensions and Properties of iVF-WEC-FB

Components	Symbols	Value
Floating Breakwater	Length, L_{fb} (m)	50.00
	Width, W_{fb} (m)	3.00
	Draft, d_{fb} (m)	2.00
Wave Energy Converter	Number, N_{wec}	25
	Length, L_{wec} (m)	2.00
	Thickness, d_{wec} (m)	0.40
	Width, W_{wec} (m)	1.25
	PTO damping B_{pto} (N.m.s/rad)	5,913
Regular Wave	Wave frequency, ω (rad/s)	1.0 to 6.0
	Wave direction, θ (°)	180, −135
	Water depth, H (m)	10
Irregular Wave	Peak Period, T_p (s)	3,4,5,6
	Significant wave height, H_s (m)	2.0

5.4.2.1 Effect of Motion of Floating Breakwater

The box-like FB, as shown in Figure 5.12, is moored by mooring lines by assuming that only the heave motion of the FB is allowed. In this analysis, the energy generation from the iVF-WEC-FB when the FB is free to heave (termed heave-free) is compared with its counterpart when it is heave-constrained. The heave-constraint is applied in ANSYS AQWA by imposing fixed motion to the heave dof. This is analogous to having the FB moored by mooring lines with very high stiffness. The pitch RAO of the single VF when it is hinged to the heave-constrained and heave-free FBs is shown in Figure 5.13. The length and width of the VF are $L_{wec} = W_{wec} = 1$ m. The result shows that the effect of the motion of the FB is negligible towards the pitch RAO when subjected to regular waves of high frequency. While the iVF-WEC-FB shows a slightly greater pitch RAO at a low ω between 1.0 rad/s to 1.7 rad/s, the heave-constrained FB results in a higher pitch RAO for the ω between 1.7 and 2.0 rad/s, i.e., wave period $T\sim$ 3 to 4 s, which is close to a typical tropical sea state in Singapore. This is in agreement with the results for the iHR-WEC-FB in Section 5.3.2, where the heave-constrained FB allows the greater energy generation from a raft-type WEC due to the greater relative pitching motion when attached to a heave-constrained FB. Figure 5.13 also shows that the pitch RAO of the VF reduces with the increase of ω. This is because the λ is short at high ω, thus resulting in

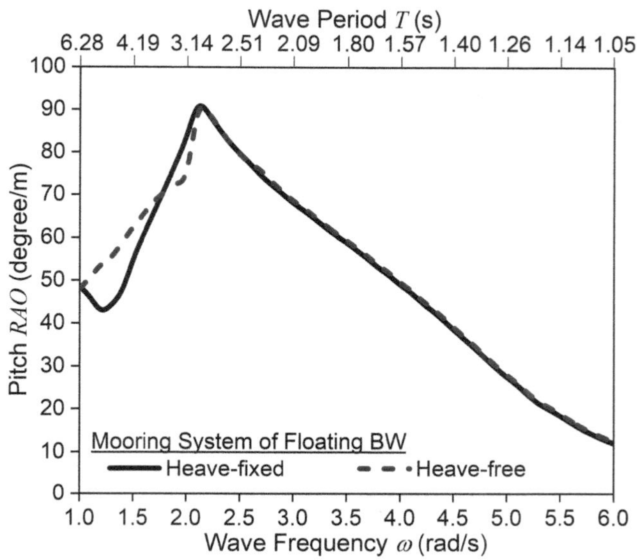

FIGURE 5.13
Pitch RAO for VF when integrated to heave-constrained and heave-free FB. $W_{wec} = 1\,\text{m}$, $L_{wec} = 1\,\text{m}$.

smaller wave energy resources. The pitch RAO is also found to peak when the ω is around 2.0 rad/s, which coincides with the most occurrence sea state of the Singapore sea at around $T \sim 2$ to 4 s.

Figure 5.14 shows P_a and CWR of the single VF with $L_{wec} = W_{wec} = 1\,\text{m}$ when subjected to regular wave. The P_a is represented by the solid line (—), whereas the CWR is represented by the broken line (- - - -). The absorbed power is found to increase with the increase of ω and maintains an almost constant peak power absorption when the wave frequency is around $\omega \sim 2.0$–3.5 rad/s, corresponding to the wave period $T \sim 2$ to 3 s. The P_a then reduces with the increase in ω due to the short λ – i.e., smaller wave energy resource. Figure 5.14 also shows that the single VF can achieve CWR above 15% for a range of wave frequency between 2.0 and 5.0 rad/s, and reaches a peak of 30% at $\omega = 4.0$ rad/s. The comparison of the performance of the VF with heave-free and heave-constrained FB shows that the motion of the FB does not significantly affect the power absorption and CWR of the WEC. However, the heave-constrained FB shall be taken in the subsequent analyses due to its relatively better performance when operating at the sea condition of the tropical sea by taking Singapore as a case study. A FB with minimal motion is important to ensure the safety and comfortability of the crew working on the structure besides generating fewer radiated waves under wave action.

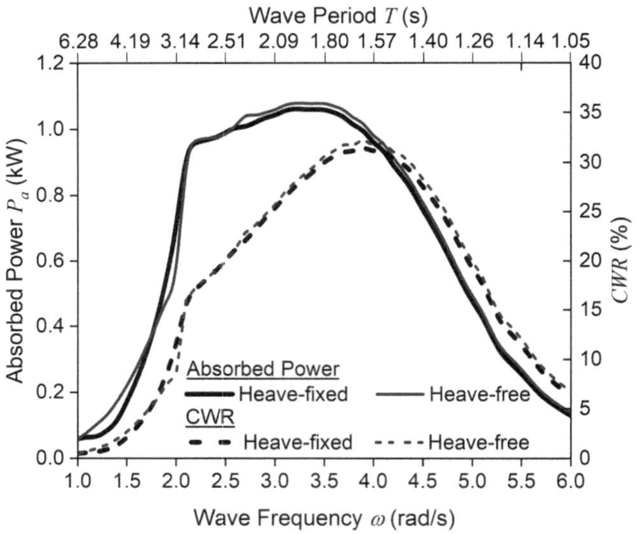

FIGURE 5.14
Absorbed power for single VF when integrated to heave-constrained and heave-free FB. $W_{wec} = 1\,\text{m}$, $L_{wec} = 1\,\text{m}, B_{pto} = 164$ N.m.s/rad.

5.4.2.2 Effect of Width of WEC

By assuming a heave-constrained FB, five different VF with varying widths, i.e., $W_{wec} = 1.00, 1.25, 1.50, 1.75$, and 2.00 m, are considered to study the effect the width has on the power generation of the WEC. The pitch RAO of the five different VFs, when attached to the heave-constrained FB under head sea condition, is presented in Figure 5.15. The length of the VF is taken to be $L_{wec} = 1.00$ m. Figure 5.15 shows that the pitch RAO for all the considered VFs has the same trend as the VF with $W_{wec} = 1.00$ m considered in Figure 5.13, i.e., the pitch RAO increases with ω and decreases after reaching a peak at $\omega = 2.0$ rad/s. It can be seen that the pitch RAO increases significantly with the increase in the width of the WEC with $W_{wec} = 2.00$ m producing a relatively larger pitching motion for all the wave frequencies considered. It should be noted that the peak pitch RAO is about 90°/m, indicating that the VF can swipe a wide angle of 45° from its vertical neutral position (see Figure 5.12b) as the wave hits on the structure.

The P_a derived from Equation (3.57) is plotted in Figure 5.16(a). It is noted that P_a depends on the B_{pto}, which differs for various geometries of the VF (varying W_{wec} and L_{wec}), as optimal B_{pto} depends on the added inertia A, radiated damping B, and hydrostatic stiffness k_y, which are distinct for VF of different geometries. The corresponding optimal B_{pto} for the VF with different W_{wec} is labelled in Figure 5.16(a). It shows that the B_{pto} is the largest for VF with $W_{wec} = 1.25$ m. The value of B_{pto} has a profound effect on the power absorption,

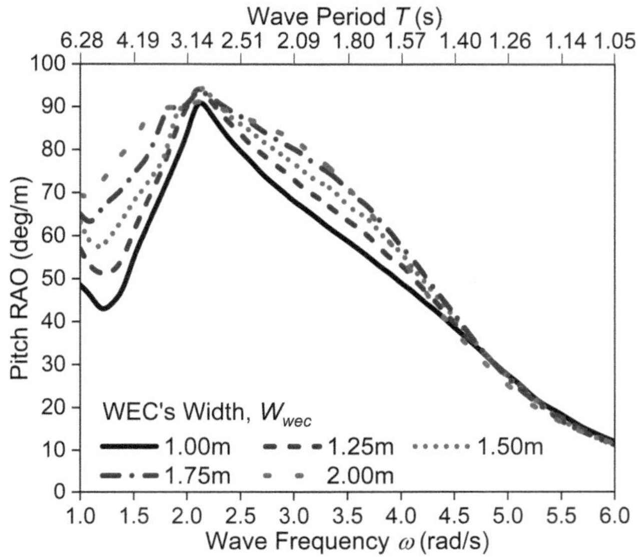

FIGURE 5.15
Pitch RAO for single iVF-WEC-FB with different W_{wec}. $L_{wec} = 1\,\text{m}$, heave-constrained FB.

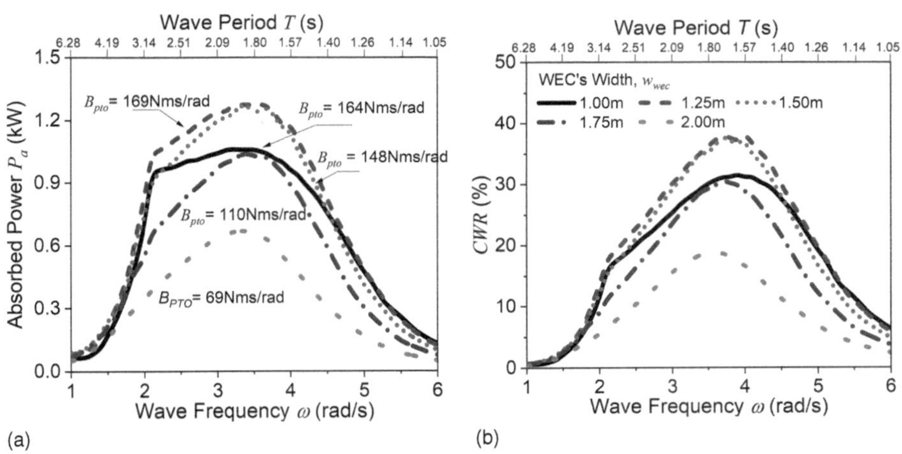

FIGURE 5.16
(a) Absorbed power P_a and (b) *CWR* for single iVF-WEC-FB with varying width W_{wec}. $L_{wec} = 1.00$ m, heave-constrained FB.

as shown in Figure 5.16(a), where the largest P_a is produced by the VF with $W_{wec} = 1.25$ m. It is interesting to see that VF with $W_{wec} = 2.00\,\text{m}$ generates the smallest P_a due to the small optimal $B_{pto} = 64\,\text{N.m.s/rad}$, even though its pitch RAO is the highest, as shown in Figure 5.15. Figure 5.16(a) shows that the P_a increases as the W_{wec} increases from 1.00 to 1.50 m with $W_{wec} = 1.25$ m generating the highest power. However, the power generated significantly decreases for W_{wec} larger than 1.50 m. This phenomenon can be explained by referring to the effect of B_{pto} where it can be seen from Figure 5.16(a) that the optimal B_{pto} for $W_{wec} \leq 1.50$ m is larger than its counterpart for $W_{wec} > 1.50$ m, thus generating more energy according to Equation (3.57).

In the parametric study to investigate the effect of W_{wec}, Figure 5.16(b) shows that under optimal $W_{wec} = 1.25$ m, a *CWR* up to 40% could be achieved at $\omega = 4.0$ rad/s, and the *CWR* is above 20% for wave frequency ranging from 2.0 to 5.0 rad/s, i.e., higher than the *CWR* presented in Figure 5.14, thereby suggesting a wider bandwidth for VF with $W_{wec} = 1.25$ m could be achieved.

5.4.2.3 Effect of Length of WEC

The effect of L_{wec} has on the pitch motion, power absorption, and *CWR* is studied when the $W_{wec} = 1.25$ m and FB are heave-constrained. Five different L_{wec} are considered, i.e., $L_{wec} = 1.00, 1.25, 1.50, 1.75$, and 2.00 m. The pitch RAO is first presented in Figure 5.17 for the varying L_{wec}, where the pitch RAO is found to increase with the decrease of the L_{wec}. This phenomenon can be explained by referring to the moment of inertia of the VF where a shorter VF has a smaller I_y, as compared to the longer VF. As a result, the VF

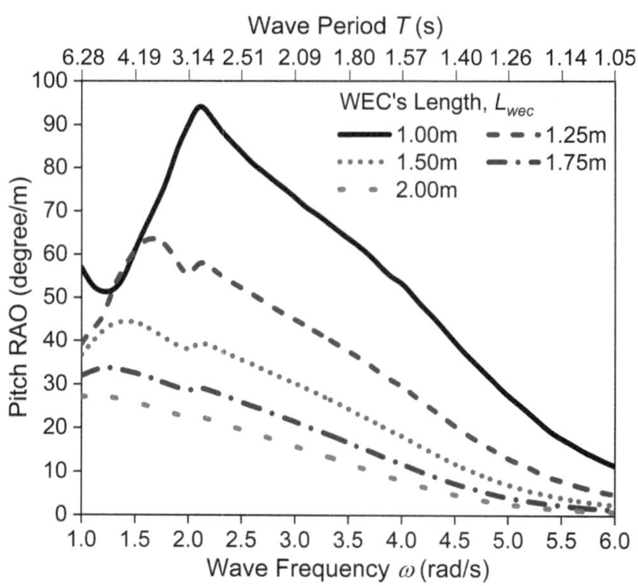

FIGURE 5.17
Pitch RAO for single iVF-WEC-FB with different length L_{wec}.$W_{wec} = 1.25$ m, heave-constrained FB.

with a smaller I_y pitches with a larger θ_y, and therefore the θ_y for the largest $L_{wec} = 2.00$ m is significantly small, as shown in Figure 5.17.

Similar to the effect of varying W_{wec} studied in Figure 5.16(a), the power absorption also depends on the optimal B_{pto} of the VF, which varies for VF with different L_{wec}. The P_a of the VF with varying L_{wec} is plotted in Figure 5.18(a). The effect of length has a greater influence compared to the width of the VF (plotted in Figure 5.16a), where the P_a differs with a larger magnitude for varying L_{wec} as compared to the influence of varying W_{wec}. Also, when keeping the WEC width at its optimal value of $W_{wec} = 1.25$ m, Figure 5.18(a) shows that the P_a of the VF could be further increased by increasing the L_{wec}, where the P_a for $L_{wec} = 2.00$ m is twice the value when $L_{wec} = 1.00$ m, as considered in Figure 5.16(a). The *CWR* of the VF for varying L_{wec} is also given in Figure 5.18(b), where it can be seen that the *CWR* is greater than 20% for ω ranging from 2.0 to 5.0 rad/s, and maximum *CWR* = 60%. The overall energy absorption bandwidth increases with the increase of L_{wec}.

5.4.3 iFB-WEC Under Regular Wave

A full-scale iVF-WEC-FB is considered, where 25 VFs with optimal $W_{wec} = 1.25$ m and $L_{wec} = 2.00$ m are hinged connected to the 50-m heave-constrained FB. The B_{pto} is 5,913 Nms/rad, taken from Figure 5.18 for

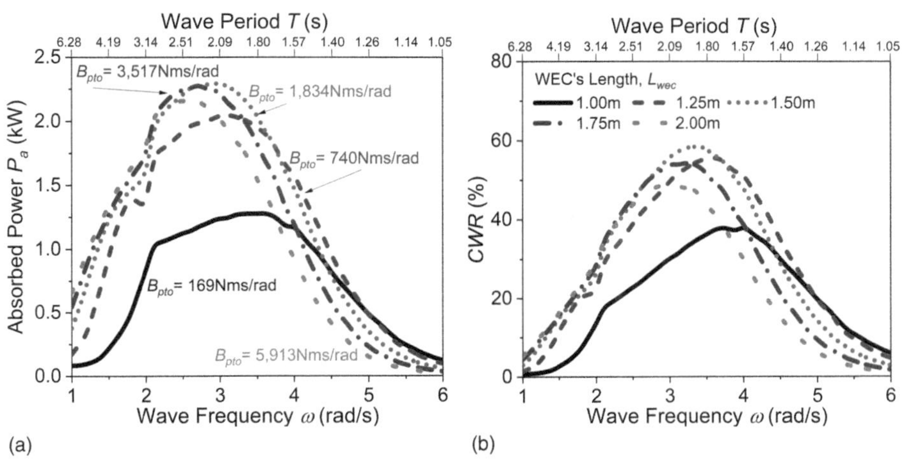

FIGURE 5.18
(a) Absorbed power P_a and (b) *CWR* for single iVF-WEC-FB with different L_{wec}. $W_{wec} = 1.25$ m, heave-constrained FB.

Note: Legend and B_{pto} are the same for Figures 3.9(a) and 3.9(b)

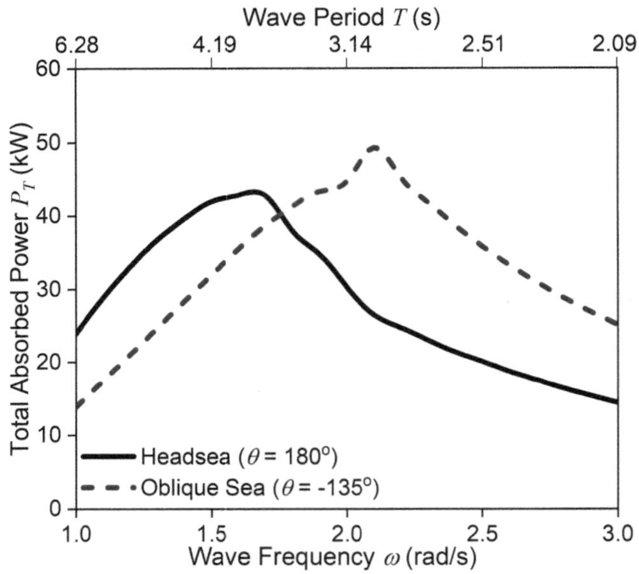

FIGURE 5.19
Total absorbed power of full-scale iVF-WEC-FB ($N_{wec} = 25$) under head sea and oblique sea conditions. Regular waves. $W_{wec} = 1.25\,\text{m}$, $L_{wec} = 2.00\,\text{m}$, $B_{pto} = 5{,}913\,\text{Nms/rad}$, heave-constrained FB.

$L_{wec} = 2.00$ m. The iVF-WEC-FB is subjected to regular and irregular waves under head sea and oblique sea to study the power generation of the VF and wave attenuation effectiveness of the FB. The total absorbed power P_T (see Equation (3.74)) of the 25 VFs under the head sea is presented in Figure 5.19. Figure 5.19 shows that the iVF-WEC-FB has a wide energy generation bandwidth where it can generate a substantial amount of energy for a wide range of frequencies. The iVF-WEC-FB under head sea is more effective at a smaller wave frequency, i.e., $\omega \leq 1.75$ rad/s, whereas iVF-WEC-FB under the oblique sea could generate more energy at $\omega > 1.75$ rad/s. The largest P_T occurs when the wave period is between $T = 2$ to 4 s.

Figure 5.20 shows the wave climate surrounding the iVF-WEC-FB under head sea. Four different wave periods are considered, i.e., $T = 3$, 4, 5, and 6 s and plotted in the sub-figure (a), (b), (c), and (d), respectively. Figure 5.20 shows that the FB is very effective in attenuating the wave force when the λ is short, i.e., $T = 3$ s, where the wave amplitude downstream of the FB is negligibly small. This is due to the relatively small wavelength to FB width ratio, i.e., $\lambda / W_{fb} = 4.68$, as compared to its counterpart with higher λ / W_{fb}.

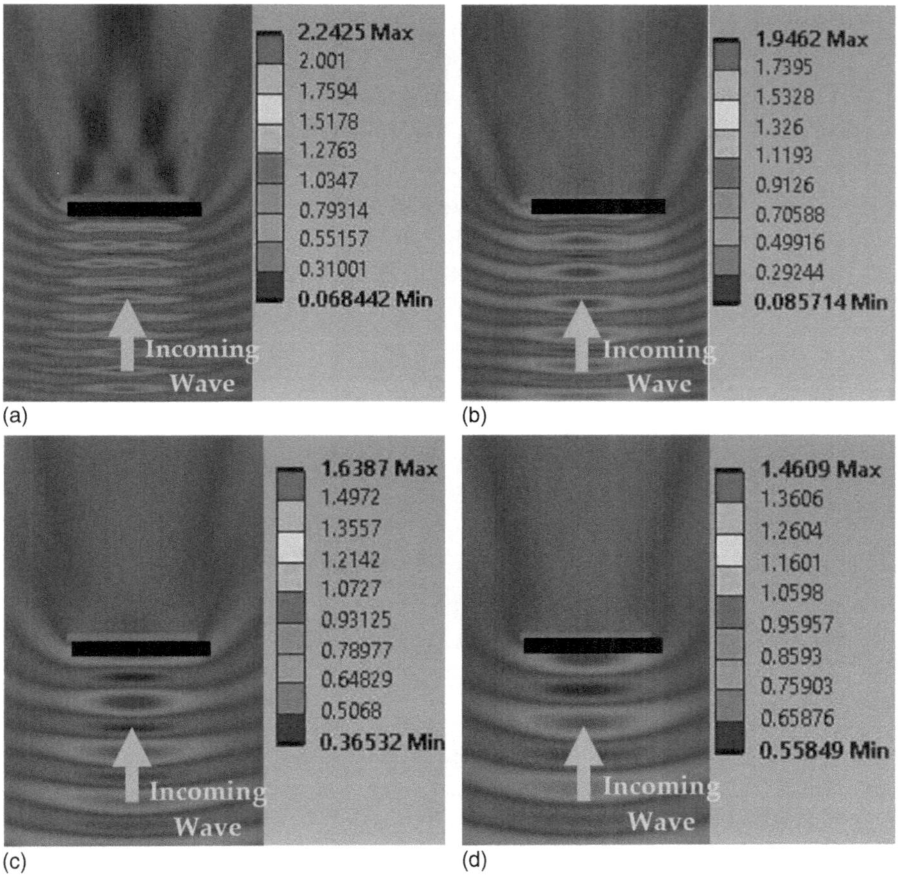

FIGURE 5.20
Wave elevation amplitude surrounding full-scale iVF-WEC-FB for wave period (a) $T = 3s, \lambda / W_{fb} = 4.68$, (b) $T = 4\,s, \lambda / W_{fb} = 8.23$, (c) $T = 5s, \lambda / W_{fb} = 12.20$, and (d) $T = 6s, \lambda / W_{fb} = 16.14$. Head sea. $W_{wec} = 1.25\,m$, $L_{wec} = 2.00\,m$, $B_{pto} = 5{,}913\,Nms/rad$, heave-constrained FB.

5.4.4 iVF-WEC-FB under Unidirectional Irregular Wave

The performance of the iVF-WEC-FB in terms of power generation under irregular waves is studied next. The BS wave spectrum given in Equation (2.104) is used to model the unidirectional waves, by taking the significant wave height as $H_S = 2$ m and peak wave period T_p ranging from 3 to 6 s, with an interval of 1 s.

The power response spectrum can be computed from (3.70) and is denoted as $S_R(\omega)$ here. The total average power $\left(\overline{P_a}\right)_T$ for the head sea can be computed from Equation (3.75) and is presented in Figure 5.21. Figure 5.21 shows that the $S_R(\omega)$ is larger when the wave frequency is small, which agrees with the finding on the P_T (under regular wave) presented in Figure 5.19, where the P_T has a greater value when the $\omega \leq 1.75$ s (i.e., $T \geq 3.6$ s) and gradually reduces when $\omega > 1.75$ rad/s (i.e., $T < 3.6$ s). The iVF-WEC-FB under head sea has the highest $\left(\overline{P_a}\right)_T$ when subjected to irregular waves with $T_p = 4$ and 5 s.

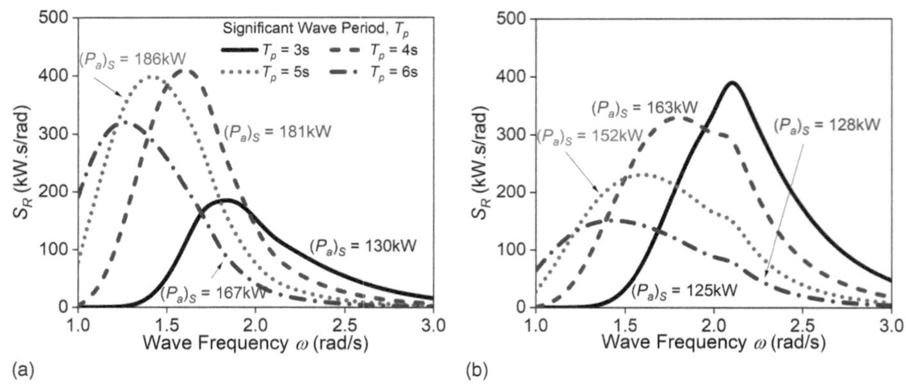

FIGURE 5.21

Power response spectrum $S_R(\omega)$ for full-scale iVF-WEC-FB subjected to irregular wave under (a) head sea (b) oblique sea. $H_s = 3$ m. $W_{wec} = 1.25$ m, $L_{wec} = 2.00$, $B_{pto} = 5{,}913$ Nms/rad, heave-constrained FB.

The $S_R(\omega)$ under the oblique sea plotted in Figure 5.22 has a different trend as compared to the head sea. It can be seen that the iVF-WEC-FB generates the most energy when $T_p = 3$ s, influenced significantly by the behaviour of the P_T (see Figure 5.19), where P_T is the largest when $\omega \approx 2.0$ rad/s (i.e., 3.1 s) and gradually decreases when $\omega < 2.0$ rad/s and $\omega > 2.0$ rad/s.

The $\left(\overline{P_a}\right)_T$ for the head sea and oblique sea is summarised in Figure 5.23. It can be seen that the iVF-WEC-FB produces more energy when waves approach from the head sea but the difference between the head sea and oblique is not significant. Also, the overall largest $\left(\overline{P_a}\right)_T$ occurs when $T_p = 4$ s. The $\left(\overline{P_a}\right)_T$ for all T_p is substantially large, i.e., $\left(\overline{P_a}\right)_T > 100$ kW implying that the iVF-WEC-FB has a wide absorption bandwidth from $T_p = 3$–6 s.

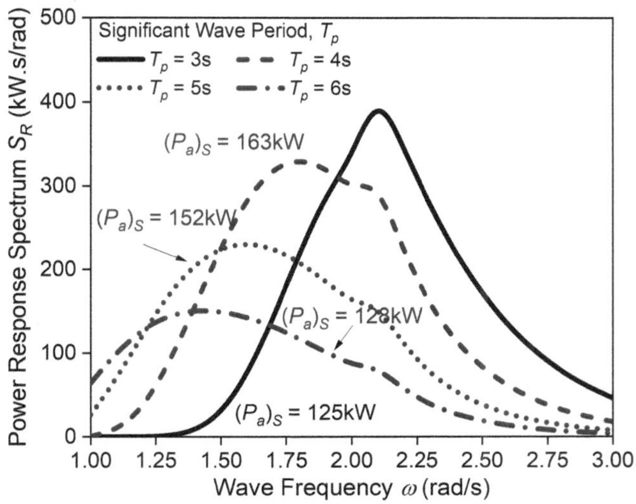

FIGURE 5.22
Power response spectrum $S_R(\omega)$ for full-scale iVF-WEC-FB under varying peak wave T_p. Oblique sea ($\theta = 45^\circ$), $H_S = 2\,\text{m}$. $W_{wec} = 1.25\,\text{m}$, $L_{wec} = 2.00\,\text{m}$, $B_{pto} = 5{,}913\,\text{Nms/rad}$, heave-constrained FB.

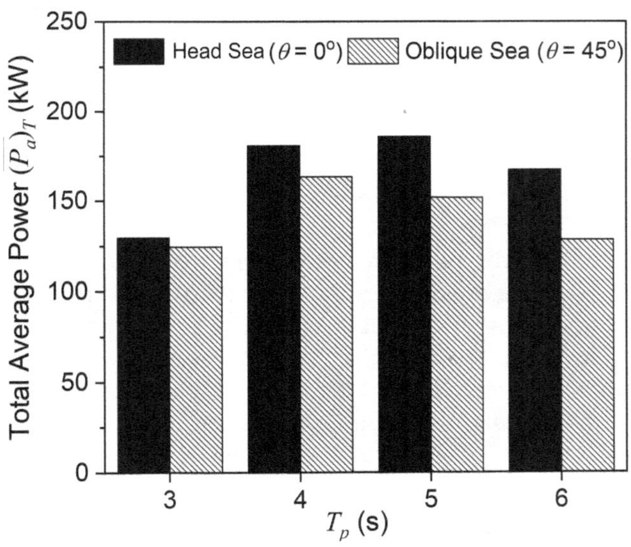

FIGURE 5.23
The total average power of full-scale iVF-WEC-FB under irregular waves. $H_S = 2\,\text{m}$. $W_{wec} = 1.25\,\text{m}$, $L_{wec} = 2.00\,\text{m}$, $B_{pto} = 5{,}913\,\text{Nms/rad}$, heave-constrained FB.

6

Power Enhancement of WEC Arrays via Sloshing Resonance and Wave Run-Up

This chapter explores the advanced methods to enhance the performance of point absorber wave energy converters (PA-WECs) through integration with floating platforms. Specifically, it investigates the deployment of WEC arrays within moonpool configurations and stepped channel platforms to maximise energy generation. The resonance effects within moonpools and the hydrodynamic interactions between WECs are shown to significantly amplify power output, particularly under optimal design conditions. Furthermore, the chapter examines the semi-enclosed moonpool platform, where wave amplification due to resonance and reflection from the moonpool's wall improve the efficiency of WEC arrays, while also serving as a floating breakwater. Another key focus is the stepped channel floating platform, which leverages wave run-up effects to enhance energy capture. The analysis demonstrates that varying the channel's width and depth optimise energy absorption, with the potential for significant increases in power output.

6.1 Background

Chapter 3 highlights the performance of WECs by evaluating their energy generation capabilities in both regular and irregular wave conditions. It has been shown that the energy output of WECs can be increased when deployed in arrays.[29,31] However, the 2015 Quadrennial Technology Review[185] noted a point of diminishing returns for WEC arrays, where adding more rows of WECs results in only slight gains in energy absorption. Despite this, several commercial-scale wave energy farms are making promising advancements such as the WaveRoller WEC, as well as the wave energy farms demonstrated by MOCEAN and AWS Ocean Energy.

As described in Chapter 4, wave energy farms can be made more cost-effective by integrating the WEC array into existing or new infrastructure such as coastal protection structures, floating breakwaters,[173,174] or floating platforms.[167,168,186] This integration can help to optimise the utilisation of sea space, especially when sea spaces are scarce and need to co-exist with other functionalities such as maritime traffic routes, recreation, or tourism. There

DOI: 10.1201/9781003387855-6

are several advantages of having a floating platform to support the WEC array, among them are the following:

(i) The floating platform could be used to support the energy storage system and substation before the electricity is transmitted to the shore.
(ii) Having a floating platform in a remote area also allows for ease of maintenance of the WECs.
(iii) The WEC array could be deployed easily by towing the platform without the need for cable mooring removal.
(iv) Utilisation of the floating platform causes minimal destruction to the seabed as compared to an array of WECs moored to the seabed.
(v) The floating platform provides shelter for marine mammals, birds, and fish.
(vi) The floating platform supports an integrated hybrid renewable energy farm with wind turbines or tidal turbines. Instead of operating as a stand-alone floating energy farm, the proposed design concept could also be integrated into an FB or floating pier, or attached to another floating platform utilised as a floating city. The integrated concept could utilise sea space efficiently and reduce the overall cost of the floating system.

In this chapter, two new means of wave power enhancement are proposed via the following:

(i) Power enhancement via resonance effect in moonpool
(ii) Power enhancement via wave run-up in channel

The moonpool and channel could be located in infrastructure such as FBs and platforms, as shown in Figure 6.1.

Various researchers have proposed this concept, including Zhao et al.[159] who work on OWC integrated with breakwaters, Zhang et al.[187,188] who considered WECs integrated into FBs, and Liu et al.,[189] who also discussed the integration of WECs into floating platforms. The impact of resonance within the moonpool and channel on the WEC array has been found to be significant.[190,235] The deployment site of the WEC array can be optimised by positioning it at the anti-node of the sloshing modes corresponding to the array's exciting natural frequency. Maximising energy output from the WEC array in the moonpool could be achieved by considering the resonance frequencies and anti-node locations of both the moonpool and channel. These resonance frequencies and anti-node positions can be determined analytically, while the array's performance in its optimised position can be evaluated numerically using the panel method.

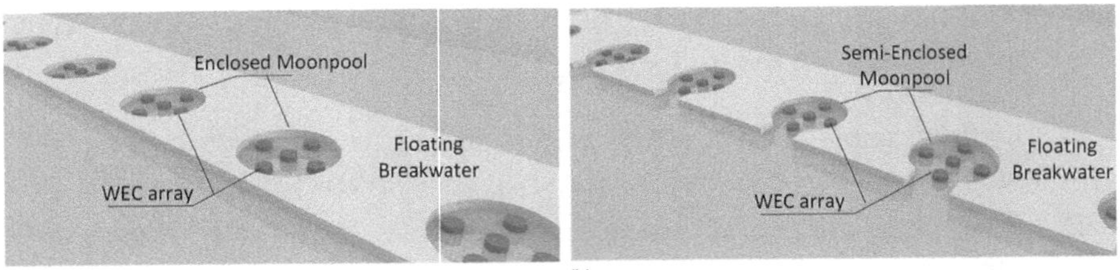

(a) (b)

FIGURE 6.1
Schematic diagram of deployment of WECs in (a) enclosed moonpool, (b) semi-enclosed moonpool, and (c) channel.

The BEM as presented in Chapter 3 is used to assess the performance of the WECs and obtain the wave elevation. Here, the higher-order panel method in WAMIT (www.wamit.com) is used due to its higher computational efficiency and requirement for coarser meshes in the modelling.[191] In the next chapter, we shall present the analytical method to obtain the natural frequencies and modes for a circular moonpool and rectangular channel. The analytical method will be used to first validate the model in WAMIT before further studies are carried out.

6.2 Natural Frequencies and Modes (Analytical Method)

6.2.1 Circular Moonpool

The natural modes and natural frequencies inside a circular moonpool are presented here. A moonpool with radius r_{mp} and draft d_{mp} floating in a water depth of D is considered in Figure 6.2.

The velocity potential inside the circular moonpool (φ^{mp}) can be decomposed as a Fourier series in the azimuthal angle ϑ, expressed in the $r-z-\vartheta$ cylindrical coordinate axis as

$$\varphi^{mp}(r,z,\vartheta)=\sum_{m=0}^{\infty}\varphi_m(r,z), \tag{6.1}$$

where the discrete set of natural modes are associated with each m value in Equation (6.1). It is noted that the lowest natural frequencies encountered by the heave, surge, and pitch motions occur when $m=0$ and $m=1$ according to Molin et al.[192]

The natural modes $\mathcal{Z}_{mn}$ can be categorised into the axisymmetric modes, i.e., $m=0$, and the antisymmetric modes, i.e., $m\geq 1$. The natural frequencies of the modes are given as $\omega_{mn}=2\pi/T_{mn}$, where T_{mn} is the natural period with

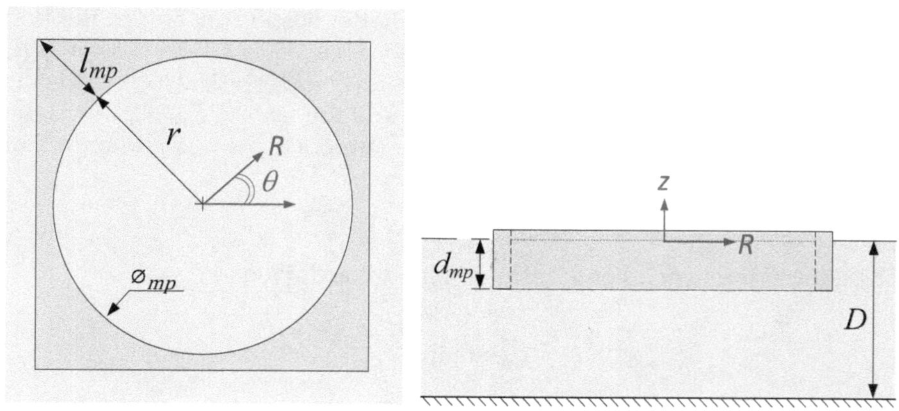

(a) (b)

FIGURE 6.2
(a) Plan view of moonpool and (b) elevated view of moonpool.

the subscript m representing the nodal line (also known as the nodal diameter) and n the nodal circle of the vibration mode.

The axisymmetric mode $\mathcal{Z}_{mn}$ is given as

$$\mathcal{Z}_{0n} \propto J_0\left(k_n r\right), \tag{6.2}$$

where J_0 is the Bessel function of the first kind of order 0, whereas the antisymmetric modes are given as

$$\mathcal{Z}_{1n} \propto J_1\left(\mu_n r\right), \tag{6.3}$$

where k_n are the roots of $J_0^{'}\left(k_n r\right) = -J_1\left(k_n r\right) = 0$, i.e., $k_1 r = 3.8317$; $k_2 r = 7.0156$, $k_3 r = 10.1735$, etc.; and μ_n are the roots of $J_1^{'}\left(\mu_n r\right) = 0$, i.e., $\mu_1 r = 1.8412$, $\mu_2 r = 5.3314$, $\mu_3 r = 8.5363$, etc.

6.2.1.1 Axisymmetric Modes $\mathcal{Z}_{mn}$ ($m = 0$)

The axisymmetric modes could be decomposed into the piston (P) mode and the symmetric (S) sloshing mode.[192] In axisymmetric mode, the indices (m, n) in ω_{mn} indicate that there are m number of nodal lines and n number of nodal circles, respectively.

6.2.1.1.1 Piston Mode $\mathcal{Z}_{00}$ ($m = 0, n = 0$)

The natural frequency of the piston mode of the circular moonpool is given as,[192]

$$\left(\omega_N\right)_{00}^2 \cong \frac{g}{d_{mp} + 4\sum_{q=1}^{\infty} \dfrac{J_1^2\left(\lambda_q r\right)}{\lambda_q^3 b^2 \tanh\left(\lambda_q \bar{D}\right) \cdot J_1^2\left(\lambda_q b\right)}}, \tag{6.4}$$

where g is the gravitational acceleration, J_γ the Bessel function of the first kind of order γ. λ_q are the roots of $J_0(\lambda_q b)=0$, i.e., $\lambda_1 b=2.4048$, $\lambda_2 b=5.5201$, $\lambda_3 b=8.6537$, etc. The radius of the moonpool is $(r=\varnothing_{mp})/2$, $\bar{D}=D-d_{mp}$ is the clearance between the moonpool platform and the seabed, $L_{mp}=l_{mp}+r$ is the distance measured from the centre of the moonpool to the outer boundary of the moonpool platform.

6.2.1.1.2 Symmetric Sloshing Mode Z_{0n} ($m=0, n\geq 1$)

The natural frequencies of the sloshing modes are given as[192]

$$(\omega_N)_{0n}^2 \cong g k_n \frac{1+\alpha_n \tanh k_n d_{mp}}{\alpha_n + \tanh k_n d_{mp}} \text{ for } n \geq 1 \tag{6.5}$$

with

$$\alpha_n = \frac{4k_n}{L_{mp}^2} \sum_{q=1}^{\infty} \frac{\lambda_q J_1^2(\lambda_q r)}{\left(\lambda_q^2 - k_n^2\right)^2 \tanh\left(\lambda_q \bar{H}\right) \cdot J_1^2\left(\lambda_q b\right)} \tag{6.6}$$

6.2.1.2 Antisymmetric Modes Z_{1n} ($m=1, n\geq 1$)

The natural frequency $(\omega_N)_{mn}$ of the single-mode approximation $(m=1)$ for the antisymmetric (AS) sloshing mode is given as[192]

$$(\omega_N)_{1n}^2 \cong g \mu_n \frac{1+\beta_n \tanh \nu_n d_{mp}}{\beta_n + \tanh \nu_n d_{mp}} \quad \text{for} \quad n \geq 1 \tag{6.7}$$

with

$$\beta_n = -\frac{4\mu_n J_1^2(\mu_n r)}{L_{mp}^2\left[J_1^2(\mu_n r) - J_0(\mu_n r)\cdot J_2(\mu_n r)\right]} \times \sum_{q=1}^{\infty} \frac{\nu_q J_1'^2(\nu_q r)}{\left(\nu_q^2-\mu_n^2\right)^2 \tanh\left(\nu_q \bar{D}\right)\cdot J_0\left(\nu_q L_{mp}\right)\cdot J_2\left(\nu_q L_{mp}\right)}, \tag{6.8}$$

where ν_q are the roots of $J_1(\nu_q L_{mp})=0$, i.e., $\nu_1 L_{mp}=3.8317$, $\nu_2 L_{mp}=7.0156$, $\nu_3 L_{mp}=10.1735$, etc.

It is noted here that the S mode represents the mode which is symmetrical about the vertical and horizontal axes of the circular moonpool whereas AS mode represents the mode which is antisymmetric about the vertical axis and symmetrical about the horizontal axis of the moonpool. In AS mode, the indices (m,n) in ω_{mn} indicate that there are m number of nodal lines and $(n-1)$ number of nodal circles, respectively.[191]

6.2.2 Rectangular Channel

Figure 6.3 illustrates the point absorber WEC arrays arranged in a channel with constant-depth d_{chn} and width W_{chn}. The global coordinate system (x,y,z)

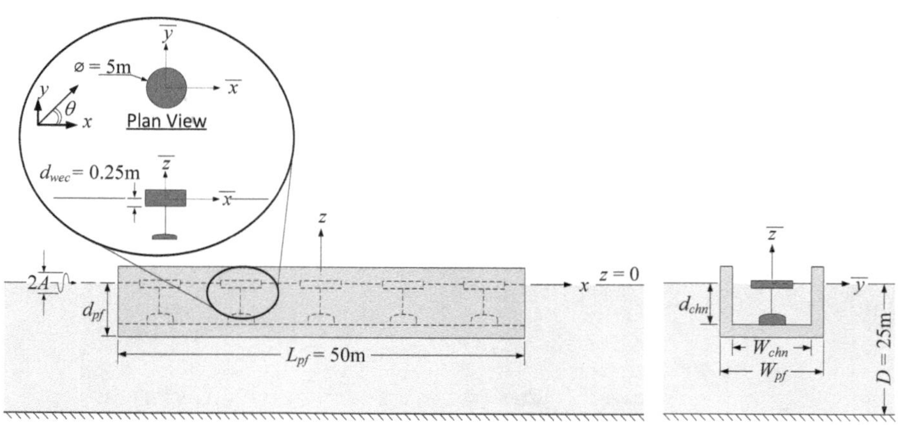

(a) (b

FIGURE 6.3
Array of PA-WECs deployed in floating platform shown in (a) elevation view $(x-z)$ and (b) side view $(y-z)$. $d_o = 5$, 10, and 15 m are considered.

defines the position, with the centre of the arrays located at (0, 0, 0), and the free surface at $z=0$. The channel consists of two vertical walls on a floating platform with length L_{pf}, width W_{pf} and draft d_{pf}. The wave farm is considered to be nearshore, where the water depth is a constant $D = 25$ m. The coordinate $(\bar{x},\bar{y},\bar{z})$ defines the local positions of individual WEC, while (x,y,z) defines the global system coordinate. Each WEC encounters an incoming wave with period T and wave height $2A$, striking the structure at an angle θ relative to the negative x-axis (as shown in Figure 6.3). The PA-WEC has a cylindrical shape with a diameter $\varnothing$ and draft d_{wec}, and is connected to a power take-off (PTO) system with damping B_{pto}. The floating platform is turret-moored, allowing it to rotate and align with varying metocean conditions so that only the head-sea scenario is considered in this context.

The natural frequency along the channel is obtained by solving an eigenvalue problem. The velocity potential φ^{chn} inside the gap formed by two vertical walls can be decomposed as a Fourier series expansion of displacement functions given by Molin et al.[194,195]

$$\varphi^{chn}(x,y,z) = \sum_{m=1}^{\infty} \left(A_m \cosh \nu_m z + B_m \sinh \nu_m z\right) \cdot \cos \mu_m z, \tag{6.9}$$

where $\mu_m = m\pi / L_{pf}$ and $\nu_m^2 = k^2 + \mu_m^2$. The expression in Equation (6.9) ensures that the Laplace equation and the no-flow condition, i.e., $\varphi^{chn} = 0$ at the vertical wall, are fulfilled. By using Garett's method[194] to match the velocity potentials on the boundary between the subdomain defined by the gap and its surrounding, Molin et al.[192,193] arrived at an eigenvalue problem where the eigenfrequencies $(\omega_N)_m$ and the associated eigenmodes A_m and B_m describe waves that are both travelling along the channel and reflected by the side walls. The eigenvectors are restricted to the dominant A_0 and B_0, where the waves appear as propagating waves in the longitudinal direction (with some amplitude modulation in the transverse direction).

Molin et al.[194,195] show that the natural frequencies in the gap by single-mode approximations as

$$(\omega_N)_m^2 \cong g k_m \frac{1 + J_m \tanh k_m d_{pf}}{\tanh k_m d_{pf} + J_m}, \tag{6.10}$$

with

$$J_m(r) = \frac{2}{n\pi^2 r} \left\{ \begin{array}{l} \displaystyle\int_0^1 \frac{r^2}{u^2\sqrt{u^2+r^2}} \left[1 + 2u + (u+1)\cos(n\pi u) - \frac{3}{n\pi}\sin(n\pi u) \right] du \\ \displaystyle -\frac{1}{\sin\theta_0} + 1 + 2r\ln\left(\frac{1+\cos\theta_0}{1-\cos\theta_0} \right) \end{array} \right\}, \tag{6.11}$$

where $k_m = m\pi / L_{pf}$, $r = W_{chn} / L_{pf}$ and $\tan\theta_0 = r^{-1}$. The natural period is given as $(T_N)_m = 2\pi / (\omega_N)_m$.

6.3 Case Study I: Deployment of WECs in Moonpool

6.3.1 Problem Definition

Consider an arrangement of WECs placed inside the closed and semi-enclosed moonpools, which are built into a FB, as illustrated in Figure 6.4(a) and (b). Each WEC is a standard cylindrical PA-WEC functioning at a constant water depth, D.

Two types of moonpools – a fully enclosed circular moonpool and a semi-enclosed moonpool – are analysed to investigate how sloshing resonance affects the power output of the WEC array, as illustrated in Figure 6.4. The semi-enclosed moonpool's opening (o) allows incident waves to enter, which is expected to increase power generation due to wave reflections from the moonpool walls. The moonpool has a diameter $\varnothing_{mp}$ and a draft $d_{mp} = d_{fb}$, with N_{wec} cylindrical PA-WECs arranged inside. The WECs in the array are spaced by a distance s_p, measured from the centre-to-centre of the WEC.

6.3.2 Optimal Dimensions of Moonpool

The natural frequencies of a moonpool are influenced by its size, making it a crucial factor for optimising power output in WEC arrays. When deciding the moonpool's diameter, the available sea space at the deployment site and

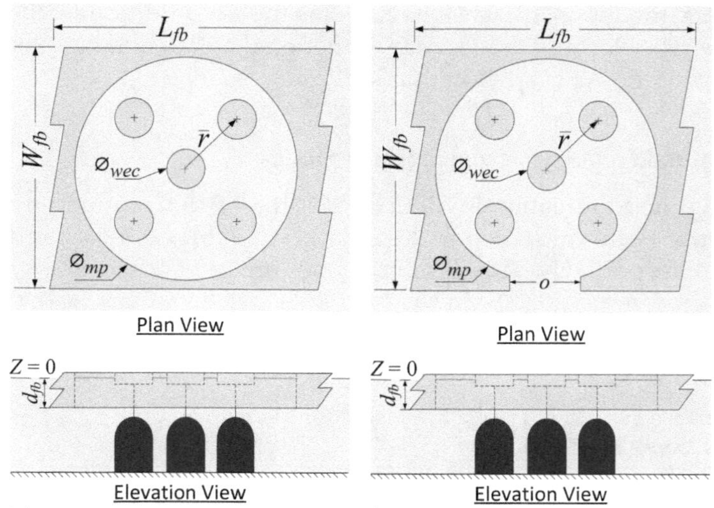

FIGURE 6.4
An array of N_{wec} WEC in (a) closed moonpool and (b) semi-enclosed moonpool.

the number of WECs N_{wec} to be installed must be considered. A larger moonpool increases the width of the breakwater, consuming valuable sea space and adding cost, while a smaller moonpool may not efficiently reduce wave forces and restricts the number of WECs that can be deployed. In this context, we aim to optimise the WEC array configuration by treating the moonpool diameter as the main constraint. A diameter range of $2\text{ m} \leq \varnothing_{mp} \leq 5\text{ m}$, drafts of $d_{mp} = 2$, 5, and 10 m, and water depths between 15 and 25 m are examined, as outlined in Table 6.1. The moonpool's resonance frequencies are calculated using the equations from Section 6.2.1, which helps identify

TABLE 6.1
Limiting Factors Considered in Determining WEC Array Configurations

		Values
Floating Breakwater	Length L_{fb}	50 m
	Width W_{fb}	$4\text{ m} \leq W_{fb} \leq 7\text{ m}$
	Draft d_{fb}	5 m
Moonpool	Diameter $\varnothing_{mp}$	$2\text{ m} \leq \varnothing_{mp} \leq 5\text{ m}$
	Draft d_{mp}	5 m
Regular Wave	Water Depth D	≥ 15 m
	Wave Amplitude A	1 m

the optimal moonpool dimensions based on the sea conditions at the deployment site. The sloshing modes of the moonpool will define the feasible configuration of the WEC array within it.

6.3.3 Natural Frequencies and Sloshing Modes

Following the formulation in Section 6.2.1, the natural sloshing periods for the moonpool are calculated and displayed in **Tables 6.2–6.4** for $d_{mp} = 2$, 5, and 10 m, respectively. The sloshing modes are categorised into symmetric (S) and antisymmetric (AS) in the tables. As per Equations (6.4), (6.6), and

TABLE 6.2

Resonance Period $(T_N)_{mn}$ in a Moonpool for Moonpool Depth $d_{mp} = 2$ m: Water Depth $D \geq 15$ m

	Resonance Period $(T_N)_{mn}$ (s)							
$\varnothing_{mp}$ =	2m		3m		4m		5m	
Mode n	S	AS	S	AS	S	AS	S	AS
1 (P)	3.32	1.47	3.32	1.79	3.7	2.06	3.87	2.28
2	1.02	0.87	1.02	1.06	1.45	1.23	1.62	1.37
3	0.76	0.69	0.76	0.84	1.07	0.97	1.2	1.09
4	0.63	0.59	0.63	0.72	0.89	0.83	0.99	0.93
5	0.55	0.52	0.55	0.64	0.78	0.74	0.87	0.82
6	0.49	0.47	0.49	0.58	0.7	0.67	0.78	0.75
7	0.45	0.44	0.45	0.53	0.64	0.62	0.72	0.69

Notes: S denotes Symmetric Mode ($m = 0$), AS denotes Antisymmetric Mode ($m = 1$), P denotes Piston Mode.

TABLE 6.3

Resonance Period $(T_N)_{mn}$ in a Moonpool for Moonpool Depth $d_{mp} = 5$ m: Water Depth $D \geq 15$ m

	Resonance Period $(T_N)_{mn}$ (s)							
$\varnothing_{mp}$ =	2m		3m		4m		5m	
Mode n	S	AS	S	AS	S	AS	S	AS
1 (P)	4.81	1.48	4.94	1.81	5.08	2.09	5.2	2.34
2	1.02	0.87	1.26	1.06	1.45	1.23	1.62	1.37
3	0.76	0.69	0.93	0.84	1.07	0.97	1.2	1.09
4	0.63	0.59	0.77	0.72	0.89	0.83	0.99	0.93
5	0.55	0.52	0.67	0.64	0.78	0.74	0.87	0.82
6	0.49	0.47	0.61	0.58	0.7	0.67	0.78	0.75
7	0.45	0.44	0.55	0.53	0.64	0.62	0.72	0.69

Notes: S denotes Symmetric Mode ($m = 0$), AS denotes Antisymmetric Mode ($m = 1$), P denotes Piston Mode.

TABLE 6.4

Resonance Period $(T_N)_{mn}$ in a Moonpool for Moonpool Depth $d_{mp} = 10$ m: Water Depth $D \geq 15$ m

	Resonance Period $(T_N)_{mn}$ (s)							
$\varnothing_{mp}$ =	2m		3m		4m		5m	
Mode n	S	AS	S	AS	S	AS	S	AS
1 (P)	6.57	1.48	6.68	1.81	6.78	2.09	6.89	2.34
2	1.02	0.87	1.26	1.06	1.45	1.23	1.62	1.37
3	0.76	0.69	0.93	0.84	1.07	0.97	1.2	1.09
4	0.63	0.59	0.77	0.72	0.89	0.83	0.99	0.93
5	0.55	0.52	0.67	0.64	0.78	0.74	0.87	0.82
6	0.49	0.47	0.61	0.58	0.7	0.67	0.78	0.75
7	0.45	0.44	0.55	0.53	0.64	0.62	0.72	0.69

Notes: S denotes Symmetric Mode ($m = 0$), AS denotes Antisymmetric Mode ($m = 1$), P denotes Piston Mode.

(6.8), the influence of water depth is linked to the hyperbolic tangent function, i.e., $\tanh(\lambda_q \bar{D}) \approx 1.0$ when $\lambda_q \bar{D} \geq 7.6$. Therefore, the natural periods remain fairly stable for intermediate water depths considered, i.e., $D \geq 15$ m, where $\tanh(\lambda_q \bar{D}) \approx 1.0$.

The resonance periods in Tables 6.2–6.4 are divided into three categories, distinguished by different colour codes: red for $(T_N)_{mn} < 1.00$ s, green for $1.00\text{ s} \leq (T_N)_{mn} \leq 3.00$ s and blue for $(T_N)_{mn} > 3$ s. Red indicates that $(T_N)_{mn}$ values are too short to be effective in energy generation for the considered sea state. Green and blue represent suitable sea state wave periods, with green indicating sloshing modes and blue for the piston mode. The piston mode has the highest $(T_N)_{mn}$, followed by the 1st AS mode and 1st S mode.

Figure 6.5 illustrates how the $\varnothing_{mp}$ and d_{mp} impact the $(T_N)_{mn}$ for P, S, and AS modes. For the P mode, Figure 6.5(a) demonstrates that increasing the d_{mp} is the most effective way to raise the $(T_N)_{mn}$, while changes in the $\varnothing_{mp}$ have minimal influence. On the other hand, the sloshing mode resonance frequencies (S and AS modes), as seen in Figure 6.5(b) and 6.5(c), are heavily affected by the $\varnothing_{mp}$ but remain largely unaffected by changes in d_{mp}. Additionally, the concept of placing the WEC array in a moonpool is only beneficial in low sea state conditions, as the maximum $(T_N)_{mn}$ for a moonpool with a $\varnothing = 5$ m and $d_{mp} = 10$ m is about 7 seconds. This suggests that the resonance mode will not be triggered by $T > 7$ s. A larger $\varnothing_{mp}$ is required for higher sea states with longer T, implying a need for a wider breakwater.

The sloshing modes in the moonpool for different (m, n) values are shown in Figure 6.6. Practically, the most efficient way to boost energy generation in the WEC array is by leveraging the P mode (0, 0), first AS mode (1, 0), and

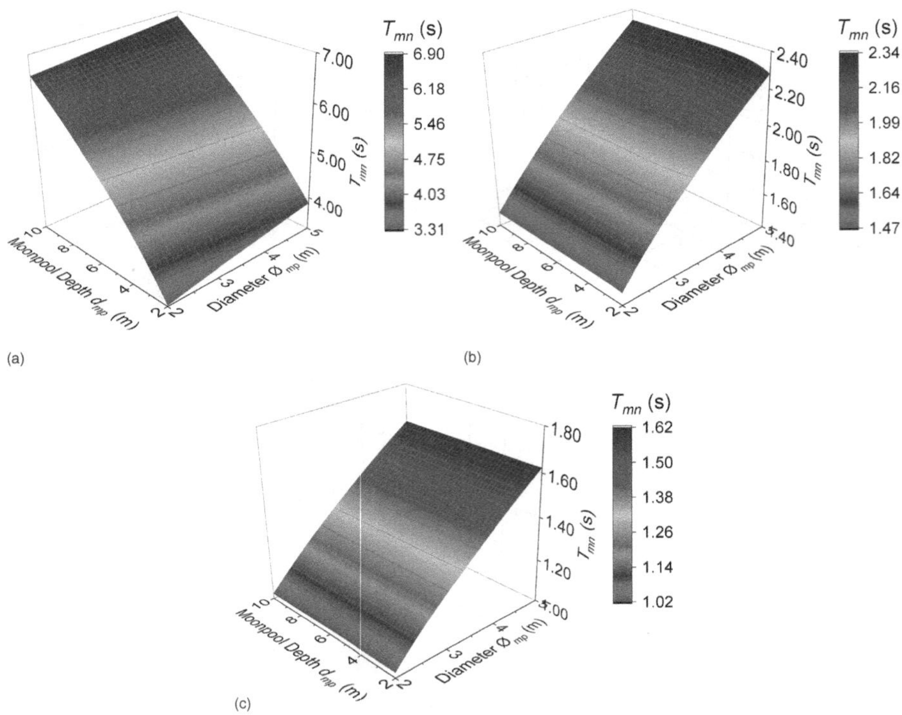

FIGURE 6.5
Moonpool natural frequencies for different modes.

first S mode (0,1). This is because the $(T_N)_{mn}$ for $n \geq 2$ are small and may not match the wave period of the given sea state.

The anti-node position during moonpool resonance for various $(T_N)_{mn}$ can be determined by finding the roots of the first derivative of Equations (6.2) and (6.3) for the S and AS modes, respectively. The first derivative represents the slope, which must be zero at the anti-node. By differentiating Equation (6.2), the slope of the S mode, m_S, is obtained.

$$(m_S)_{0n} = -k_n J_1(k_n r) \tag{6.12}$$

And taking the derivative of Equation (6.3), the slope for AS mode m_{AS} is

$$(m_{AS})_{1n} = \frac{1}{2} k_n \left[J_0(\mu_n r) - J_2(\mu_n r) \right]. \tag{6.13}$$

By setting Equations (6.12) and (6.13) to zero, we can find the values for m_S and m_{AS}. To determine the anti-node positions from the moonpool's centre, divide these values by the k_n and μ_n for the nth S and AS modes, respectively. These positions are then normalised using the moonpool radius, r, resulting in the normalised radial position $\overline{r}$. Figure 6.4 displays

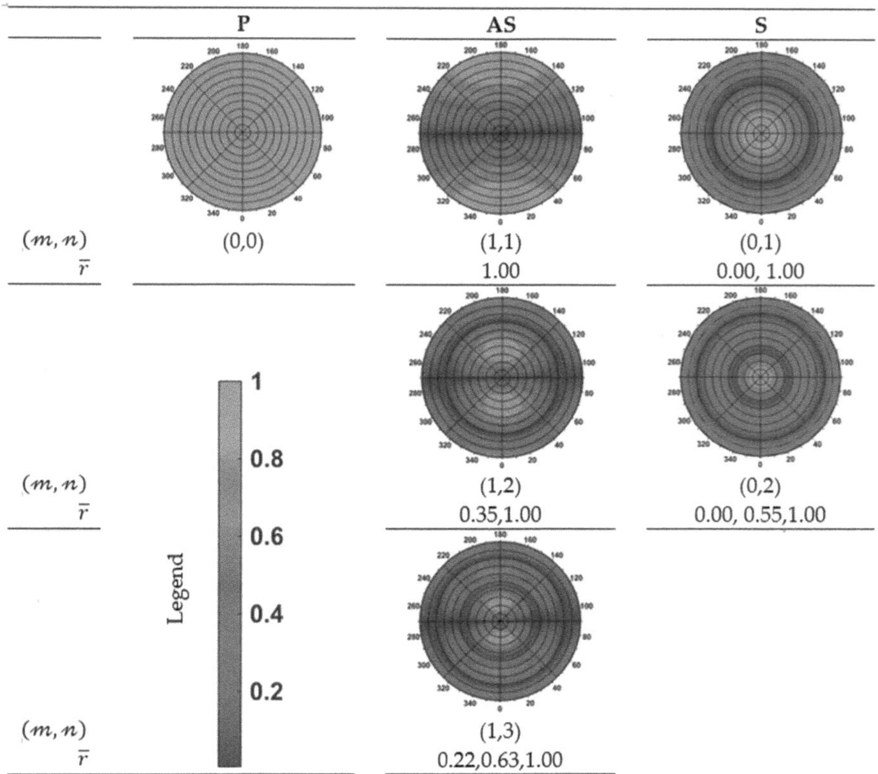

FIGURE 6.6
Elevation (absolute value) of sloshing mode in moonpool.

these $\bar{r}$ values. Figure 6.6 illustrates that for S mode (0,1), the anti-nodes are at the centre ($\bar{r}=0$) and the moonpool boundary ($\bar{r}=1.00$). For S mode (0,2), the anti-nodes are at $\bar{r}=0,0.55$, and 1.00. In AS mode, anti-nodes vary with the mode (m,n) but are all located at the moonpool boundary ($\bar{r}=1.00$). Since placing point absorber WECs at the boundary is impractical, it is better to position them closer to the boundary to leverage the resonance effect for the next highest T_{mn}, i.e., $(m,n)=(1,1)$.

6.3.4 Optimal Array Configuration

To assess how different WEC array configurations perform in a moonpool under resonance, we examine several setups. For this study, we use a wave period of $T=6$ s in regular wave conditions. As shown in Figure 6.5(a), the potential $\varnothing_{mp}$ is 5 m, and d_{mp} is 7.2 m. Four PA-WEC array configurations are analysed, as illustrated in Figure 6.7. These

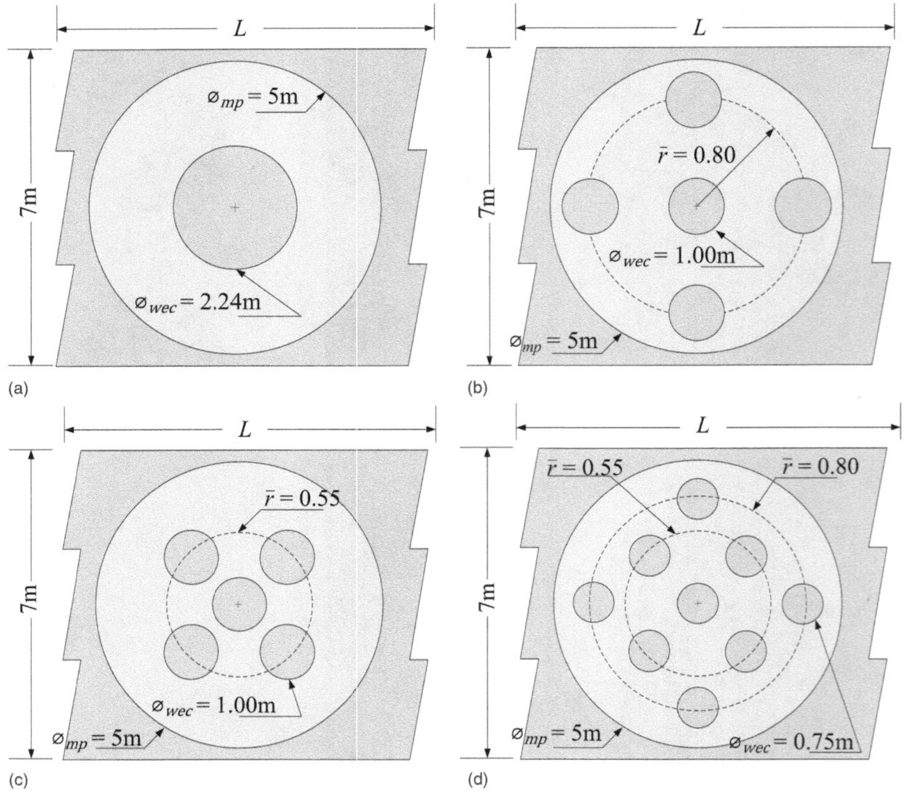

FIGURE 6.7
Plan view of PA-WEC array arranged in configuration (a) configuration 1 (C1), $N_{wec} = 1$, $\varnothing_{wec} = 2.40$ m; (b) configuration 2 (C2), $N_{wec} = 5$, $\varnothing_{wec} = 1.00$m; (c) configuration 3 (C3), $N_{wec} = 5$, $\varnothing_{wec} = 1.00$m; and (d) configuration 4 (C4) $N_{wec} = 9$, $\varnothing_{wec} = 0.75$m, and $d_{mp} = 7.2$ m.

configurations are based on significant resonance modes in the moonpool, as detailed in Figure 6.6:

- C1 for P mode
- C2 for P and first AS mode
- C3 for P and first S Mode
- C4 for P, first AS, and first S Mode.

To ensure a fair comparison, the total A_{wp} of the WECs within the moonpool is maintained at approximately 1.25π m^2. Since the number N_{wec} varies across configurations C1 to C4, the $\varnothing_{wec}$ differs accordingly: $\varnothing_{wec} = 2.70$ m for $N_{wec} = 1$ (C1), $\varnothing_{wec} = 0.75$ m for $N_{wec} = 13$ (C2), $\varnothing_{wec} = 0.66$ m for $N = 17$ (C3), and $\varnothing_{wec} = 0.59$ m for $N_{wec} = 21$ (C4).

6.3.5 Performance of WEC Arrays

The performance of the WEC array in the moonpool is assessed by using WAMIT, which utilises the panel method, as described in Section 2.2. The WEC array deployed in a moonpool is first considered, followed by a line array of moonpools embedded in a floating breakwater. A moonpool with an opening is then considered to further enhance the effect of resonance.

6.3.5.1 Closed Moonpool

Figure 6.8 shows the P mode, the first S mode, and the first AS mode in the moonpool computed from WAMIT. The results show that the $(T_N)_{mn}$ and mode shapes agree well with the analytical solution, thereby validating the numerical model in software.

The performance of the WEC array in the closed moonpool was examined, with the average heave RAO displayed in Figure 6.9 for different modes: P mode, P-AS mode, P-S mode, and P-AS-S mode, corresponding to C1 through C4. Figure 6.9 reveals that the piston mode is the most effective in boosting power generation, especially when a single cylindrical PA-WEC with a larger A_{wp} (C1) or an array near the anti-node of P mode and the first S mode (C3) is used. Conversely, C2 and C4, where some WECs are at the anti-node of the AS mode, are less effective in increasing energy generation. Additionally, the resonance of the floating platform around 4 s significantly improves energy generation for the WEC array. It is worth noting that this resonance mode of the moonpool due to the floating platform's resonance is similar to the P mode.

The *CW* of the array can be easily calculated from the heave RAO using Equation (3.59). Figure 6.10 illustrates the *CW* for configurations C1 through C4. The *CW* indicates that resonance effects in both the moonpool and the floating platform can boost energy generation, with *CW* values

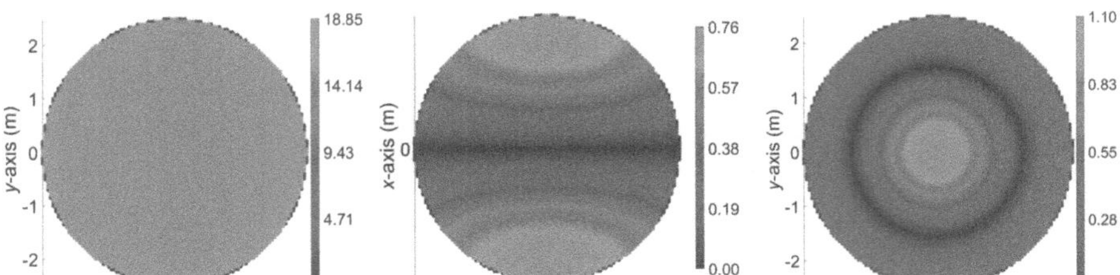

FIGURE 6.8
Mode shape for (a) $(T_N)_{mn} = 6.04$ s, P mode; (b) $(T_N)_{mn} = 2.34$ s, first AS mode; and (c) $(T_N)_{mn} = 1.62$ s, first S mode.

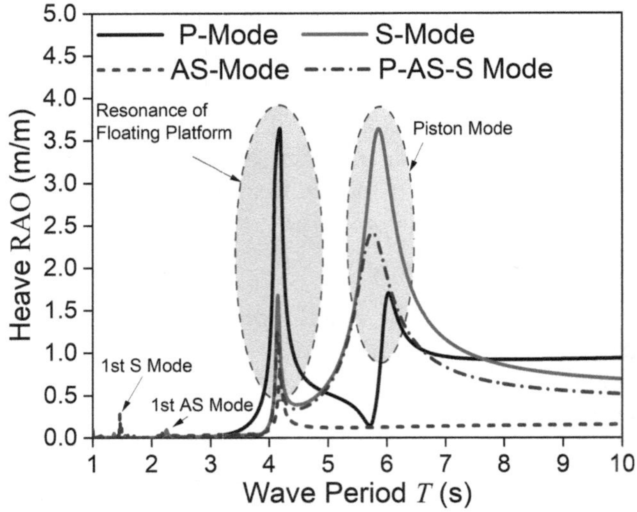

FIGURE 6.9
Average heave RAO for WEC array in moonpool.

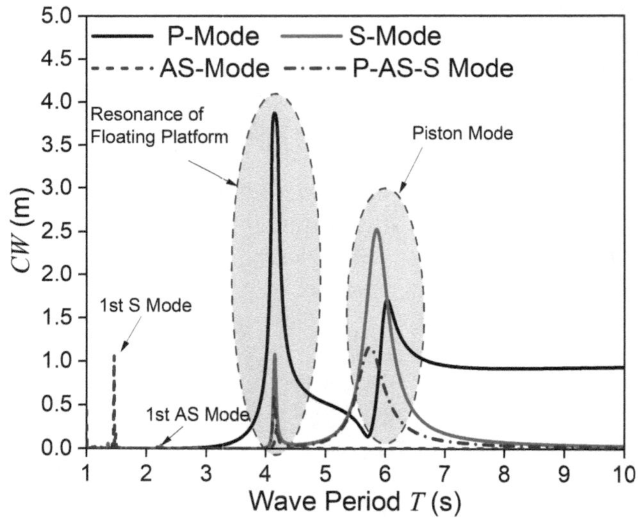

FIGURE 6.10
Average *CW* for WEC array in moonpool.

much higher compared to scenarios without resonance. Between the two most effective configurations for energy production, C3 is preferred over C1 because the WECs in C3 capture additional energy from the first AS mode and first S mode.

6.3.5.2 Semi-Enclosed Moonpool

The performance of the WEC array placed in a semi-enclosed moonpool, as illustrated in Figure 6.4(b), is next examined. The *CW* values for the four configurations under regular wave conditions are shown in Figure 6.7 and depicted in Figure 6.11. According to Figure 6.11, the C2 configuration, where the WECs are positioned at the anti-node of the first AS mode, performs better than those in the P mode and S mode configurations. Specifically, the *CW* for the semi-enclosed moonpool in the C2 set-up is notably higher than the highest *CW* observed in the closed moonpool under configurations C1 and C3. This indicates that the semi-enclosed moonpool's openings, which allow incident waves to enter, positively impact WEC power generation by enhancing wave reflection from the moonpool's walls. Additionally, Figure 6.11 shows that the influence of the P mode is minimal, likely due to the wave interference effects when the incident wave aligns with the P mode.

6.3.5.3 Large Array of Moonpools in Floating Breakwater

A case study with a large array of closed and semi-enclosed moonpools integrated with an FB is considered here, as shown in Figure 6.12. The FB has a length $L_{fb} = 49$ m and width of $W_{fb} = 7$m.

Figure 6.13 illustrates the *CW* for the optimal set-up derived from a single moonpool: C3 for a closed moonpool and C2 for a semi-enclosed moonpool

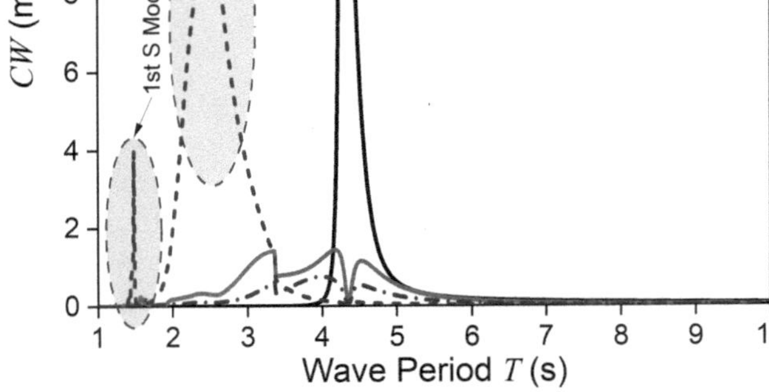

FIGURE 6.11
Average capture width for WEC array in moonpool with opening.

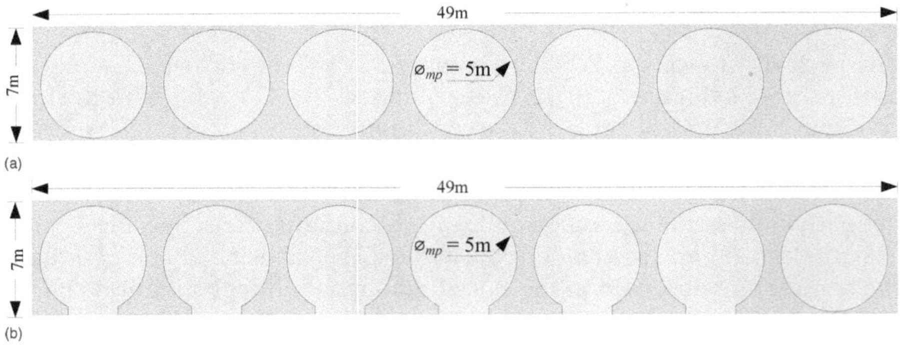

FIGURE 6.12
FB with (a) closed moonpools and (b) semi-enclosed moonpools. C2 and C3 are considered (WECs not shown in figures).

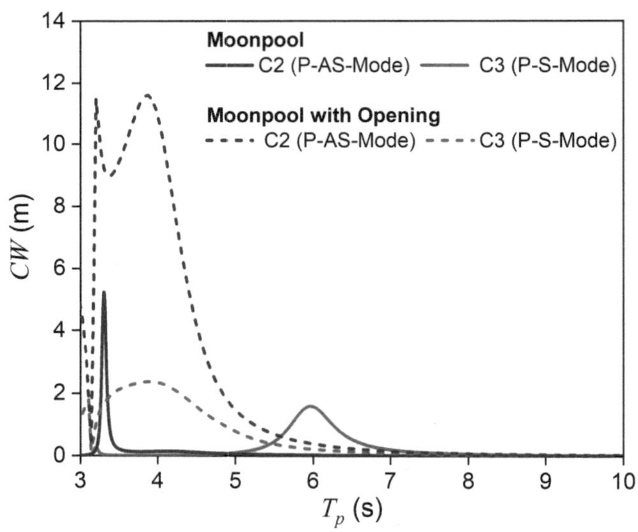

FIGURE 6.13
Average capture width for moonpool with opening breakwater.

in irregular wave conditions. The data show that C3 performs well for energy generation at higher wave periods, while C2 is more effective at lower wave periods. Thus, choosing the type of moonpool should be based on the prevailing sea conditions at the site of the FB.

Figure 6.14 examines how well the integrated FB combined with the WEC array reduces wave forces. Figure 6.14 shows a notable decrease in

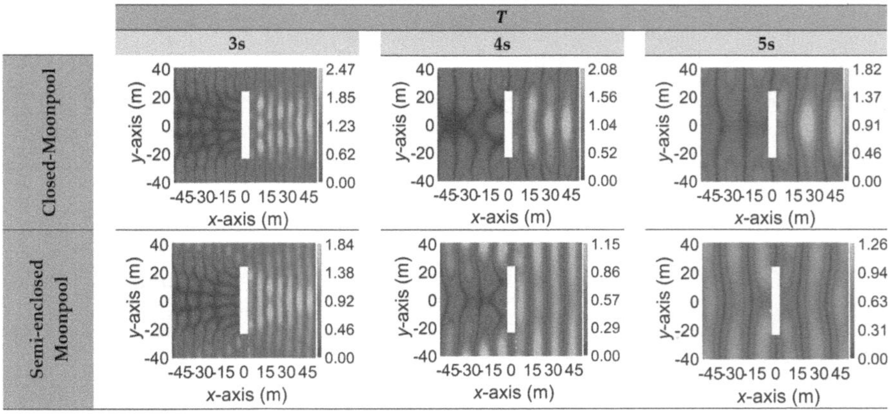

FIGURE 6.14
Wave elevation surrounding FB. Wave direction from right to left.

wave height behind the FB. Conversely, there is a constructive interference between incoming waves and those reflected by the breakwater, leading to an increased wave height in front of the breakwater. Additionally, the wave heights for the semi-enclosed moonpool counterparts are lower, indicating that the FB with a semi-enclosed moonpool is more effective at reducing waves compared to the closed moonpool counterparts.

6.4 Case Study II: Deployment of WECs in Channel

6.4.1 Problem Definition

In this case study, it is assumed that an array of PA-WEC is deployed in a platform with channel formed by two vertical walls, as shown in Figure 6.3. The channel in the floating platform has constant-depth d_{chn} and width W_{chn}. Three d_{chn} are considered, i.e., $d_{chn} = 5$, 10, and 15 m. The PA-WEC has a cylindrical geometry with $\varnothing_{wec} = 5\,\text{m}$ and $d_{wec} = 0.25\,\text{m}$. Two problem analyses are considered, and the principal dimensions of the WEC and floating platforms are summarised in Table 6.5.

6.4.2 Channel Resonance Properties

According to Equation (6.10), wave run-up is analysed for waves travelling in a channel between two side walls, similar to the set-up between two barges.

TABLE 6.5

Principal Dimensions and Properties for Point Absorber WEC and Floating Platform

		N_{wec}		
		5	**13**	**23**
Point Absorber WEC	**Diameter $\varnothing_{wec}$**		5 m	
	Draft d_{wec}		0.25 m	
	Mass m		4 Metric-Tonne	
	Mass Moment of Inertia I		6,770.8 kg.m2	
	PTO Damping B_{pto}		29,150 N.m.s	
Floating Platform Default Dimensions	L_{pf}		50 m	
	W_{pf}		$w_o + 4\,\text{m}$	
	W_{chn}		35 m	
	d_{pf}		$d_o + 2\,\text{m}$	
1st Problem Analysis	d_{chn}		5, 10 15 m	
2nd Problem Analysis	W_{chn}	55 m	35 m	55 m
Water	**Water Depth D**		25 m	
	Wave Period T		2 to 22 s (0.2 s interval)	

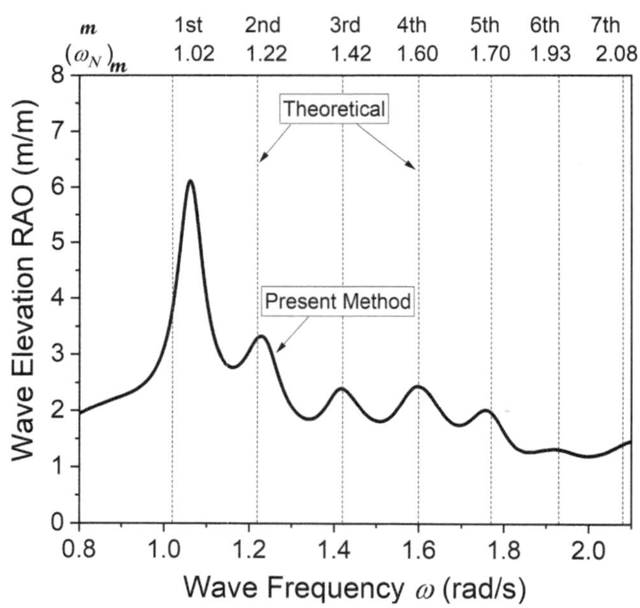

FIGURE 6.15

Validation of natural frequencies of wave run-up along channel formed by two barges. $\theta = 0^\circ$.

Figure 6.15 illustrates the natural frequencies of the first seven modes in such a channel, with barges measuring 50 m in length, 30 m in width, and a draft of 5 m, separated by a 5 m gap. The theoretical natural frequencies (eigenvalues) from Equation (6.10) are represented by a dashed blue line, while the natural frequencies from the BEM are shown by a solid black line. Figure 6.15 demonstrates that wave run-up at these natural frequencies decreases as the wave frequency increases.

The number of waves along the channel, denoted as N_{wave}, is determined by the mode $\mathcal{Z}_{mn}$, with N_{wave} equalling $m/2$. It is important to note that for odd m, the wave elevation is symmetric about $x = 0$, while for even m, it is axisymmetric. According to the theoretical natural frequencies provided in Equation (6.10), these frequencies along the channel increase as the channel width W_{chn}, length L_{pf}, and depth d_{chn} decrease. This information can help in designing the channel dimensions to align with the sea state where the wave farm will be set up, aiming to optimise wave energy production.

6.4.3 Constant-Depth Channel Floating Platform

For the problem analysis, we consider the constant-depth channel floating platform as presented in Table 6.5. Two problem analyses will be considered:

i. Problem Analysis 1: Effect of varying channel depth
ii. Problem Analysis 2: Effect of varying channel width

In the first problem analysis, three depths—$d_{chn} = 5$, 10, and 15 m – are examined, with the channel width kept constant at $W_{chn} = 15$ m. In the second problem analysis, various channel widths are explored to assess their impact on wave farm energy production. The schematic for this analysis is shown in Figure 6.16, where the WECs are arranged in a staggered pattern with spacing $s_p = \varnothing_{wec}$. As channel width increases, so does the N_{wec} deployed along the channel. Specifically, for channel widths of W_{chn} = 15, 35, and 55 m, the corresponding N_{wec} values are 5, 13, and 23, respectively.

6.4.3.1 Resonance Frequencies of Wave Run-Up in Channel

The resonance wave frequencies and periods for wave run-up in the channel of the floating platform are initially determined using BEM. It should be noted that the frequencies obtained through this numerical approach are approximation and not as precise as those calculated from Equation (6.10). Table 6.6 provides a summary of the first three near-resonance wave periods in the channel without the WEC. Generally, the resonance wave period increases with W_{chn} and d_{chn}, aligning well with the trend predicted by the analytical solution in Equation (6.10).

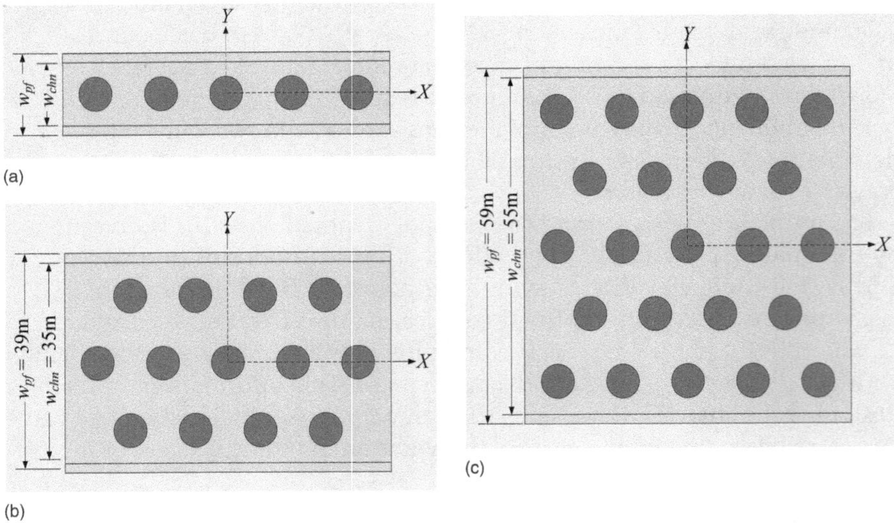

FIGURE 6.16
Array of PA-WECs arranged in floating platform with different W_{chn}. $d_{chn} = 10$ m.

TABLE 6.6
Natural Period $(T_N)_m$ in the Channel in the Floating Platform

		$(T_N)_m$ (s)		
d_{chn}	W_{chn}	$m = 1$	2	3
5 m	15 m	6.8	5.9	3.9
	35 m	7.2	6.1	5.3
	55 m	7.9	7.5	6.1
10 m	15 m	7.3	6.1	4.2
	35 m	7.0	6.5	5.0
	55 m	8.5	7.0	6.2
15 m	15 m	8.1	6.3	4.8
	35 m	9.1	6.9	5.0
	55 m	10.5	7.0	5.8

Figure 6.17 illustrates the wave run-up along the channel (without WECs) for different channel depths – namely, $d_{chn} = 5$, 10, and 15 m (as analysed in Problem Analysis 1), under wave periods of $T = 4.0$ and 6.5 s. It is important to note that $T \approx 6.5$ s represents the natural period for wave run-up in the channel. The wave elevation $\hat{\eta}$ along the channel, considering both the effects of resonance (i.e., $T = 6.5$ s) and no resonance, increases with decreasing channel depth. This occurs because a smaller depth leads to a greater

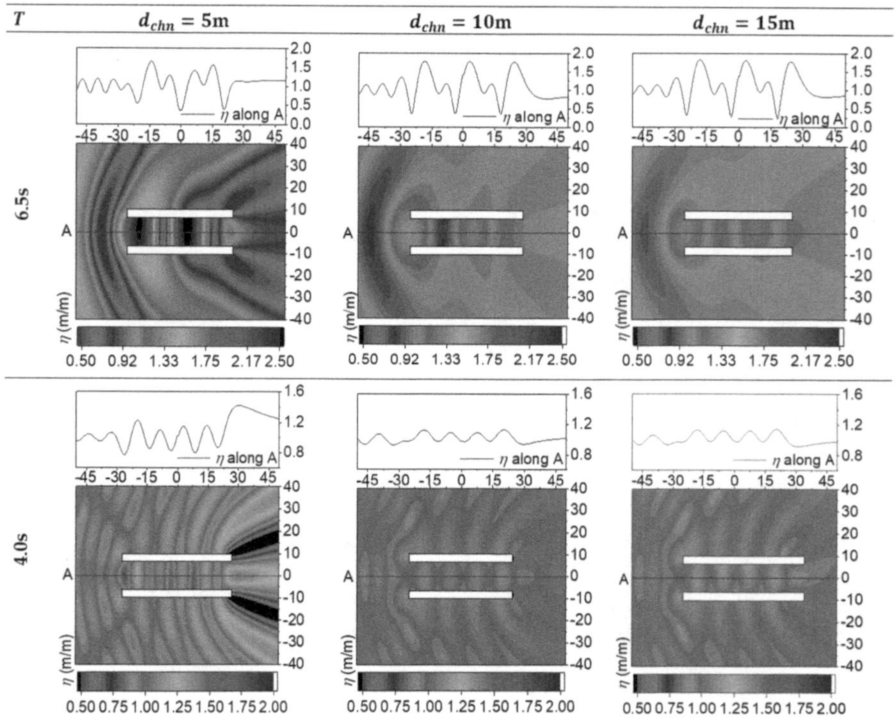

FIGURE 6.17
Normalised wave elevation $\hat{\eta}$ surrounding floating platform with different d_{chn} under wave period $T = 6.5$ and 4.0 s. $W_{chn} = 15$ m.

water volume surge, with the wave having less room to dissipate compared to when the channel depth is larger.

Figure 6.18 illustrates wave run-up along the channel in the platform (without WECs) for different channel widths: W_{chn} = 15, 35, and 55 m (refer to Figure 6.16 for the platform used in Problem Analysis 2) with wave periods T = 4.0 and 6.5 s ($T \approx 6.5$ s is the channel's natural wave run-up period). The wave elevation $\hat{\eta}$ along cross-section A in each case shows that $\hat{\eta}$ decreases as the W_{chn} increases. This is because wider channels dissipate more wave energy. For W_{chn} = 55 m, standing waves become more prominent due to wave reflections from the sidewalls, compared to a narrower channel like W_{chn} = 15 m. Results for W_{chn} = 55 m also reveal that the resonance mode includes both symmetric and axisymmetric patterns along the x- and y-axes, which is a result of the increased channel width (as opposed to the 5 m narrow channel), causing resonance in two-dimensional directions.

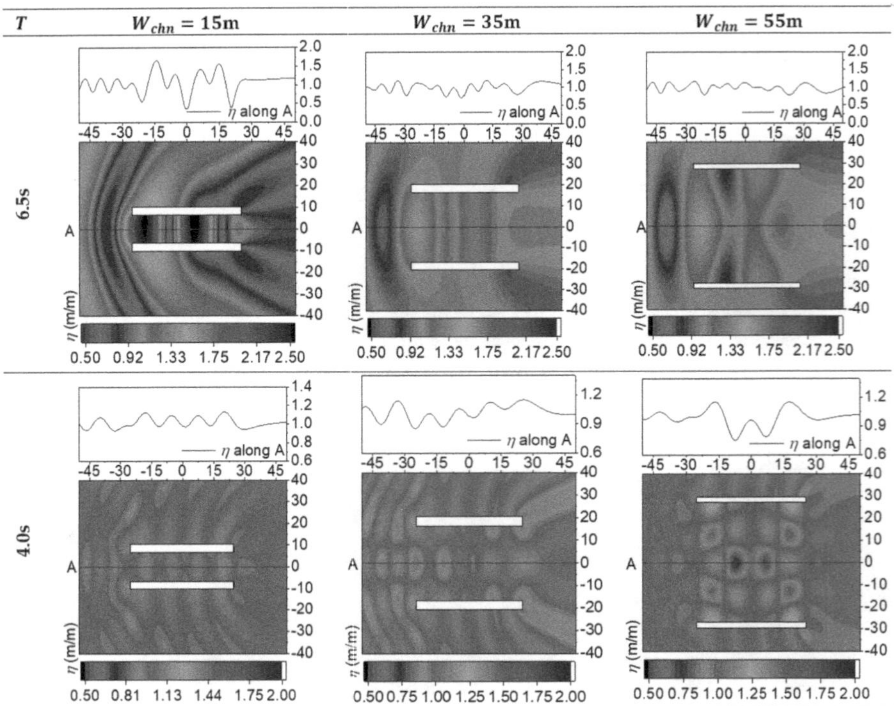

FIGURE 6.18
Normalised wave elevation $\hat{\eta}$ surrounding floating platform with different w_o under wave period $T = 6.5$ and 4.0 s. $d_{chn} = 10$ m.

It should be noted here that the wave elevation along the channel increases with the wavelength, i.e., wave periods, due to increasing energy in longer wavelength, as shown in Figure 6.19. The floating platform considered has a width $W_{chn} = 15$ m and a depth $d_{chn} = 10$ m for demonstration purpose.

An array of N_{wec} number of PA-WECs arranged in staggered configuration is then deployed in the floating platform for Problem Analyses 1 and 2, presented in the subsequent chapters.

6.4.3.2 Problem Analysis 1: Effect of Varying Channel Depth d_{chn}

In the first problem analysis, the impact of different depths ($d_{chn} = 5$, 10, and 15 m) on the wave run-up along the channel and its effect on array performance are examined. Figure 6.20 compares the average heave RAO ($\overline{RAO}$), which relates to absorbed power via Equation (3.57), for the three

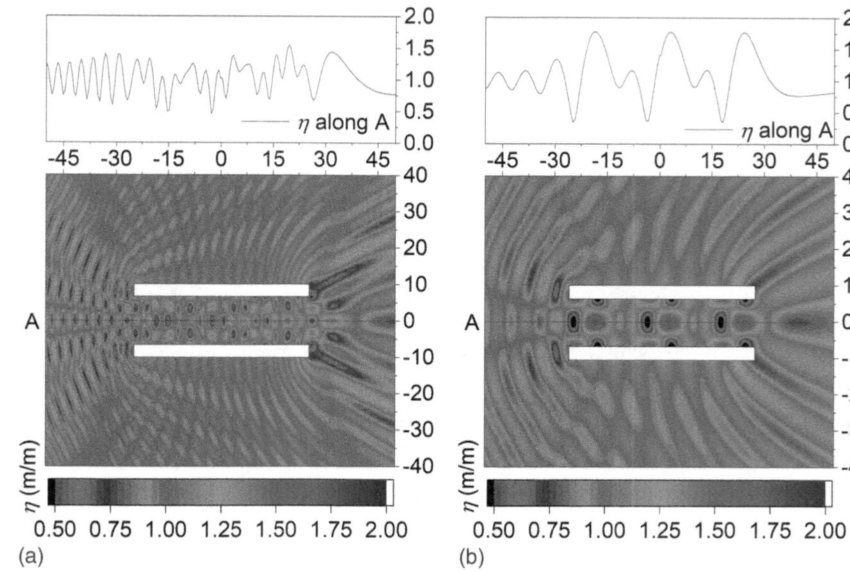

FIGURE 6.19
Normalised wave elevation $\hat{\eta}$ surrounding floating platform under different wave period T. $w_{chn} = 15\,\text{m}$ and $d_{chn} = 10\,\text{m}$.

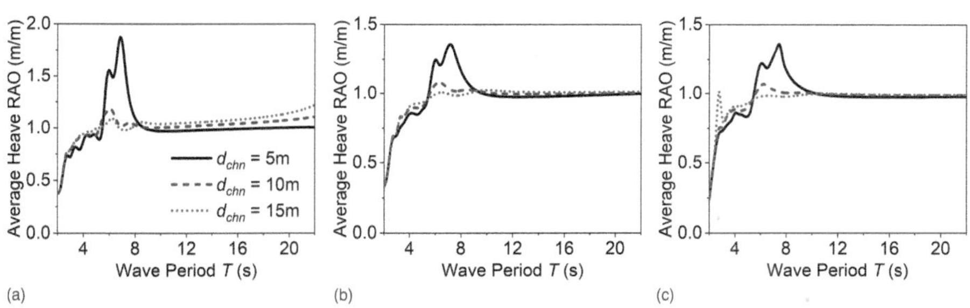

FIGURE 6.20
Comparison of average heave RAO of PA array deployed in floating platform with different $d_{chn} = 5$, 10, and 15 m when (a) $W_{chn} = 15\,\text{m}$ and (b) $W_{chn} = 35\,\text{m}$ (c) $W_{chn} = 55\,\text{m}$.

d_{chn} with constant $W_{chn} = 15$ m (Figure 6.20a), 35 m (Figure 6.20b), and 55 m (Figure 6.20c). Since the number of deployed N_{wec} varies with different W_{chn} values (refer to Table 6.1), the $\overline{RAO}$, which normalises the total heave RAO by the N_{wec} in the array, is used to assess the array's performance. The $\overline{RAO}$ is calculated as follows:

$$\overline{RAO} = \left(\sum_{n=1}^{N_{wec}} RAO_n \right) / N_{wec}. \tag{6.14}$$

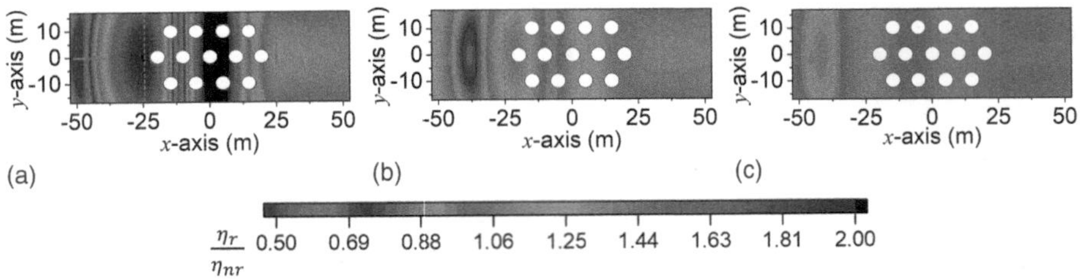

FIGURE 6.21
Wave elevation (η_r / η_{nr}) surrounding WEC array deployed in floating platform for (a) $d_{chn} = 5\,\text{m}$, (b) $d_{chn} = 10\,\text{m}$, and (c) $d_{chn} = 15\,\text{m}$. $W_{chn} = 35\,\text{m}, \theta = 0^{\circ}, T = 6.5\,\text{s}$ (near-resonance wave period). Two grey dash lines represent boundaries of channel.

Figure 6.20 illustrates that the array performs best when the channel depth, d_{chn} , is at its minimum. This is evident from Figure 6.21, which depicts wave run-up along a 35-m-wide channel for various d_{chn} values under resonance conditions, specifically when $T = 6.5$ s. The run-up for d_{chn} = 5 m is much higher compared to $d_{chn} = 10$ and 15 m. This increase in fluid volume as waves enter the shallower channel results in a higher wave elevation ratio (η_r / η_{nr}), as shown in Figure 6.21(a). Here, η_r represents wave elevation in the channel due to run-up, while η_{nr} is the elevation in the open sea with the same sea space as the channel. Conversely, for a deeper channel such as $d_{chn} = 15$ m, the wave run-up is closer to the incident wave amplitude of $A = 1.0$ m, resulting in an $\overline{RAO}$ near 1.0. This is because the WECs oscillate in phase with the wave's orbital motion. Figure 6.20 also indicates that the first and second mode resonance occurs with wave periods around $T \approx 6$ s (refer to Table 6.6 for resonance periods), leading to an increased in power absorption in this period range.

A modified interaction factor, known as q_{array}-factor, is introduced and can be calculated using Equation (6.15). It is derived by comparing the total power harvested from the array on the floating platform with that from the open sea. If the q_{array}-factor is greater than 1.0, it indicates that the wave run-up in the channel enhances energy harvesting for the array compared to the open sea.

$$q_{array} = \sum_{n=1}^{N_{wec}} \left(P_a\right)_n / \sum_{n=1}^{N_{wec}} \left(P_{open\,sea}\right)_n, \tag{6.15}$$

where $P_{open\,sea}$ is the absorbed power of the WECs in the open sea.

Figure 6.22 shows that the array's performance in the channel improves when the wave period exceeds $T = 6$ s. The $d_{chn} = 5$ m yields the highest q_{array}-factor, followed by d_{chn} = 10 and 15 m. Additionally, the q_{array} -factor remains nearly constant at 1.0 for a wide channel (W_{chn} = 55 m), indicating that the wide channel does not significantly enhance the array's performance compared to deployment in open sea.

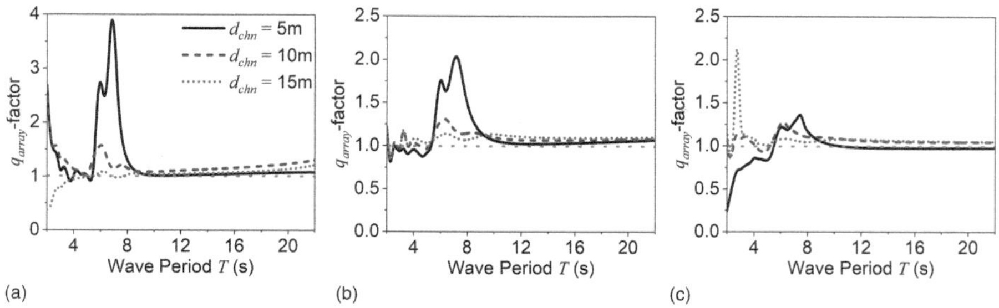

FIGURE 6.22
Comparison of q_{array}-factor of PA array deployed in floating platform with different $d_{chn} = 5, 10$, and 15 m when (a) $W_{chn} = 15$ m, (b) $W_{chn} = 35$ m, and (c) $W_{chn} = 55$ m.

6.4.3.3 Problem Analysis 2: Effect of Varying Channel Width w_o

Figure 6.23 illustrates how W_{chn} affects the q_{array}-factor for three different depths: $W_{chn} = 15$, 35, and 55 m. It shows that with a channel depth of $W_{chn} = 10$ m, the average heave *RAO* is higher in narrower channels due to increased wave run-up. This is supported by Figure 6.24, which indicates that $\eta_r / \eta_{nr} > 1.0$ means higher wave elevation in the channel compared to open sea conditions. Figure 6.24 also reveals that the maximum wave run-up occurs at W_{chn} = 55-m-wide, aligning with the findings of the resonance mode in wider channels in Figure 6.18. However, when 23 WECs are placed in the 55-m-wide channel, some of the WECs encounter waves of very low wave elevation. In contrast, in a 15-m-wide channel, five WECs encounter higher average wave elevations compared to those in the 35-m and 55-m channels. Therefore, when designing a channel, one must balance between a narrower channel that offers better wave farm performance but limited space for a large array and a wider channel that can accommodate more WECs but provides lower average wave elevations.

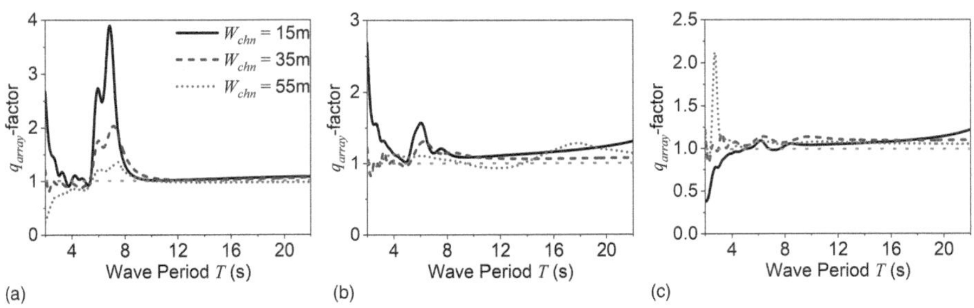

FIGURE 6.23
Comparison of q_{array}-factor of PA array deployed in floating platform with different $W_{chn} = 15, 35$, and 55 m when (a) $d_{chn} = 5$ m, (b) $d_{chn} = 10$ m, and (c) $d_{chn} = 15$ m.

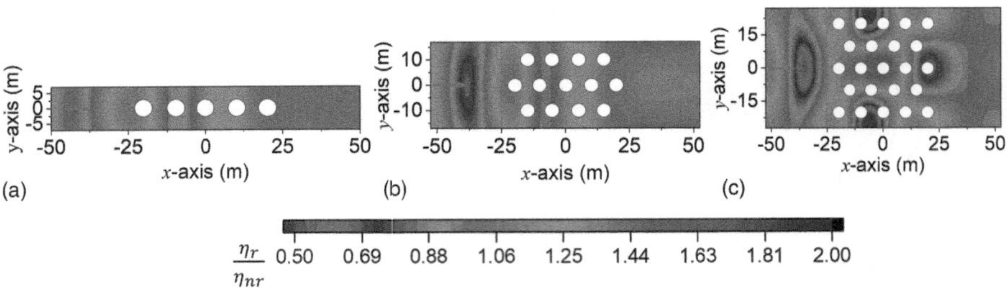

FIGURE 6.24
Wave elevation $\left(\frac{\eta_r}{\eta_{nr}}\right)$ surrounding WEC array deployed in the floating platform for (a) $W_{chn} = 15\,\text{m}$, (b) $W_{chn} = 35\,\text{m}$, and (c) $W_{chn} = 55\,\text{m}$. $d_{chn} = 10\,\text{m}$, $\theta = 0^\circ$, $T = 6.5\,\text{s}$ (resonance wave period). Two grey dash lines represent boundaries of channel.

7

Optimisation Scheme for Power Enhancement of WEC

This chapter presents the enhancement of power generation from the WEC via optimisation of the WEC designs and layout configuration. The genetic algorithm optimisation technique is used to demonstrate the enhancement of the energy generation from two case studies – namely (i) power enhancement via spacing optimisation for point absorber, attenuator, and terminator type WECs and (ii) power enhancement via variable PTO.

7.1 Genetic Optimisation Scheme

Genetic algorithm (GA) is a class of optimisation algorithm inspired by the principle of natural selection and genetics. Since their introduction by John Holland in the 1970s,[197] GA has become a popular optimisation tool due to their flexibility and robustness in solving complex problems. The GA uses the search heuristic approach which is inspired by the process of natural selection. It is used to solve optimisation and search problems by iteratively improving a set of candidate solutions. The framework of a general GA optimisation scheme is given in Figure 7.1 and starts with the initial population followed by an iteration process of fitness function, selection, crossover, mutation, replacement and finally termination when stopping criteria is met. Below is the mathematical framework for a GA.

7.1.1 Step 1: Initial Population

The GA starts with an initial population of potential solutions. Each solution, called an individual or chromosome, is typically represented as a string of numbers (binary, real, or other types).

$$\mathcal{P}(0)=\left\{x_1(0),x_2(0),\ldots,x_{N_{\mathcal{P}}}(0)\right\}, \tag{7.1}$$

where $\mathcal{P}(0)$ is the initial population, $x_i(0)$ represents the i-th individual at generation 0, and $N_{\mathcal{P}}$ is the population size.

DOI: 10.1201/9781003387855-7

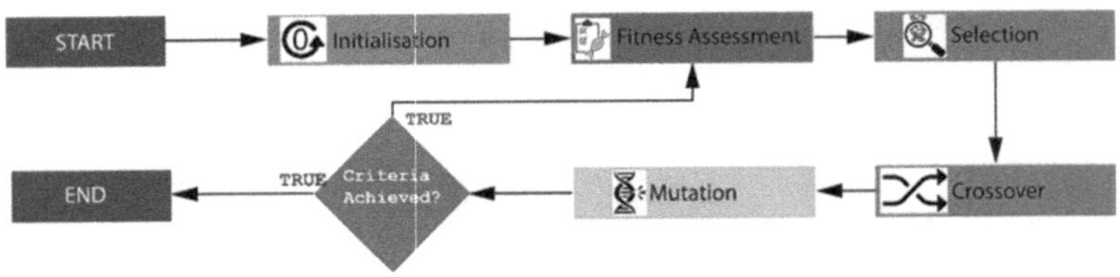

FIGURE 7.1
GA optimisation scheme framework.

7.1.2 Step 2: Fitness Function

Each individual in the population is evaluated using a fitness function, which measures the quality of the solution with respect to the optimisation objective.

$$f(x): \mathbb{R}^n \rightarrow \mathbb{R}, \tag{7.2}$$

where $\mathbb{R}^n$ represents an n-dimensional real-valued vector space (the solution space) and $\mathbb{R}$ represents the real numbers (the fitness value). The fitness function $f(x)$ is to be maximised (or minimised).

7.1.3 Step 3: Selection

Selection is the process of choosing individuals from the current population to create offspring for the next generation. Individuals with higher fitness are more likely to be selected. Common selection methods include the following:

- **Roulette Wheel Selection**: Probability of selection is proportional to fitness.
- **Tournament Selection**: Randomly select a subset of individuals and choose the best among them.
- **Rank Selection**: Rank individuals by fitness and assign selection probabilities based on rank.

7.1.4 Step 4: Crossover (Recombination)

Crossover combines two parent individuals to produce offspring. The goal is to mix the parents' genetic material to create potentially better solutions. Common crossover methods include the following:

- **Single-Point Crossover**: A single crossover point is chosen, and the genetic material is swapped between the two parents.
- **Multi-point Crossover**: Multiple crossover points are chosen, and genetic material is swapped at each point.
- **Uniform Crossover**: Each gene is chosen randomly from one of the parents with equal probability.

Mathematically, if x_i and x_j are two parents, the offspring x_k and x_l might be

$$x_k = \left(x_i^1, x_i^2, \ldots, x_i^c, x_j^{c+1}, \ldots, x_j^n\right),$$

$$x_l = \left(x_j^1, x_j^2, \ldots, x_j^c, x_i^{c+1}, \ldots, x_i^n\right), \tag{7.3}$$

where c is the crossover point and n the number of crossover points.

7.1.5 Step 5: Mutation

Mutation introduces random variations into the offspring to maintain genetic diversity within the population and to explore new solutions. Each gene in the offspring is altered with a small probability p_m. Note that for a binary representation, if $x_k = \left(x_k^1, x_k^2, \ldots, x_k^n\right)$, a mutation might change from x_k^i to $\left(1 - x_k^i\right)$ with probability p_m.

7.1.6 Step 6: Replacement

The new generation is formed by replacing some or all individuals in the current population with offspring. Common replacement strategies include:

- **Generational Replacement**: The entire population is replaced by the offspring.
- **Steady-State Replacement**: Only a few individuals are replaced in each generation.

7.1.7 Step 7: Termination

The algorithm terminates when a stopping criterion is met. Common stopping criteria include the following:

- A fixed number of generations is reached.
- A solution with a satisfactory fitness level is found.
- The population has converged (i.e., little or no improvement over generations).

7.2 Case Study I: Power Enhancement of OWSC from Spacing Optimisation

To maximise power generation from an array of oscillating wave surge converters (OWSC), it is important to arrange the devices at their ideal distances. In the GA optimisation method, the spacing between the WECs is treated as a variable, with the goal of maximising the q-factor as the objective function. This study examines three array configurations: single, double, and triple arrays. The hydrodynamic software WAMIT, an industry standard, is used to assess the pitch RAO of each device in the array from which the power output and q-factor are calculated. The effect of varying wave periods and their directions on the optimal spacing and q-factor of the arrays is also analysed.

7.2.1 Problem Definition

Figure 7.2 displays three configurations of OWSC arrays. Each layout features 12 OWSCs arranged differently: in a single row (single-array, Figure 7.2(a)), two rows (double-array, Figure 7.2(b)), and three rows (triple-array, Figure 7.2(c)). In the double-array set-up, the 12 OWSCs are evenly split into two rows of six devices each. In the triple-array set-up, there are four devices per row. The arrays use three spacing variables: s_p^{y1}, s_p^{x}, and s_p^{y2}, as shown in Figure 7.2, which are applied in the GA optimisation process to maximise the q-factor. Each OWSC features a rectangular flap hinged at the bottom to a fixed foundation, allowing only pitching motion about the hinge (refer to

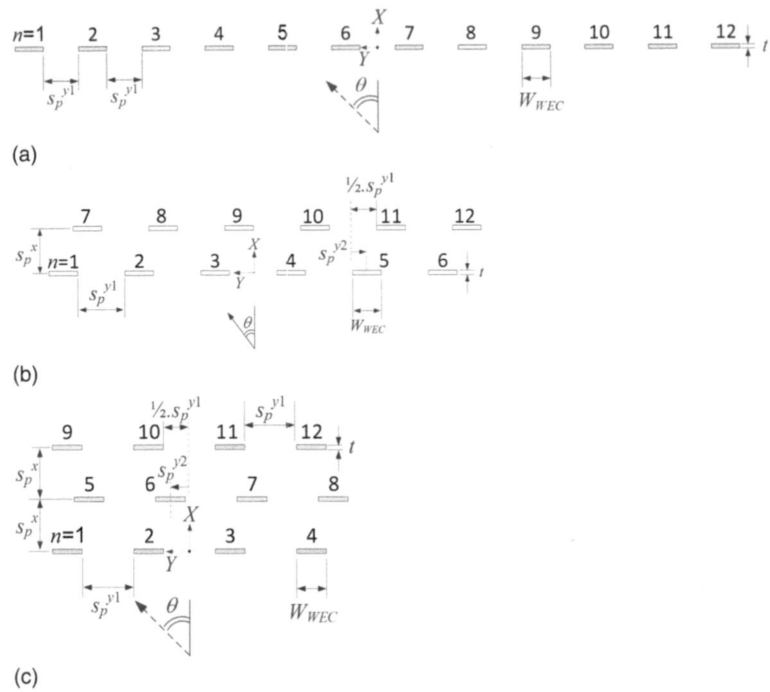

FIGURE 7.2
Array layout for OWSCs arranged in (a) single-array, (b) double-array, and (c) triple-array.

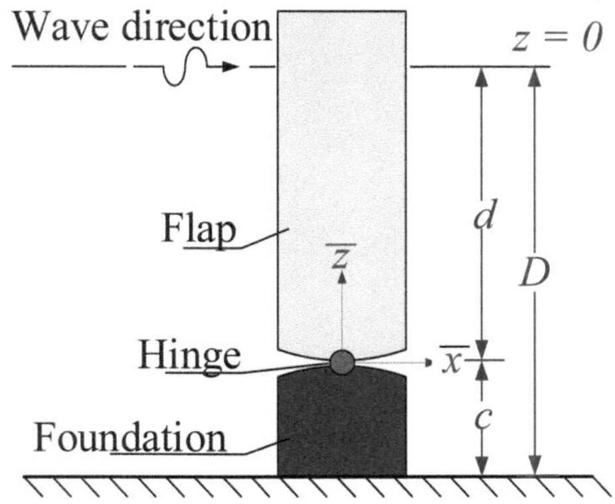

FIGURE 7.3
Schematic diagram of OWSC.

Figure 7.3) for waves of amplitude A. This flap is linked to a PTO to convert kinetic energy into electricity.

7.2.2 GA Optimisation Scheme

The modified GA optimisation method for this problem is shown in Figure 7.4. To achieve the highest q-factor, which results in the maximum power generation from the wave array, the horizontal and vertical distances between OWSCs are kept constant at $s_p^{y_1}$ and s_p^x, respectively. The variable $s_p^{y_2}$ represents the offset of the device's centre of gravity in the second row (for double- and triple-arrays) relative to $s_p^{y_1}$. When $s_p^{y_2} = 0$, the layout is staggered. It is important to limit the range of these variables during optimisation since placing devices too far apart would use more sea space, while keeping them too close could complicate the installation process. Therefore, the dimensionless spacing variables $s_p^{y_1}$, s_p^x, and $s_p^{y_2}$ must fall within specific ranges:

$$\frac{1}{2} \leq \frac{s_p^{y_1}}{a} \leq 2 \tag{7.4a}$$

$$1 \leq \frac{s_p^x}{a} \leq 2 \tag{7.4b}$$

$$0 \leq \frac{s_p^{y_2}}{a} \leq \frac{s_p^{y_1}}{a} + \frac{1}{2} \tag{7.4c}$$

The total number of possible combinations for the three variables specified in Equations (7.4a) to (7.4c) is represented as N_T. Initially, a population of size $N_{1/10}$, which is $1/10$ of N_T, is generated using a biased distribution and processed in WAMIT. From this initial population, N individuals with the

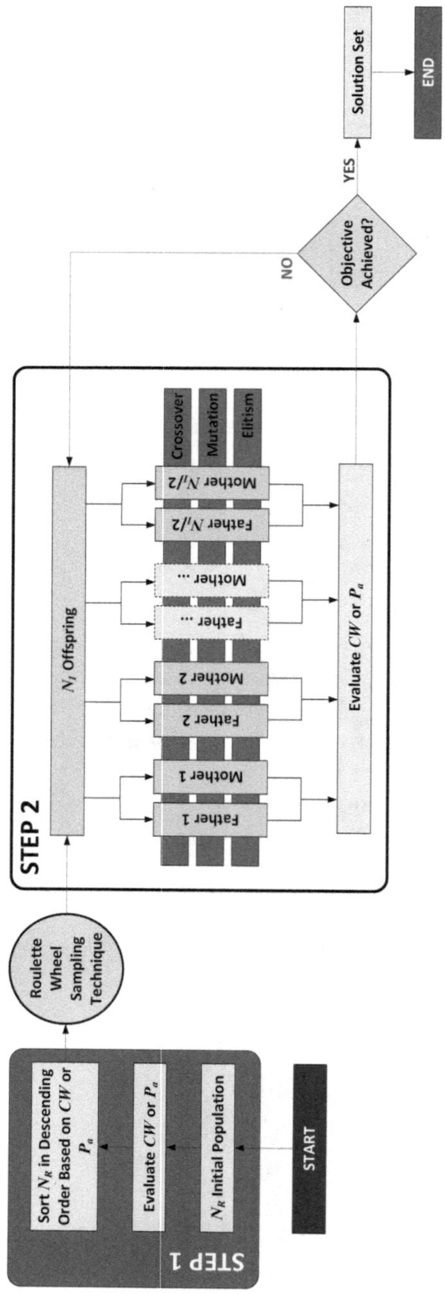

FIGURE 7.4
Flow chart of GA optimisation scheme for Case Study I.

highest fitness values based on the q-factor are selected using the roulette wheel sampling method. Two parents are then chosen from these N individuals to perform crossover and mutation, and this is repeated N times to produce N offspring for the next generation. The individual with the highest fitness value, known as the elite child, is retained for the next generation. This cycle of crossover, mutation, and elitism continues until the objective function is achieved, i.e., the maximum q-factor has converged. Convergence is defined as the change in the maximum q-factor from one generation to the next being less than 0.01%. Note that increasing N will lengthen computation time but lead to faster convergence, and vice versa. For this study, N is set to 40. The crossover p_c and mutation p_m probabilities are taken as 1.000 and 0.015, respectively.

The crossover processes are performed by using three random distributions that add up to p_c, i.e. $r_1 + r_2 + r_3 = p_c = 1.000$ according to the following formulation.

$$\left(s_p^{y_1}\right)_1^{\text{crossover}} = \left(s_p^{y_1}\right)_2 - r_1\left|\left(s_p^{y_1}\right)_2 - \left(s_p^{y_1}\right)_1\right| \tag{7.5a}$$

$$\left(s_p^{x}\right)_1^{\text{crossover}} = \left(s_p^{x}\right)_1 + r_2\left|2a - \left(s_p^{x}\right)_1\right| \tag{7.5b}$$

$$\left(s_p^{y_2}\right)_1^{\text{crossover}} = \left(s_p^{y_2}\right)_2 - r_2\left|\left(s_p^{y_2}\right)_1\right| \tag{7.5c}$$

and

$$\left(s_p^{y_1}\right)_2^{\text{crossover}} = \left(s_p^{y_1}\right)_1 + r_1\left|2a - \left(s_p^{y_1}\right)_1\right| \tag{7.6a}$$

$$\left(s_p^{x}\right)_2^{\text{crossover}} = \left(s_p^{x}\right)_2 - r_2\left|\left(s_p^{x}\right)_2 - \left(s_p^{x}\right)_1\right| \tag{7.6b}$$

$$\left(s_p^{y_2}\right)_2^{\text{crossover}} = \left(s_p^{y_2}\right)_1 + r_3\left|\frac{\left(s_p^{y_1}\right)_1^{\text{crossover}} + W_{wec}}{2} - \left(s_p^{y_2}\right)_1\right|, \tag{7.6c}$$

where the two initial populations are, respectively, denoted by $\left[\left(s_p^{y_1}\right)_1, \left(s_p^{x}\right)_1, \left(s_p^{y_2}\right)_1\right]$ and $\left[\left(s_p^{y_1}\right)_2, \left(s_p^{x}\right)_2, \left(s_p^{y_2}\right)_2\right]$, whereas the two crossover populations are, respectively, $\left[\left(s_p^{y_1}\right)_1^{\text{crossover}}, \left(s_p^{x}\right)_1^{\text{crossover}}, \left(s_p^{y_2}\right)_1^{\text{crossover}}\right]$ and $\left[\left(s_p^{y_1}\right)_2^{\text{crossover}}, \left(s_p^{x}\right)_2^{\text{crossover}}, \left(s_p^{y_2}\right)_2^{\text{crossover}}\right]$. The crossover populations (in decimal numbers) are then converted into binary numbers represented in terms of "1s" and "0s" where the mutation process is then performed. The crossover and mutation processes are depicted in the GA flow chart (Figure 7.4). It is noted here that the crossover process is performed in the decimal format following Equations (7.5a) to (7.5c) as the spacings with decimal place could not be expressed in the binary format. The formulation in the modulus $|\blacksquare|$ of Equations (7.5a) to (7.5c) will ensure that the crossover populations generated are in the range stipulated in Equations (7.4a) to (7.4c).

7.2.3 Results and Discussions

The performance of the OWSC is obtained from WAMIT and the GA is developed in MATLAB®. Validation of the numerical model and GA code is given in paper by Tay.[29] The results are plotted against the scattering parameter kW_{wec} ranging from 1.00 to 3.05, which corresponds to the wave frequency ranging 0.42 to 0.90 rad/s for the OWSC ($W_{wec} = 26$ m).

7.2.3.1 Optimal Spacing

Figure 7.5 illustrates an example of the optimal spacing for the OWSC triple-array under head sea conditions with a wave period $T = 10$ s. Figures 7.5(a) to 7.5(c) display the q-factors for all possible optimal spacings s_p^{y1}, s_p^x, and s_p^{y2} determined through the GA optimisation method. These spacings are normalised by the OWSC width W_{wec}. For this case, the optimal spacings are $s_p^{y1} = 2W_{wec}$, $s_p^x = 1.27W_{wec}$ and $s_p^{y2} = 0$. The optimal layout based on these spacings is shown in Figure 7.5(d).

Figures 7.6(a) to 7.6(c) show the results for optimal spacings in single-, double-, and triple-arrays, respectively. For the single-array, both s_p^x and s_p^{y2} are set to 0, while for the double- and triple-arrays, only s_p^{y2} is 0. The optimal spacings are found to be closely related to the scattering parameter kW_{wec}. This suggests that the optimal spacings indicated in Figure 7.6 can be applied to OWSC arrays with varying widths W_{wec}. The key observations regarding these spacings are as follows:

For spacing s_p^{y1}:

i. In general, the spacing s_p^{y1} is determined to be $2W_{wec}$ (at its maximum allowable spacing) when the scattering parameter is low, which occurs under long wavelength conditions.
ii. For a single array, s_p^{y1} decreases as kW_{wec} increases (see Figure 7.6(a)). However, in the case of double- and triple-arrays, s_p^{y1} ranges from $2W_{wec}$ to $1.5W_{wec}$ when the scattering parameter is between 1.80 and 2.60 (see Figures 7.6(b) and 7.6(c)).

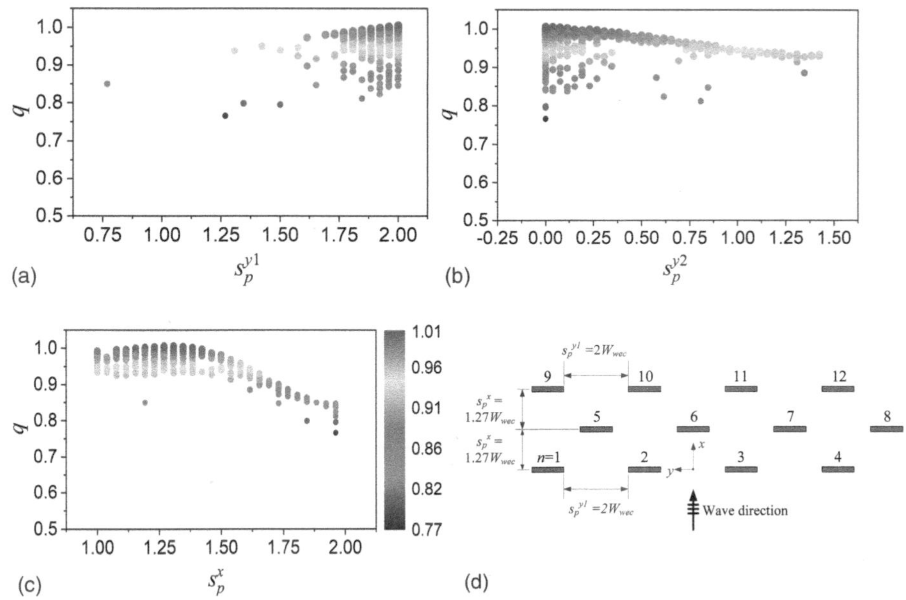

FIGURE 7.5
Example of q -factor for OWSC triple-array with respect to (a) s_p^{y1}, (b) s_p^x, (c) s_p^{y2}, and (d) triple-array layout with optimal spacing. $T = 10s$, $A = 1m$, $\theta = 0^\circ$.

For spacing s_p^x:

i. The ideal spacing s_p^x varies between $1.0W_{wec}$ and $1.5W_{wec}$ for $kW_{wec} \leq 2.0$ in the double- and triple-array configurations, as illustrated in Figures 7.6(b) and 7.6(c).
ii. For higher values of kW_{wec}, the optimal spacing s_p^x is seen to exceed $1.5W_{wec}$.

A noticeable increase in the magnitude of s_p^{y1} and s_p^x is seen when kW_{wec} exceeds 2.0, i.e., for when wavelength is short. This implies that the optimal array spacings are heavily influenced by the diffracted and radiated waves. In every case, s_p^{y2} must be zero, indicating that the best arrangement (i.e., the highest q -factor) is achieved with a staggered layout. This layout results in constructive hydrodynamic interactions when $q > 1.0$ and minimal destructive interactions when $q \leq 1.0$.

7.2.3.2 Pitch RAO

The power output of the OWSC is greatly influenced by the pitch *RAO*, as described in Equation (3.57). Figure 7.7 shows the average pitch *RAO* for

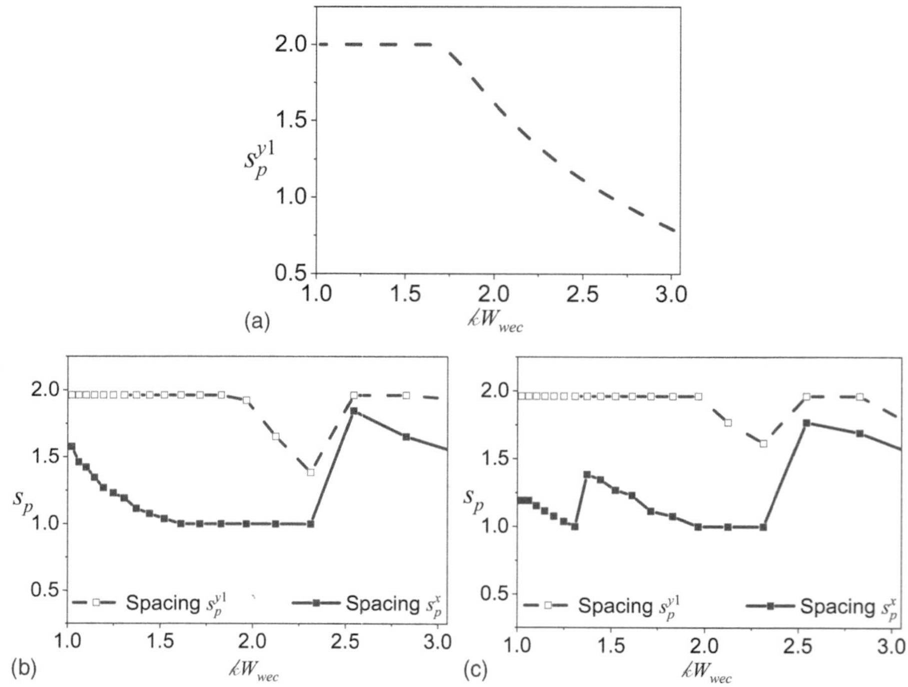

FIGURE 7.6
Optimal spacing for (a) single-array $s_p^x = 0$ and $s_p^{y2} = 0$, (b) double-array $s_p^{y2} = 0$, and (c) triple-array $s_p^{y2} = 0$ under regular wave. $\theta = 0^\circ$.

each row in the single-, double-, and triple-array set-ups, considering the optimal spacing and wave direction $\theta = 0^\circ$. It also includes the average pitch RAO for arrays with spacings that yield the lowest q-factor (referred to as worst spacing), obtained through a GA optimisation scheme aimed at minimising the q-factor. The mean pitch RAOs for each row are calculated as the average of $\left(\sum_i^N RAO_i\right) / N_d$, where N_d represents the total number of devices in the row.

7.2.3.3 q-Factor

The q-factor, obtained using Equation (3.74) for the single-, double-, and triple-arrays are presented in Figure 7.8. Comparing the q-factors shows that only the single-array demonstrates constructive interference for the range of kW_{wec} considered while the q-factors for the double- and triple arrays are below 1.0 when the kW_{wec} is lesser than approximately 1.25 and 1.75, respectively.

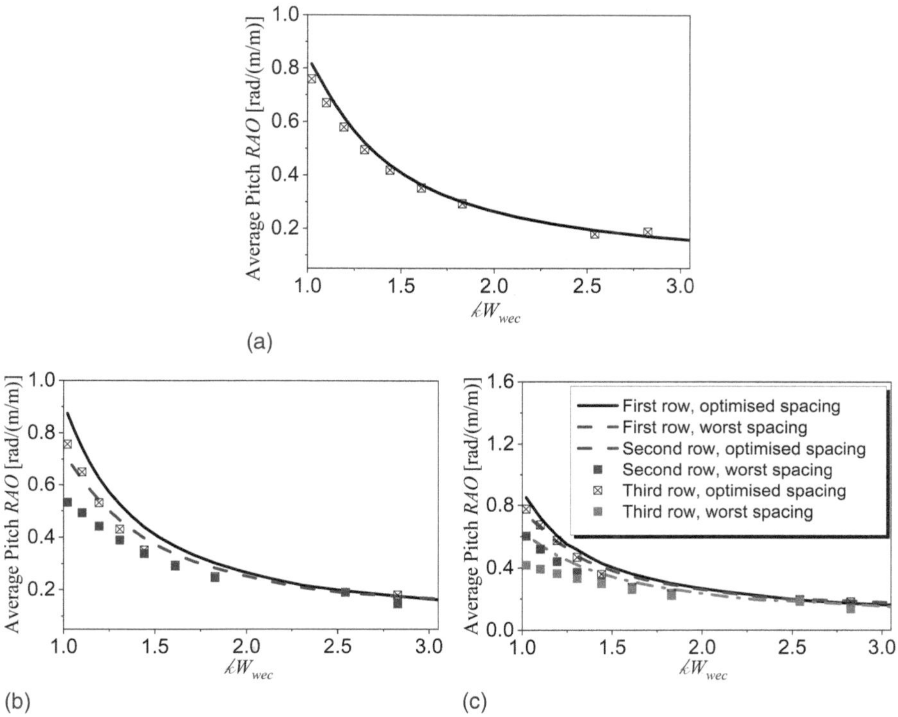

FIGURE 7.7
Comparison of average pitch RAO for (a) single-array, (b) double-array, and (c) triple-array with and without optimal spacing under $\theta = 0°$.

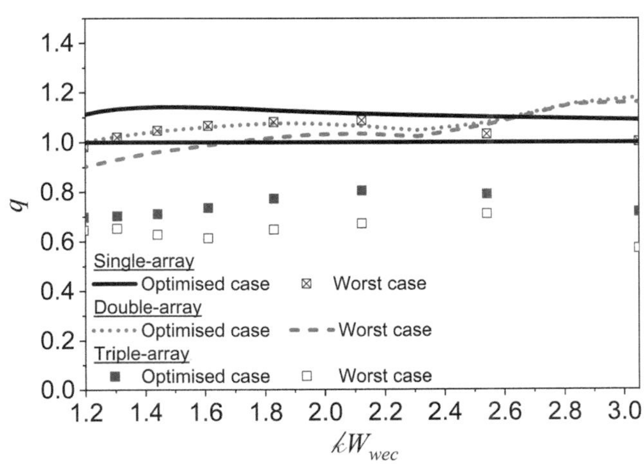

FIGURE 7.8
Comparison of q-factor between single-array, double-array, and triple-array for OWSC.

7.2.3.4 Wave Elevations

The wave elevation surrounding the single array under optimal spacing condition is compared with their counterpart under worst spacing conditions in Figure 7.9. The comparison shows that the value for the former case is always lower than that of the latter case for most of scattering parameters kW_{wec} considered. It can be deduced that in the process of seeking the optimal spacing, the OWSC should be arranged such that it allows the waves to propagate or pass through the array without resulting in a large reflection or trapping of waves.

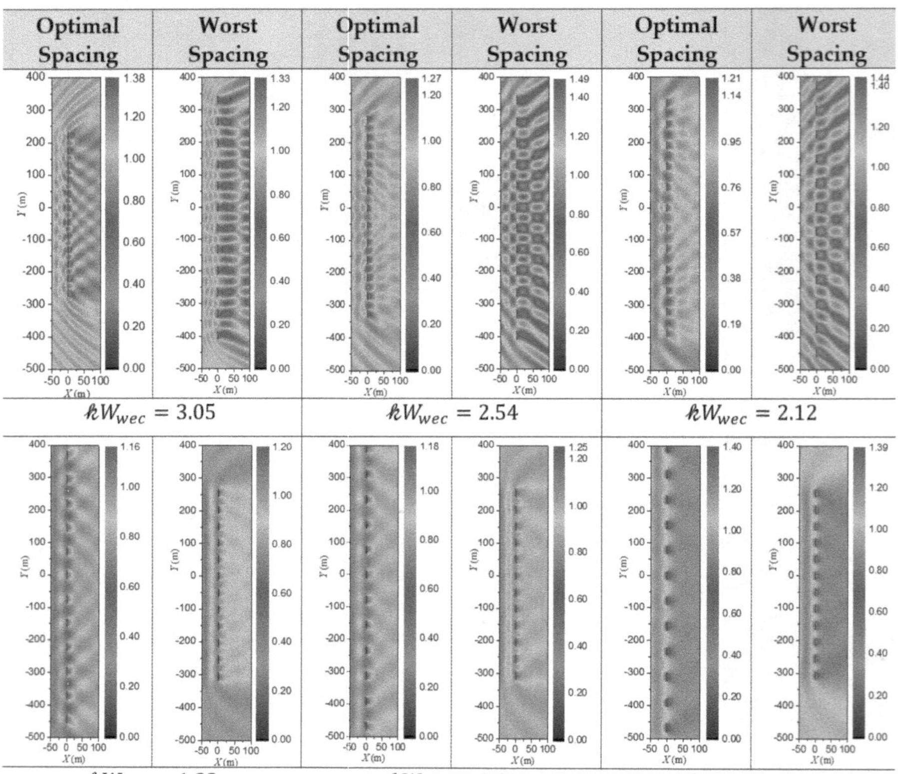

FIGURE 7.9
Comparison of wave elevation surrounding OWSC single-array under optimal and worst spacing conditions.

7.3 Case Study II: Power Enhancement from Hinged-Connected Modular Units

7.3.1 Problem Definition

The pontoon-type WEC shown in Figure 4.15 is used to demonstrate the power enhancement of the hinged-connected WEC by using the GA optimisation method. The descriptions of the interconnected modules have been presented in Chapter 4. The properties of the different pontoon-type WECs and wave conditions can be found in Section 4.6.

The GA optimisation technique will be applied to seek for the optimal PTO damping following the scheme presented in Section 7.1. Two cases are considered in the GA optimisation to seek the optimal variable B_{pto}, i.e.,

- Case 1: The GA optimisation scheme is applied to all the B_{pto} in the pontoon-type WEC.
- Case 2: The GA optimisation scheme is applied to the B_{pto} attached to the line hinge connector only.

It is noted here that for the variable PTO system, the B_{pto} are assumed to vary along the x-axis direction but are kept constant for each line connector (along the y-axis direction) in order to reduce the computation time.

The pontoon-type WEC is subjected to a series of regular waves with a constant wave amplitude $2A$. The wave frequencies ω range from 0.1 to 1.6 rad/s with an interval of 0.025 rad/s where the regular waves approach the WEC at the head sea direction. The WEC is assumed to operate in a deepwater condition where the effect of seabed on the structural motion is negligible.

7.3.2 GA Optimisation Scheme

The ideal constant and variable PTO damping for the pontoon-type WEC are determined using a custom GA optimisation code in MATLAB®. The goal is to maximise the absorbed power P_a (Equation (3.57)) or CW (Equation (3.59)), which represents the highest power capture from the waves. The optimisation involves adjusting the B_{pto}, which ranges from 0 MN.s/m to $66\frac{2}{3}$ MN.s/m with 40 intervals, allowing for 41 possible values for each PTO system. The upper limit of $66\frac{2}{3}$ MN.s/m is chosen because beyond this point, the effect of B_{pto} on the hydroelastic response is minimal.

Without the GA optimisation scheme, the total possible combination of the B_{pto} is denoted by N_T, where N_T is given as

$$N_T = \left(\frac{\left(B_{pto}\right)_{max} - \left(B_{pto}\right)_{min}}{\left(B_{pto}\right)_{\text{interval}}} + 1 \right)^{N_{pto}}, \quad (7.7)$$

where N_{pto} represents the number of PTOs in the pontoon-type WEC, with $N_{pto} = 99$. Equation (7.7) would result in 41^{99} potential PTO damping combinations for this WEC, which is computationally impractical due to its vast number of possibilities. To simplify the optimisation, B_{pto} is varied along the y-axis but kept fixed along the x-axis, reducing N_{pto} in Equation (7.7) to 11 for the pontoon-type WEC being examined.

In order to reduce the computational time to seek for the optimal B_{pto}, the GA optimisation scheme is utilised where it is divided into two steps as depicted in Figure 7.4:

(i) Step 1: Generating an initial population of N_0
(ii) Step 2: Applying GA optimisation scheme to the initial population to seek the optimal B_{pto}

In Step 1, a starting population of $N_0 = 500{,}000$ different B_{pto} combinations are generated using a bias distribution. Step 2 involves selecting N_i individuals with the highest fitness values based on P_a or CW, using the roulette wheel sampling method.[198] These top N_i individuals are then randomly paired as parents for crossover and mutation operations to produce N_i offspring for the next generation. The individual with the best fitness in the current generation, known as the elite child, is retained for the next generation. This process of crossover, mutation, and elitism continues until the objective function is met, which means the maximum P_a or CW has converged. The convergence threshold for the maximum P_a or CW is set at 0.01%. It is important to note that while increasing N_i lengthens computational time, it speeds up convergence for P_a or CW. For the case study, $N_i = 50$ is chosen as it provides a good balance for faster convergence with reasonable accuracy. Additionally, in the GA optimisation scheme, parents are referred to as DNA, and the N_{pto} numbers of B_{pto} are called chromosomes.

The chromosomes will be converted into binary numbers of 20 bits for the crossover and mutation processes. The crossover p_c and mutation p_m probabilities are taken as 0.2 and 1.0, respectively. This implies that the crossover process between the two DNAs applies only to 20% of the chromosomes whereas the mutation process only applies to all the chromosomes of the best fit.

7.3.3 Results and Discussion: Effect of Non-uniform Optimised PTO Damping for VLFS

The power output of the pontoon-type WEC can be improved by allowing some structural deformation, as detailed in Chapter 4. Additionally, energy production can be further increased by applying an optimal, non-uniform PTO damping, also known as a variable PTO system. To find the best PTO damping, the GA optimisation method described in Section 7.3.2 is employed. This is demonstrated using the Group III ($L = 300$ m) pontoon-type WEC with two different wavelength-to-structural length ratios, $\lambda / L = 0.2$ and 0.4.[199] An initial population of $N_0 = 500,000$, $p_c = 0.2$ and $p_m = 1.0$ is used. The evolution of B_{pto} for Group III pontoon-type WEC to achieve the optimal B_{pto} is plotted in Figure 7.10.

Figures 7.10(a) and 7.10(b) represent the WEC connected by different numbers of connection joints and subjected to regular waves of $\lambda / L =$ 0.2. The optimal damping value at each PTO system that results in the maximum CW is presented in each subfigure. It can be seen that by using the GA optimisation scheme, the optimal non-uniform distributed PTO damping at each line connector that produces the maximum power could be obtained.

Figure 7.10 indicates that for a pontoon-type WEC, the combination of the minimum B_{pto} (0 MN.s/m) and the maximum B_{pto} ($66\frac{2}{3}$ MN.s/m) can help achieve the highest CW. However, the best B_{pto} might fall between these extremes, as seen in the Type-A pontoon-type WEC, which has multiple

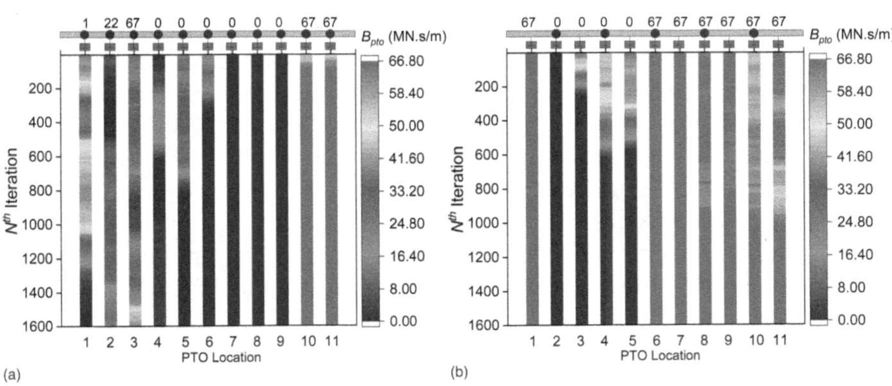

FIGURE 7.10
Evolution of PTO damping B_{pto} for Group III ($L = 300$ m) pontoon-type WEC under GA optimisation scheme (a) Type-A: 11 Hinges ($N = 12$) (b) Type-B: 5 Hinges ($N = 6$).

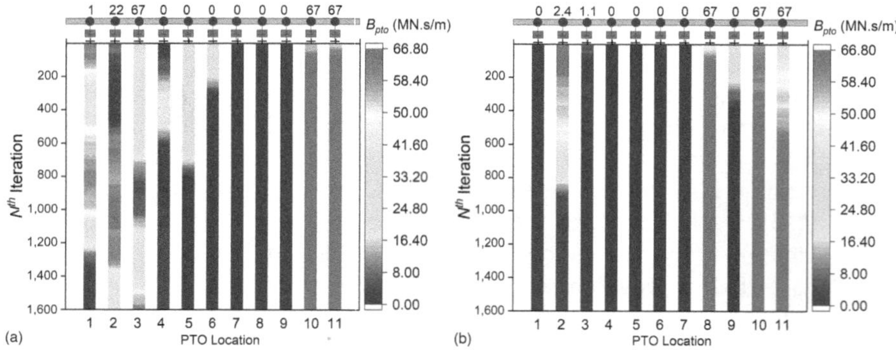

FIGURE 7.11
Evolution of PTO damping B_{pto} for Group III ($L = 300$ m) Type-A ($N = 12$ modules) pontoon-type WEC under GA optimisation scheme.

connection joints. Focusing on the Group III Type-A pontoon-type WEC, Figure 7.11 shows how B_{pto} changes for different λ / L values, specifically $\lambda / L = 0.2$ and 0.4. Figure 7.11(a) reveals that the highest CW can be reached with various B_{pto} combinations, where the optimal B_{pto} is not necessarily the minimum or maximum value. The optimal B_{pto} might be a mix of both extremes to maximise power generation under large wavelengths. Figure 7.12 illustrates the hydroelastic response of Group III pontoon-type WECs with uniform and non-uniform optimal B_{pto}. It shows that the CW with non-uniform optimal B_{pto} is generally higher than with uniform B_{pto}, especially for Type-A due to its greater rotational motion. This implies that a variable PTO system works better for waves with longer wavelength due to the increased hydroelastic response. Additionally, CW also increases with the wavelength. Figure 7.13 compares the hydroelastic responses of Type-A pontoon-type WECs to that of a continuous pontoon-type WEC (Type-D) across different λ / L values from 0.2 to 0.8. Notably, the continuous pontoon-type WEC is the most effective in generating wave energy with an optimal variable PTO system, consistent with Chapter 4 results. The optimal B_{pto} for maximum CW, as determined by GA optimisation, is in the higher range. It is to note that the computation of the GA optimisation scheme to seek for the optimal non-uniform PTO required large computation time thereby may limit the range of B_{pto} considered in the study. One way to overcome this is to consider a surrogate model developed using neural network method to predict the hydroelastic analysis as presented by Tay.[236]

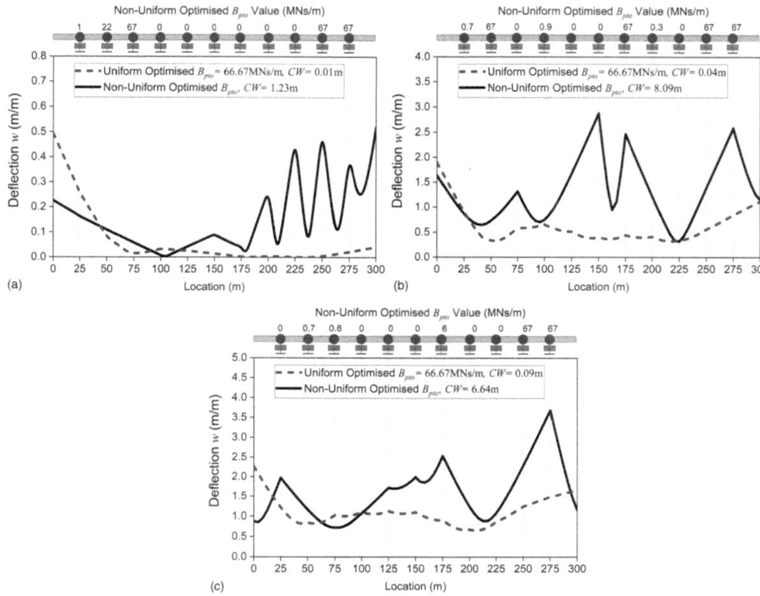

FIGURE 7.12
Hydroelastic response for Group III pontoon-type WEC under (a) $\frac{\lambda}{L}=0.2$ (b) $\frac{\lambda}{L}=0.6$ (c) $\frac{\lambda}{L}=0.8$ with uniform and non-uniform optimised PTO damping.

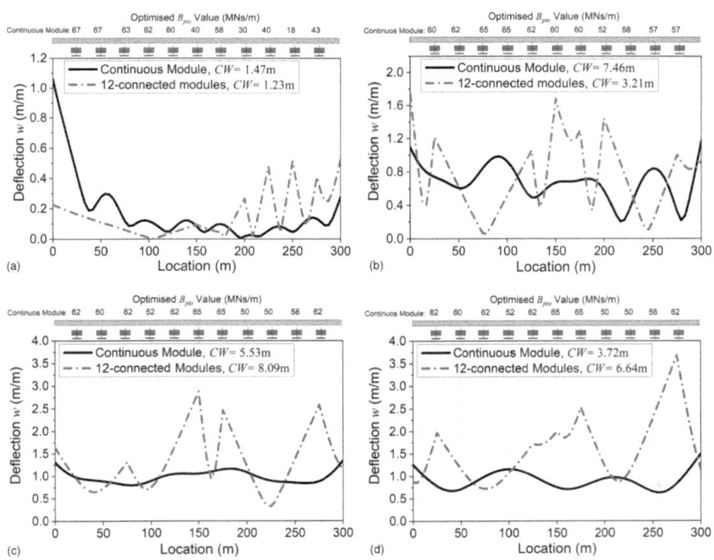

FIGURE 7.13
Comparison of hydroelastic response for Group III continuous (Type-D) pontoon-type WEC and 12-connected modules (Type-A) under their respective optimised B_{pto} for (a) $\frac{\lambda}{L}=0.2$ and (b) $\frac{\lambda}{L}=0.4$ (c) $\frac{\lambda}{L}=0.6$ (d) $\frac{\lambda}{L}=0.8$.

8

Offshore Floating Solar Photovoltaic Farm

This chapter focuses on the hydroelasticity of offshore floating solar photovoltaic (OFPV) farm and the mitigation techniques by using floating breakwater and hinged-connected articulated plate as an antimotion device. PTO system can also be attached to the floating breakwater and OFPV farm to convert the mechanical motion into electricity. Four case studies will be presented, starting with understanding the motion performance of the OFPV, followed by energy generation from the integrated floating breakwater with WECs and mitigation methods using an antimotion device. The final case study will investigate the performance of integrated OFPV with WECs. Power absorption and compliance are used to quantify the performance of the WECs and an antimotion device. For simplicity, the deflection of the plate $\hat{w}$ is denoted by w. Other symbols used in this chapter follows those that appear in the previous chapters unless otherwise specified.

8.1 Hydroelastic Response of Offshore Floating Solar Photovoltaic Farm

8.1.1 State-of-the-Art Review

The increasing global demand for energy highlights the urgent need to explore alternative, more sustainable sources to reduce the environmental impacts of fossil fuel-based electricity production.[200–203] Floating photovoltaic (FPV) systems, which place solar panels on platforms floating on water bodies such as oceans, lakes, reservoirs, and canals, have emerged as a viable solution to address land scarcity problems.[204–206] Over the years, the total installed capacity of FPV systems has grown steadily, with projects being implemented in oceans, lakes, estuaries, and natural basins.[207,208] FPV systems offer several advantages, including unobstructed sunlight, ease of installation, enhanced energy and power generation efficiency, and decreased water evaporation.[209,210] However, deploying these systems in marine environments, particularly in open seas, presents challenges such as higher environmental loadings due to wind and wave forces as well as corrosion and biofouling, which differ significantly from onshore conditions.[205,211]

The continuous reduction in the cost of solar photovoltaic (PV) panels has driven a global expansion of solar energy installations, playing a critical role

DOI: 10.1201/9781003387855-8

in decreasing dependence on hydrocarbons. Some of the largest FPV systems are situated in lakes and reservoirs across Asia. For example, Thailand's Sirindhorn Reservoir hosts an FPV system that covers an area comparable to 70 soccer fields, with a power generation capacity of 45 megawatts (MW). China houses three of the world's largest FPV installations: the Dingzhuang system in Dezhou (320 MW), the Three Gorges FPV system (150 MW), and the China Energy Conservation and Environmental Protection Group's (CECEP) FPV project (70 MW) in Anhui. In 2021, Singapore introduced one of the world's most extensive FPV projects, spanning an area equal to 45 football fields on a western reservoir, with a capacity of 60 megawatt-peak (MWp). Meanwhile, South Korea is developing the Saemangeum floating solar project, which will become the largest floating solar farm in the world upon its completion in 2024, with a capacity of 2.1 gigawatts (GW).

With the encouraging energy yields from FPV panels installed on water bodies, energy companies are now investigating the potential of offshore FPV farms (OFPVs) for deployment in open ocean environments. The ocean offers a compelling opportunity, with a theoretical global PV capacity estimated at approximately 4,000 gigawatts.[212] Studies show that solar panels floating at sea outperform land-based systems by nearly 13% on average, with some months showing energy production increases of up to 18%, thanks to cooler temperatures and less cloud cover.[213] In recent years, OFPVs have grown substantially, now spanning areas measured in hectares. A prominent example is Singapore's Sunseap OFPV, with a 5 MWp capacity, recognised as one of the largest offshore solar projects globally (Figure 8.1(a)). Covering an area of five hectares, it is projected to generate up to six megawatt-hours (MWh) of electricity annually. In Europe, energy provider like SolarDuck, a Dutch-Norwegian company, is partnering with German energy firm Rheinisch-Westfälisches Elektrizitätswerk (RWE) to create a floating solar facility with a raised deck at the North Sea wind farm (Figure 8.1(b)). Moreover, Norway's Ocean Sun (Figure 8.1(c)) has introduced a floating platform where solar panels are mounted on a flexible surface that adapts to wave movements.

8.1.2 Problem Definition

A modular OFPV comprises universal units is proposed, classified as a mat-like or pontoon-type VLFS, which is buoyancy-supported with a large surface area. The structure consists of modular units connected in a grid layout, as illustrated in Figure 8.2. This configuration reduces the structural rigidity, allowing the OFPV to deform flexibly under wave forces.[214] The grid layout has a total length L_x in the x-direction, L_y in the y-direction, and depth d. Each modular unit in the platform has dimensions l_x or l_y for length and w_m for width, forming a uniform grid, as depicted in Figure 8.2. All grid units are of the same size due to the uniformity of the solar PV panels. Symbols for wave and plate properties are

FIGURE 8.1
Offshore floating solar PV (a) Sunseap, (b) Solar duck, and (c) OceanSun.

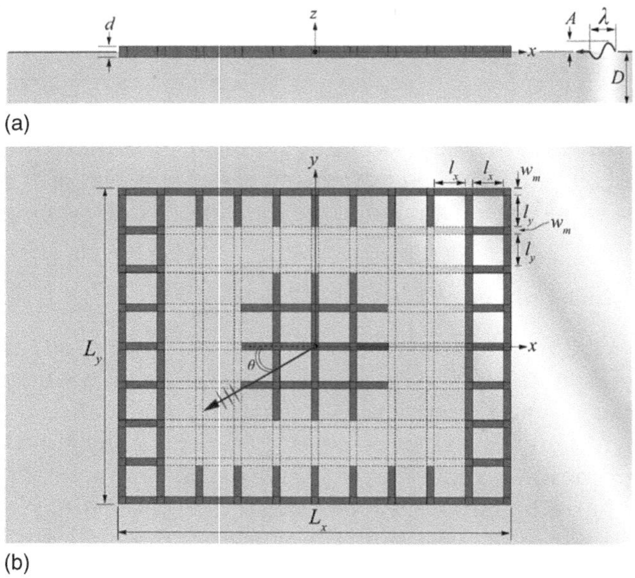

FIGURE 8.2
Schematic diagram showing (a) elevated view and (b) plan view of OFPV. Grey modules indicate that the OFPV could be expanded or reduced in size using integrated modular units.

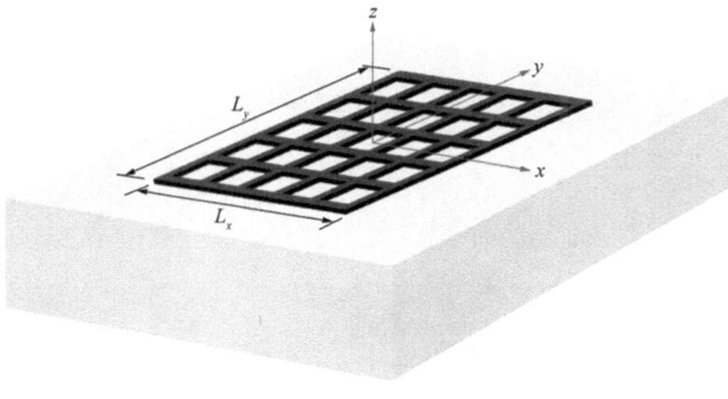

FIGURE 8.3
Example of OFPV arranged in a 5 × 5 grid configuration.

consistent with previous chapters unless otherwise specified. For visualisation, Figure 8.3 shows a three-dimensional representation of the OFPV in a five-by-five grid arrangement.

Figures 8.4(a)–8.4(c) are broken down into smaller solar farms, resulting in the layouts presented in Figures 8.4(d)–8.4(f). This segmentation facilitates the study of wave interactions between the OFPVs and their effect on reducing the hydroelastic response. The separated solar farms are placed at varying distances, denoted as s_p. In this analysis, three spacings are considered: s_p = 1, 5, and 10 m. Evaluating these different configurations helps to understand the influence of varying stiffness, which is characterised by the stiffness coefficient β. This coefficient is defined as $\beta = \mathcal{D}/\left(\rho_f g L^4\right)$, where $\mathcal{D} = \left(\mathcal{E}d^3\right)/\left[12\left(1-\nu^2\right)\right]$ represents flexural rigidity. Here, $\mathcal{E}$ is set to 534 MPa, $\rho_f = 1{,}000$ kg/m^3 is the water density, g = 9.81 m/s^2 is gravitational acceleration, and L is taken as $L = L_x$. Table 8.1 provides the detailed dimensions, properties, and parameters for the case studies.

For simplicity, Figures 8.4(a)–8.4(c) are referred to as OFPV-IA, OFPV-IB, and OFPV-IC, respectively, while Figures 8.3(d)–8.4(f) are termed OFPV-IIA, OFPV-IIB, and OFPV-IIC, respectively. The grid layout dimensions are indicated in the sub-captions of Figure 8.4, such as OFPV-IA having a 73 × 73 grid layout, similar to Figure 8.3.

8.1.3 Model Validation

The OFPV modelled by using the thin plate theory (see Appendix A) is first validated with its counterparts obtained from the finite element software ABAQUS. The natural frequencies are determined through free vibration analysis of the OFPV with free edge boundary conditions, accomplished by solving the eigenvalue problem described in Equation (4.75).

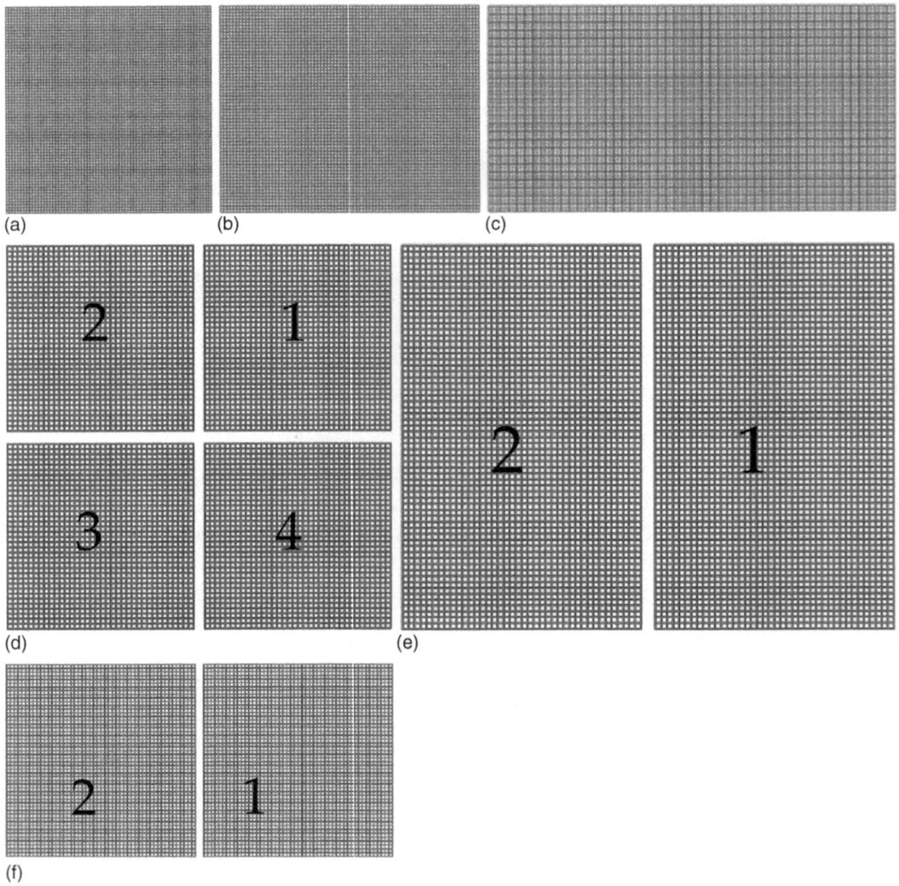

FIGURE 8.4
Layout configurations considered in case studies: (a) OFPV-IA: $\alpha = 1.00$, (b) OFPV-IB: $\alpha = 0.80$, (c) OFPV-IC: $\alpha = 0.50$, (d) OFPV-IIA: $\alpha = 1.00$, (e) OFPV-IIB: $\alpha = 1.62$, and (f) OFPV-IIC: $\alpha = 1.01$.

Note that α is for each separated OFPV.

Five different module depths are considered, as presented in Table 8.1, i.e., $d = 0.600, 0.200, 0.060, 0.030,$ and $0.006\,\text{m}$, to study the convergence of the present numerical model for the thin plate theory. The comparisons of the natural frequencies f_N of OFPV-IA, OFPV-IB, and OFPV-IC are presented in Table 8.2, respectively, where f_N is given by

$$f_N = \sqrt{\left(\frac{\overline{\omega_N}}{2\pi}\right)^2 \cdot \frac{g}{L}}, \text{unit in Hertz}(\text{hz}). \tag{8.1}$$

TABLE 8.1

Principal Dimensions, Properties, and Parameters Considered in Case Studies

		OFPV Properties					Regular Waves
		α	d (m)	$L_x \times L_y$ (m×m)	T(s)	λ(m)	θ(°)
OFPV-I	OFPV-IA	1.00	0.2 m	101.2×101.2	3, 4, 5	14, 25, 37	0, 30, 45, 60, 90
	OFPV-IB	0.800		112.4×90.0			
	OFPV-IC	0.500		141.8×71.8			
OFPV-II†	OFPV-IIA	1.00		50.8×50.8			
	OFPV-IIB	1.62		56.4×90.0			
	OFPV-IIC	1.01		70.4×71.8			

† spacing $s_p = 1$, 5, and 10 m; Young's modulus $\mathcal{E} = 534\,\text{MPa}$; mass density $\rho_p = 960\,\text{kg/m}^3$; and water depth $D = 10\,\text{m}$.

TABLE 8.2

Comparison of Natural Frequency Between Present Model with ABAQUS for OFPV-I ($h = 0.2000\,\text{m}$, $\Delta_e = 0.2\,\text{m}$)

	OFPV-IA		OFPV-IB		OFPV-IC	
L	101.2 m		112.5 m		141.8 m	
$\bar{h} = \frac{h}{L}$	1.9763 ×10^{-3}		1.7778 ×10^{-3}		1.4104 ×10^{-3}	
	f_N (hz)		f_N (hz)		f_N (hz)	
Mode	PM	ABQ	PM	ABQ	PM	ABQ
1 (RBM*)	0.0000	0.0000	0.0000	0.0000	0.0000	0.0000
2 (RBM*)	0.0000	0.0000	0.0000	0.0000	0.0000	0.0000
3 (RBM*)	0.0000	0.0000	0.0000	0.0000	0.0000	0.0000
4	0.0101	0.0094	0.0098	0.0093	0.0062	0.0060
5	0.0120	0.0120	0.0102	0.0095	0.0101	0.0092
6	0.0122	0.0121	0.0153	0.0148	0.0171	0.0165
7	0.0232	0.0224	0.0226	0.0209	0.0211	0.0193
8	0.0235	0.0224	0.0255	0.0237	0.0240	0.0232
9	0.0334	0.0332	0.0271	0.0262	0.0314	0.0296
10	0.0334	0.0332	0.0407	0.0381	0.0335	0.0320
11	0.0434	0.0411	0.0422	0.0408	0.0346	0.0323

*RBM: Rigid body motion, PM: Present Method, ABQ: ABAQUS.

The depth d is normalised with the length L in this study, where the nondimensional depth is defined as $\bar{d} = d / L$.

Table 8.2 presents the natural frequencies of OFPV-I as obtained from the present method (PM)[1] and ABAQUS (ABQ). The results indicate that the natural frequencies derived from the PM are in good agreement with those

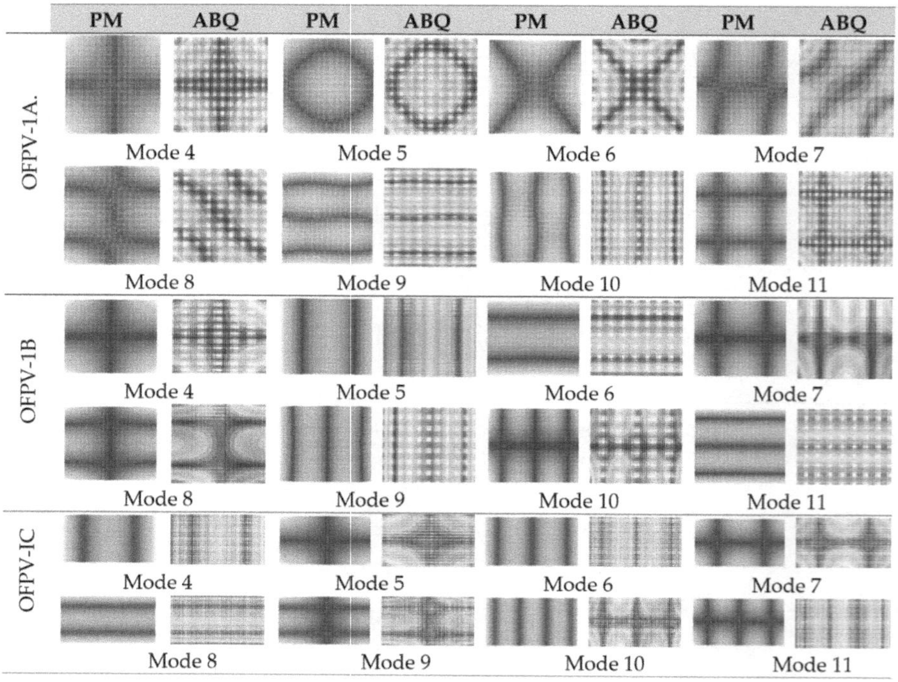

FIGURE 8.5
Vibration model for OFPV-IA, OFPV-IB, and OFPV-IC.

predicted by ABAQUS. Figure 8.5 illustrates the comparisons of vibration modes for OFPV-IA, OFPV-IB, and OFPV-IC between the two models. The similarities observed in the vibration modes further validate the accuracy of the present model.

8.1.4 Results and Discussion

The hydroelastic response of the OFPV is solved using the FE-BE method presented in Chapter 4. The deflection of the OFPV is denoted as w instead of $\hat{w}$ for simplification.

8.1.4.1 OFPV-I

8.1.4.1.1 Effect of Wave Direction

Figure 8.6 presents the hydroelastic response under regular wave conditions for OFPV-IA, OFPV-IB, and OFPV-IC. The wave periods considered range from T = 3–5 s, reflecting typical conditions in tropical regions like Singapore. Generally, the results indicate that the hydroelastic response of OFPV-I increases with longer wavelengths (larger wave periods) and decreases with shorter wavelengths (smaller wave periods).

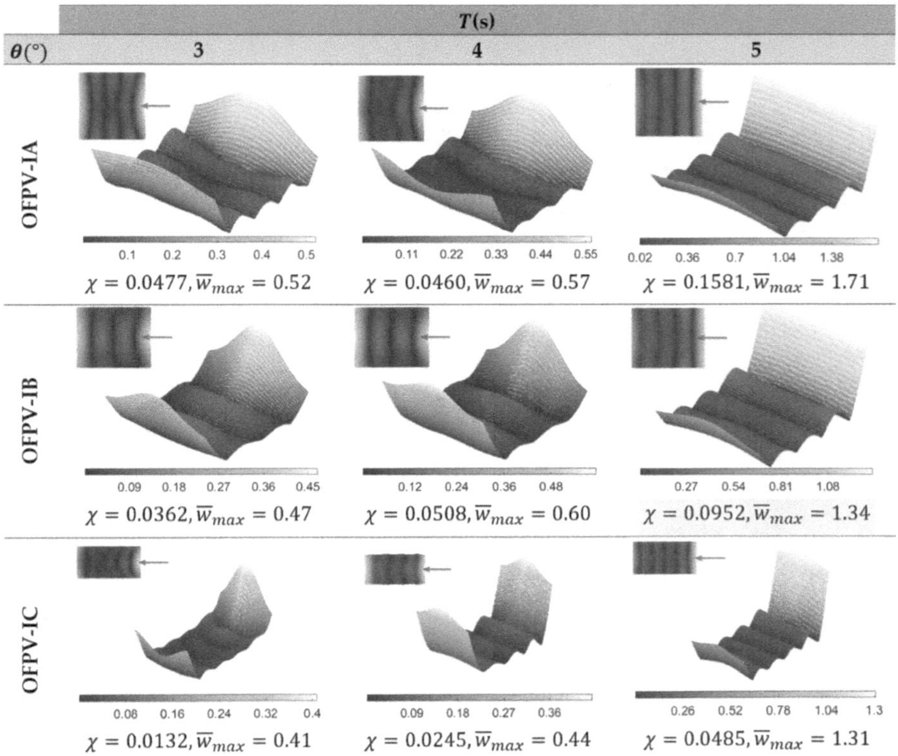

FIGURE 8.6
Hydroelastic response of OFPV-IA, OFPV-IB, and OFPV-IC under head sea and various wave periods.

Note: Wave direction is indicated by the red arrow. Small contour figure represents plan view and large contour figure represents isometric view of the OFPV.

Notably, the elastic deformation of OFPV-I is more pronounced when the λ / L is small. Additionally, the hydroelastic response is typically greater under head sea conditions ($\theta = 0°$). The data in Figure 8.6 illustrate that the hydroelastic response is highest at the forefront of the OFPVs, where incident waves first impact the structure. As the response travels towards the end of the structure, it diminishes due to energy absorption. To mitigate the hydroelastic response, it is recommended that the forefronts of OFPVs be constructed from materials with higher rigidity, functioning as wave attenuation devices. When comparing OFPV-I with varying aspect ratios α in Figure 8.6, it is observed that the maximum hydroelastic response, defined in Equation (8.2), decreases as α reduces, with OFPV-IC exhibiting the smallest response. However, the elastic deformation is greater for the longer structures due to reduced flexural rigidity in the longitudinal direction (x-axis) as the layout elongates.

$$\bar{w}_{\max} = \frac{w}{A}, \tag{8.2}$$

where w is the hydroelastic response of the structure.

In comparison to the square OFPV-IA floating farm, the hydroelastic response of the elongated OFPV-IC farm dampens more quickly, resulting in lower overall hydroelastic responses. This trend is clearly illustrated in Figure 8.7, which presents the elevated view of the hydroelastic response along the centreline of OFPV-IA, OFPV-IB, and OFPV-IC under head sea conditions. While the elongated OFPVs (OFPV-IB and OFPV-IC) exhibit greater elastic deformation, the magnitude of this deformation is smaller compared to the square OFPV-IA. This discrepancy leads to a reduced overall hydroelastic response for the elongated configurations. The compliance χ for OFPV-IA, OFPV-IB, and OFPV-IC, as shown in Figure 8.6, further supports this observation.

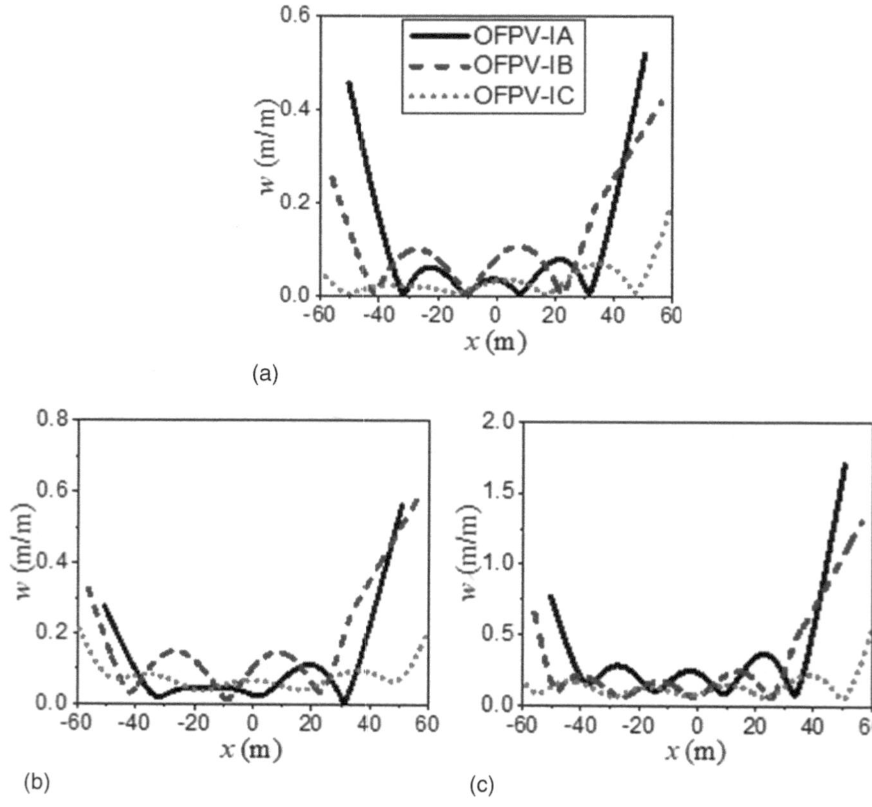

FIGURE 8.7
Elevated view of hydroelastic response along centreline of OFPV-IA, OFPV-IB, and OFPV-IC under (a) $T = 3$ s, (b) $T = 4$ s, and (c) $T = 5$ s. Incoming waves travel from right to left.

8.1.4.1.2 Effect of Layout Configuration

The hydroelastic response of the OFPV can be further reduced by using shorter modular units to enhance the overall structural stiffness. In this study, modular units with lengths of $l_x = 1.0$ m and $l_y = 0.5$ m are utilised to create OFPV-I, maintaining the same overall footprint of 10,000 m^2 as indicated in Table 8.1, where both l_x and l_y were originally 1.0 m. By using shorter l_y modular units, the longitudinal (flexural) stiffness in the x-direction of the modified OFPV-I′ increases. The prime symbol (′) distinguishes this enhanced version from the original OFPV-I, with OFPV-I′, OFPV-IA′, OFPV-IB′, and OFPV-IC′ referring to the variants with greater longitudinal stiffness.

Figure 8.8 illustrates the hydroelastic response along the centreline of OFPV-I′ and OFPV-I under a wave period of $T = 5$ s and head sea direction ($\theta = 0^{\circ}$). The reduction in hydroelastic response is significant for OFPV-IA′, with the maximum response being halved compared to OFPV-IA. While OFPV-IB′ and OFPV-IC′ show minimal reduction in maximum responses,

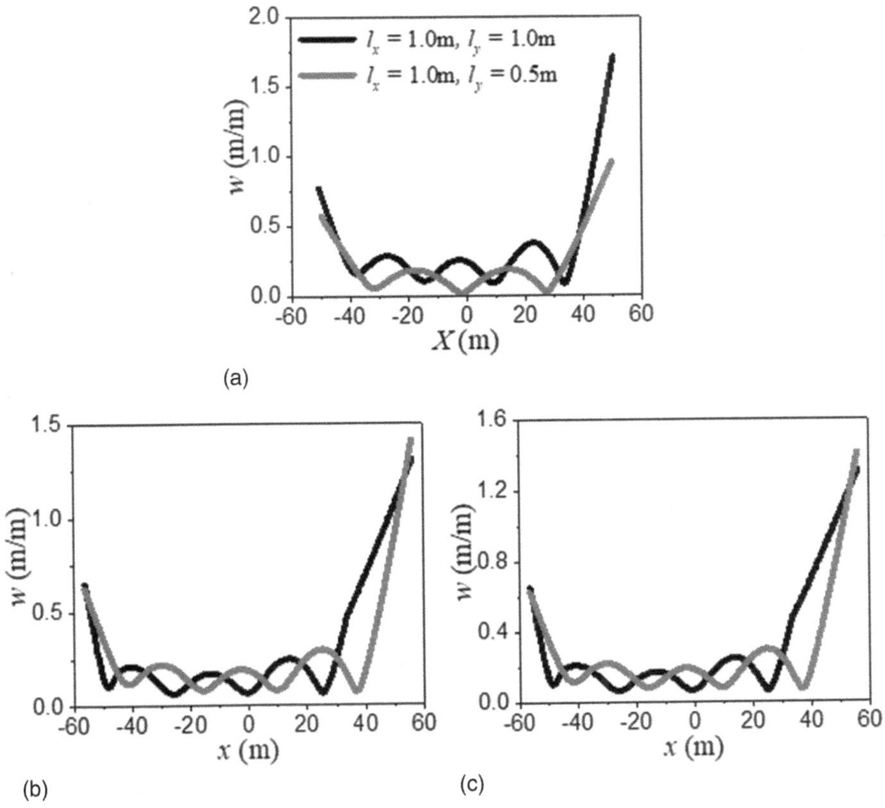

FIGURE 8.8
Elevated view of hydroelastic response along centreline of (**a**) OFPV-IA′, (**b**) OFPV-IB′, and (**c**) OFPV-1C′, under $T = 5$ s and head sea direction. Incoming waves travel from right to left.

they experience less elastic deformation due to their larger modular units in the y-direction, which enhance the longitudinal rigidity of the OFPV. Consequently, the overall hydroelastic response of OFPV-I′ is lower than that of OFPV-I. Additionally, increasing the number of rows in the grid layout adds mass to the OFPV, further contributing to the reduction in hydroelastic response.

8.1.4.1.3 Compliances

To quantify the deformation of the OFPV, the parameter compliance χ is introduced:

$$\chi = \int_{S_{HB}} w \cdot dS_{HB}. \tag{8.3}$$

The compliance χ in Equation (8.3) represents the total volume under the hydroelastic response w, where a higher χ indicates a higher overall response of the OFPV, and vice versa. The χ values for OFPV-IA, OFPV-IB, and OFPV-IC are presented in Figure 8.9, which shows that OFPV-IC has the smallest χ values among the three configurations considered. Thus, this implies that the long-ish layout configuration (small α) for the OFPV is desirable, as it has a smaller hydroelastic response.

8.1.4.2 OFPV-II

This section discusses the division of OFPV-I into multiple layouts, referred to as OFPV-II in Table 8.1. The OFPVs are spaced apart at distances of $s_p = 1\,\text{m}$,

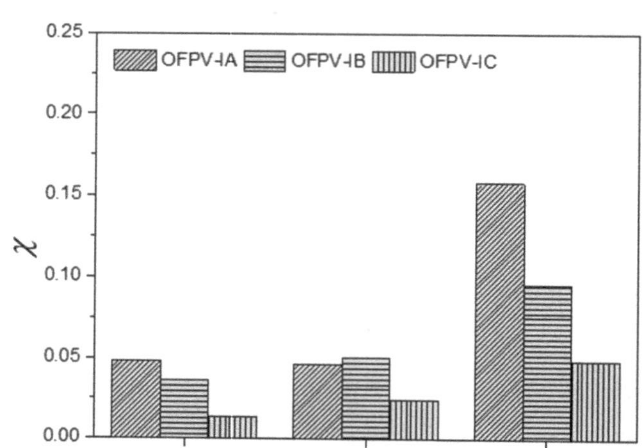

FIGURE 8.9
Compliance χ for (a) OFPV-IA, (b) OFPV-IB, and (c) OFPV-IC.

5m, and 10 m to examine the impact of gap spacing on hydroelastic response. Figure 8.10 presents a comparison of compliance χ between OFPV-II and OFPV-I at wave periods of T = 3, 4, and 5 s. The results indicate that breaking OFPV-I into smaller modules does not reduce the overall hydroelastic response. The χ value for OFPV-I (the interconnected floating farm shown by the black bar in Figure 8.10) is significantly lower under wave action compared to the separated layouts. Generally, among the three OFPV-II configurations, dividing the OFPV into more modules, such as OFPV-IIA, leads to higher compliance due to increased interference between the four separate OFPVs. Furthermore, the reduced mass of the separated farm in OFPV-II results in a greater hydroelastic response compared to OFPV-I. The findings also reveal that the hydroelastic response for the square OFPV-IIA increases with larger gap spacing s_p, while the response for the elongated OFPV-II (OFPV-IIC) decreases as s_p increases. Interested readers can refer to the publication by Tay[215] for more information on the deflection of OFPV-II.

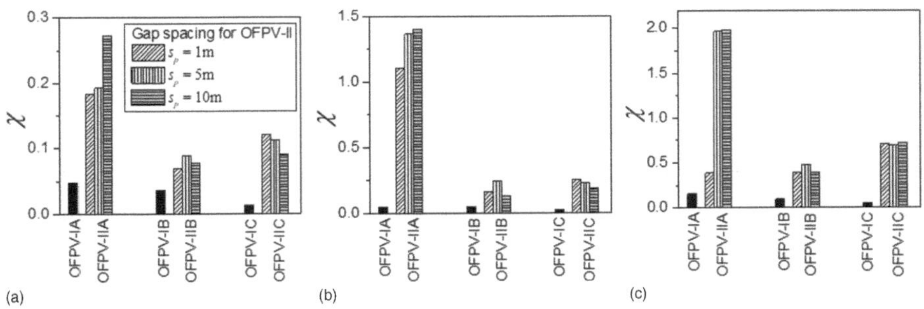

FIGURE 8.10
Comparison of compliance χ between OFPV-I and OFPV-II under (a) T = 3 s, (b) T = 4 s, and (c) T = 5 s.

8.2 OFPV Protected by Integrated Floating Breakwater and WECs

8.2.1 State-of-the-Art Review

As OFPVs are placed in the open sea, they face intense random waves, necessitating designs that consider these cyclic loads. These floating platforms are generally large but have a minimal thickness, leading to low structural stiffness.[216] As a result, when the platform is subjected to wave action, it deforms flexibly under hydroelastic response. Various strategies can help reduce the hydroelastic response of floating solar farms. A traditional method involves using breakwaters.[217,218] However, as the floating solar farm moves further offshore and the water depth increases, fixed breakwaters may become impractical. Additionally, these breakwaters can harm the environment by damaging the seafloor, destroying marine habitats, and taking up more sea space.[70,154,180] This makes FBs a more appealing option.

Floating breakwaters can be divided into seven main types[154]: box-type, pontoon-type, frame-type, mat-type, tethered float-type, horizontal plate-type, and other specialised designs. Each type has distinct features and benefits in terms of wave reduction and stability. The box-type is the most widely used reflective FB, having been effectively implemented in various locations worldwide for many years.[154] These box-type breakwaters are specifically designed to reduce wave force in calm conditions, with their effectiveness increases with the increase in the width of the structure.[167,173,190] Examples of FBs globally include the Holy Loch breakwater (Figure 8.11(a)), Fezzano FB (Figure 8.11(b)), Kan-on floating composite breakwater (Figure 8.11(c)), and the SF Marine FB (Figure 8.11(d)). Many of these breakwaters have been operational for decades, proving their ability to protect ports, marinas, and coastal areas from strong waves and erosion.

A significant example is the FB used in the Monaco Pier Hercules extension project,[219] which features a counter-jetty to improve the port's ability to

FIGURE 8.11
Examples of floating breakwater (a) Holy Loch breakwater, (b) Fezzano floating breakwater, (c) Ujina Pier and Kan-on composite breakwater, and (d) SF Marine floating breakwater.

receive cruise ships and large yachts. In 2000, a FB was constructed next to the Mega-Float to shield the prototype floating runway from strong waves.[220] Researchers have also explored hybrid FBs combined with attenuator-type WEC systems, as they can both protect the coast and harness wave energy. Additionally, innovative designs such as hybrid FBs with point absorber WECs[221,222] and OWC WECs[223] have been suggested. Placing WECs alongside FBs could optimise the use of ocean space and lower deployment costs, making WECs more cost-effective.

This section considers a hybrid FB-WEC system that serves dual purposes: as a wave attenuator and a wave energy extractor. The effect of the FB on mitigating the hydroelastic response of the OFPV will be investigated. The I-, L-, U- and box-shape breakwaters will be considered,[237] and its effectiveness in wave attenuation under monochromatic waves arriving from different directions will be studied. Additionally, due to the long structure length-to-wavelength ratio of the OFPV and FB, the hydroelastic response of the structures will be taken into consideration.

8.2.2 Problem Definition

An integrated floating breakwater, known as FB-WEC, incorporates attenuator-type WECs to safeguard a OFPV array, as shown in Figure 8.12. The OFPV comprises modular units that create a large mat-like formation, while the FB is elongated, exceeding the dimensions of the OFPV. Both structures are categorised as VLFS due to their significant horizontal area compared to their vertical size, which results in their movements being predominantly affected by hydroelastic responses (see Suzuki et al.'s[122] definition of VLFS in Equation (4.1)). The attenuator-type WECs are attached to the FB with hinges, where PTO systems on these hinges transform rotational motion into electricity. The OFPV is organised in a grid layout with dimensions, length L_{OFPV}, width W_{OFPV}, and depth d_{OFPV}. This OFPV is symmetric along the x- and y-axes. In contrast, the FB has a box-like shape with dimensions, length L_{fb}, width W_{fb}, and depth d_{fb}. Each WEC also has a box-like form, with dimensions, length L_{wec}, width W_{wec}, and depth d_{wec}. The breakwater is positioned a distance $(s_p)_{OFPV}$ away from the OFPV, and the WECs are spaced by $(s_p)_{wec}$. A regular wave with amplitude A, period T, and wavelength λ / L approaches the floating farm at an angle θ measured counterclockwise from the positive x-axis. The water depth D is assumed to remain constant.

The study examines the OFPV when it is protected by a FB and an integrated FB with a WEC, referred to as FB-WEC, as illustrated in Figure 8.13. The OFPV has a square configuration with dimensions $L_{OFPV} = 101.2$ m, depth $d_{OFPV} = 0.4$ m and occupies an area of 1 hectare, similar to the OFPV-IA discussed in Section 8.1. The structural characteristics of the OFPV are detailed in Table 8.3. In contrast, the FB has a length $L_{fb} = L_{OFPV} + 2s_p$, where

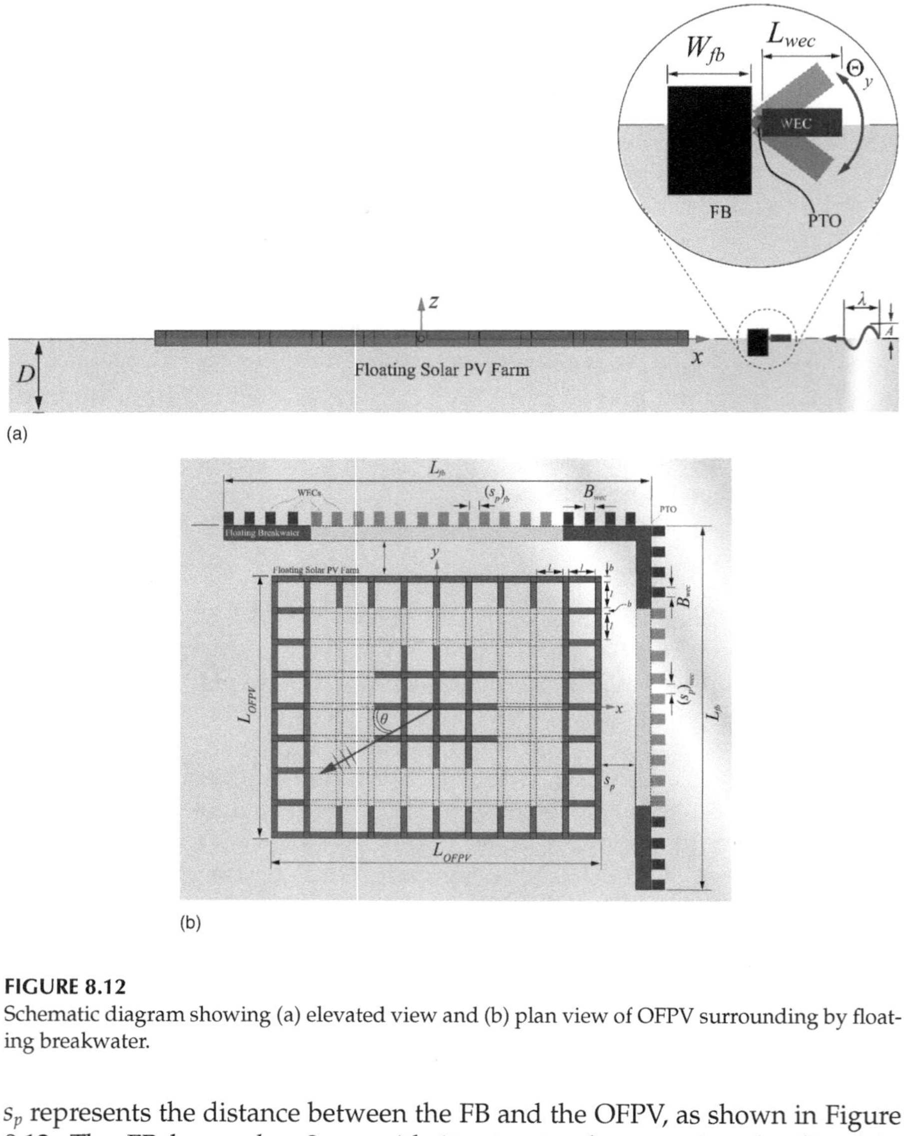

FIGURE 8.12
Schematic diagram showing (a) elevated view and (b) plan view of OFPV surrounding by floating breakwater.

s_p represents the distance between the FB and the OFPV, as shown in Figure 8.12. The FB has a $d_{fb} = 2$ m, with its structural properties also listed in Table 8.3. The dimensions, along with the structural and water properties for the cases analysed, are summarised in Table 8.3. To facilitate the discussion, FB-WEC-I, FB-WEC-L, FB-WEC-U, and FB-WEC-BOX refer to the I-shape, L-shape, U-shape, and box-shape FB-WEC, respectively.

8.2.3 Model Validation

Table 8.4 presents the comparison of the natural frequencies of FB-WEC-I, FB-WEC-L, FB-WEC-U, and FB-WEC-BOX obtained from the PM with ABQ.

TABLE 8.3

Principal Dimensions, Properties, and Parameters of Case Studies

	Principal Dimensions	Structural Parameters $\mathcal{E}$	ν	ρ_p	Water Properties T(s)	θ(°)
OFPV	$L_{OFPV} \times L_{OFPV} = 70\,m \times 70\,m$ $d_{OFPV} = 0.2$ m	534 MPa	0	960 kg/m^3	3, 4, 5	0, -30, -45
FB	$L_{fb} = L_{fb} + 2s_p$ $W_{fb} = 1$m, 2 m, 5 m $s_p = 1$m	11.9 GPa	0.2	256.25 kg/m^3		
WEC	$L_{wec} = 1, 2, 3$ m $W_{wec} = 1, 2, 3$ m $d_{wec} = 0.2, 0.4, 0.6$ m					

TABLE 8.4

Comparison of Vibration Frequency of FB-WEC Between PM with ABAQUS (ABQ)

	FB-WEC-I		FB-WEC-L		FB-WEC-U		FB-WEC-BOX	
Mode	**PM**	**ABQ**	**PM**	**ABQ**	**PM**	**ABQ**	**PM**	**ABQ**
RBM*	0		0		0		0	
1	1.25	1.24	0.85	0.84	0.21	0.15	0.35	0.26
2	3.45	3.40	1.22	1.21	0.69	0.68	0.54	0.53
3	6.76	6.65	2.74	2.70	1.17	1.08	1.13	0.98
4	11.17	10.94	3.36	3.31	1.22	1.20	1.15	1.00
5	16.69	16.25	5.70	5.57	2.44	2.40	1.22	1.21
6	23.31	22.53	6.55	6.32	3.02	2.98	2.44	2.26
7	31.03	29.77	9.71	9.86	3.52	3.38	2.71	2.66
8	39.86	37.90	10.74	10.99	5.23	5.10	2.78	2.73
9	43.91	46.88	14.64	14.60	6.32	6.11	3.86	3.56
10	49.79	56.67	15.38	15.96	6.53	6.33	4.86	4.77

*Number of rigid body mode (RBM) = Number of WECs + 3.

Table 8.4 shows that the natural frequencies obtained from PM is in good agreement with that obtained from ABQ except for Mode 1 of FB-WEC-I and FB-WEC-BOX where the accuracy differs by a little.

8.2.4 Results and Discussion

The performance of the four types of FB-WEC summarised in Figure 8.13 in mitigating the hydroelastic response under varying wave periods (i.e., varying wavelength) is first investigated. The effectiveness of the FB-WEC in mitigating the hydroelastic response of the OFPV is studied where compliance of the OFPV χ_{OFPV} is used to quantify the effectiveness of the breakwater. The power generation of the WECs integrated in the FB-WEC is then computed.

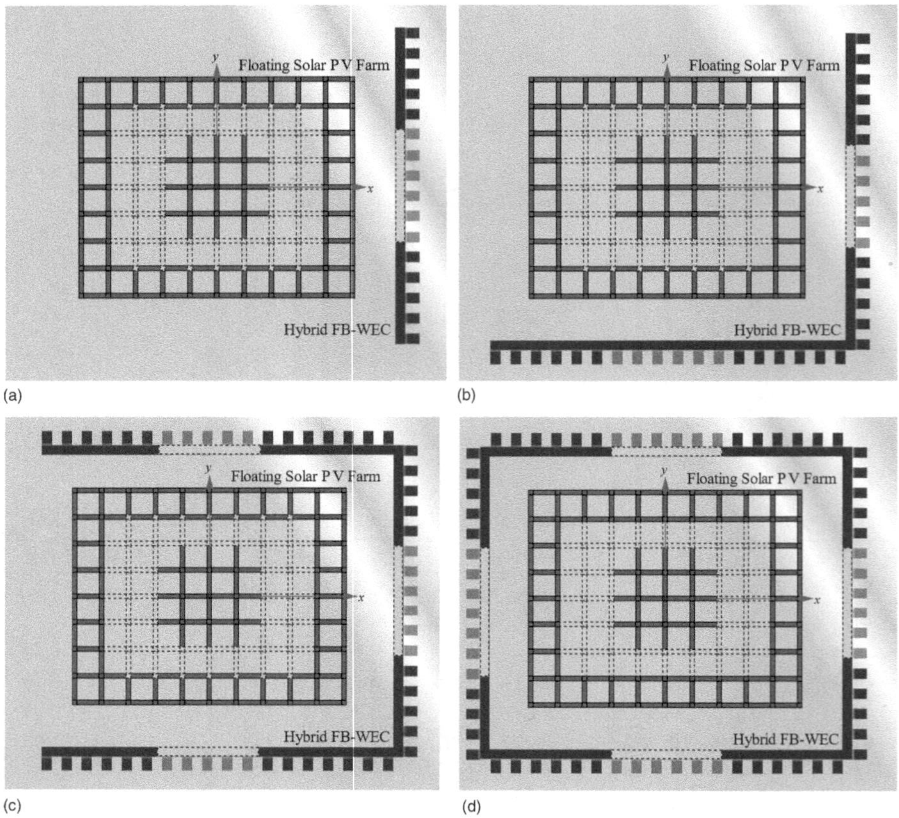

FIGURE 8.13
OFPV protected by (a) I-shape FB-WEC, (b) L-shape FB-WEC, (c) U-shape FB-WEC, and (d) box-shape FB-WEC.

8.2.4.1 Performance of FB-WEC

8.2.4.1.1 Compliance

The hydroelastic response of the OFPV protected by the integrated FB-WEC is presented in Figure 8.14. In general, the FB-WECs show a consistent trend in its performance at $T = 5$ s where it is not effective in mitigating the hydroelastic response for larger wavelength due to the longer wavelength-to-structural length ratio, i.e., λ / L_{OFPV}. However, there is no clear trend in the performance of the FB-WEC when $T = 3$ and 4 s due to the complex hydrodynamic interaction between the WECs, FB, and OFPV where the χ_{OFPV} of the OFPV is higher for FB-WEC-I, FB-WEC-L, and FB-WEC-BOX when $T = 3$ s whereas χ_{OFPV} is higher for FB-WEC-I when $T = 4$ s.

A comparison of the χ_{OFPV} of the OFPV protected by FB-WEC is summarised in Figure 8.15. In general, the hydroelastic response of the OFPVs under $T = 3$ and 4 s are significantly smaller compared to their counterparts when $T = 5$ s

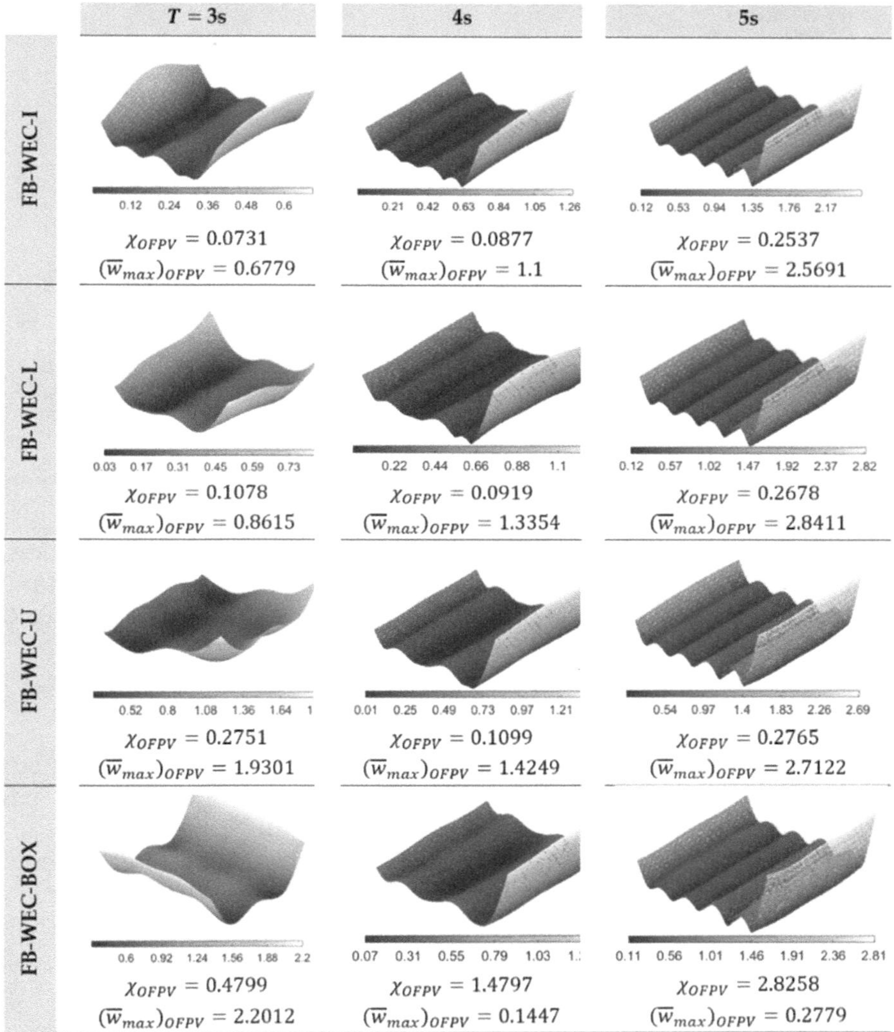

FIGURE 8.14
Hydroelastic response of OFPV protected by various FB-WEC under varying wave periods. Regular waves. Head sea.

when waves arrive from the head sea. However, the FB is not effective in mitigating the response when waves arrive from the oblique direction, i.e., $\theta = 45^{\circ}$, where the χ_{OFPV} are almost similar. While most of the hydroelastic responses of the OFPV are higher under head sea, this is not true for the case of FB-WEC-U and FB-WEC-BOX where the χ_{OFPV} is significantly greater when waves arrived from $\theta = 45^{\circ}$. This could be due to the greater interaction effect occurring at the gap formed between the OFPV and FB-WECs as

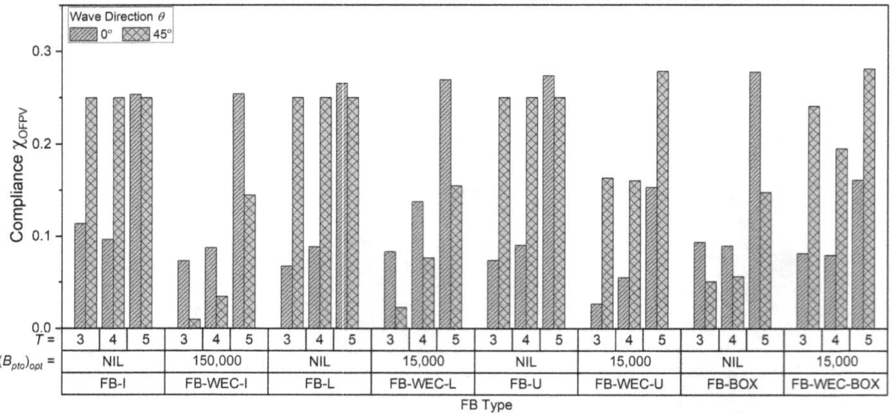

FIGURE 8.15
Comparison of compliance of OFPV protected by various FB and FB-WEC under regular waves.

a result of the interference between the WECs, FB, and OFPV. Nevertheless, the integration of WECs and FB could be used to extract marine energy from the sea to enhance the energy generation for OFPV. The energy generation for the FB-WECs will be presented in the next section.

8.2.4.1.2 Absorbed Power

As the FB-WECs are subjected to the regular waves of varying wave periods and wave directions, each WEC rotates about an angle Θ_y about the x-axis and Θ_x about the y-axis.

The power extracted/absorbed from the WECs could be computed from the rotations by using Equation (3.57). Figures 8.16 and 8.17 show the comparisons of the absorbed power of the four FB-WECs under head sea and oblique sea ($\theta = 45°$), respectively. To study the effect of PTO damping has on the energy generation of the WECs in Figures 8.16 and 8.17, five PTO damping are considered, i.e., B_{pto} = 15, 150, 1,500, 15,000, and 150,000 N.s/m. The results show that for FB-WEC-I, the P_a is the highest when the largest PTO damping, i.e., B_{pto} = 150,000 N.s/m, is employed. However, for FB-WEC-L, FB-WEC-U, and FB-WEC-BOX, the largest energy generation occurs when B_{pto} = 15,000 N.s/m. Considering the head sea direction in Figure 8.16, the FB-WEC-I generates the most energy at T = 3 s followed by FB-WEC-BOX at T = 4s. This can be shown clearly in Figure 8.14 where the WECs have the highest rotations/deflections when the maximum power absorption occurs.

To compare the maximum energy extraction from the FB-WEC, the absorbed power under head sea and oblique sea is placed side by side in Figure 8.18 for wave period 3, 4, and 5 s. Energy extraction from the

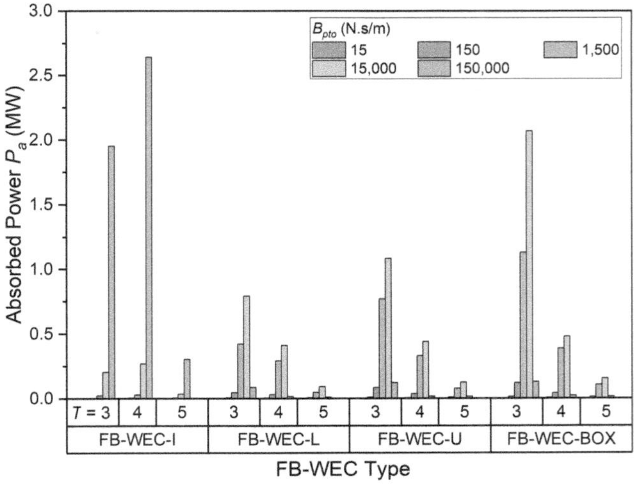

FIGURE 8.16
Comparison of absorbed power of various FB-WEC under regular waves. Head sea.

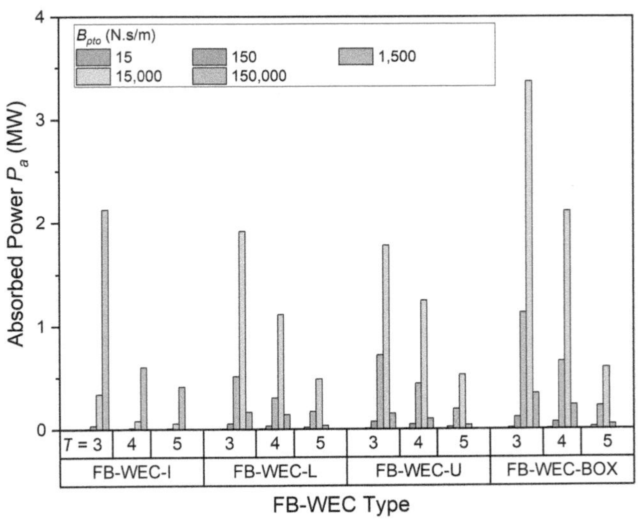

FIGURE 8.17
Comparison of absorbed power of various FB-WEC under regular waves. Oblique sea $\theta = 45^{\circ}$.

oblique sea for FB-WEC-U and FB-WEC-BOX is significantly larger than their counterparts of the head sea and at the same time, both FB-WECs have the capability to mitigate the hydroelastic response of the OFPV as evidenced by the values summarised in Figure 8.15. Therefore, to choose

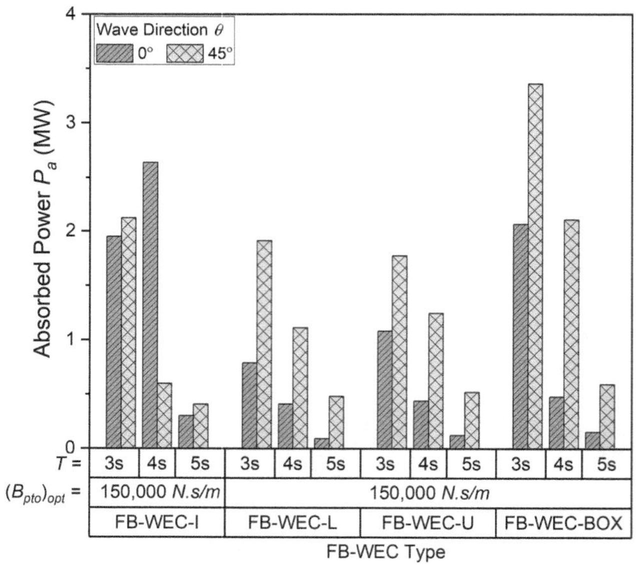

FIGURE 8.18
Comparison of absorbed power of various FB-WEC under regular waves.

the appropriate FB-WEC, a sweet spot has to be achieved between the maximising the P_a and minimising the χ_{OFPV} values.

The absorbed power P_a under oblique sea shown in Figure 8.17 shows that energy generation by the WECs in oblique sea is generally higher when compared to their counterparts under head sea in Figure 8.16. This is due to greater energy extractions by the WECs moving at different phases when waves arrive from the oblique direction. The maximum energy generation from the WECs occurs when the wave periods is small at $T = 3$ s because of the greater wave interaction between the WECs that results in constructive interferences.

8.2.4.1.3 Average Capture Width

The $\overline{CW}$ for the WECs computed from Equation (3.59) for head sea and oblique sea is plotted in Figure 8.19. The $\overline{CW}$ indicates the capability of each WEC in extracting the energy from the incoming regular wave train. The $\overline{CW}$ of the WEC for FB-WEC-I is significantly high when $T = 3$ s, denoting the effectiveness of the FB-WEC-I configuration in extracting energy when wavelength is small. Despite the smaller $\overline{CW}$ for FB-WEC-U and FB-WEC-BOX, both configurations allow a significantly large energy generation from the oblique sea.

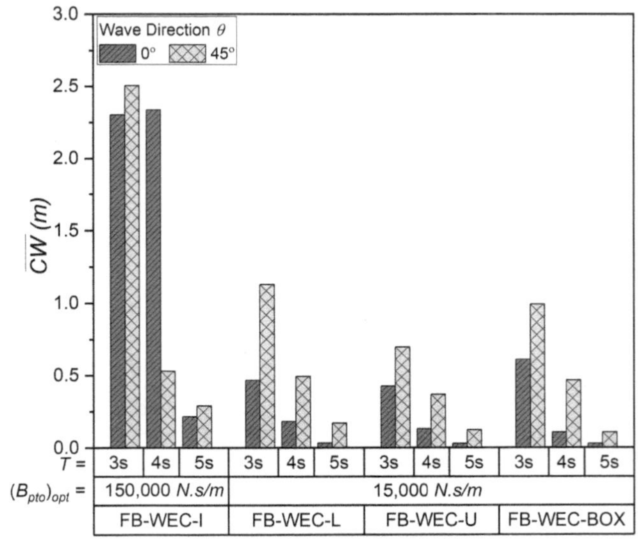

FIGURE 8.19
Comparison of average capture width $\overline{CW}$ of various FB-WEC under regular waves.

8.3 Integrated OFPV with AMDs

8.3.1 State-of-the-Art Review

An antimotion device (AMD) consisting of a multi-modular articulated plate could be attached to the OFPV. This solution could also reduce the footprint occupied by the OFPV compared to the FB solution for mitigating hydroelastic response. Khabakhpasheva and Korobkin[222] demonstrated that the hydroelastic response of a floating elastic beam could be effectively reduced using rigid, semi-rigid, and hinge connectors that connect the auxiliary beam to the elastic beam. They found that the semi-rigid mechanical joint is more effective in reducing hydroelastic responses compared to the rigid or simple hinge mechanical joints. Kim et al.[223] also examined the effects of hinged-connected auxiliary plates with varying stiffnesses and shapes on reducing the hydroelastic responses of VLFS. They discovered that a single hinged-connected auxiliary plate with a rounded shape and two hinged-connected auxiliary plates with rectangular shapes are the most effective in minimising the hydroelastic responses of the VLFS. Wang et al.[224] proposed a novel approach to reduce the hydroelastic response of an elongated VLFS by attaching auxiliary beams with appropriate rotational stiffness and strategically placing mechanical joints along the beam. They found that employing a semi-rigid connector at optimal locations in the floating beam system significantly reduces the hydroelastic responses of the main floating beam.

A comprehensive review of AMDs for VLFSs has been conducted.[70] Takagi et al.[225] proposed a simple box-shaped body attached to the edge of a VLFS as an AMD. The device's performance was investigated both theoretically and experimentally, showing good antimotion capabilities, including reducing deformation, shear force, and bending moment of the platform. The device performed well in beam sea and oblique sea conditions. Another proposed AMD is a submerged plate attached to the fore-end of the VLFS, designed to dissipate incident wave energy and reduce the incident wavelength by generating breaking waves, wave fission, and vortices. Ohta et al.[226] proposed a submerged vertical plate, which is found to significantly decrease hydroelastic response with decreasing wavelength. Masanobu et al.[227] proposed an innovative submerged AMD attached to the VLFS using a curtain wall (a submerged vertical plate with slits) and an inverted-L type AMD to reduce the hydroelastic response. They found that the wave energy could be dissipated via eddies generated from the slits of the curtain wall. Hybrid-type AMDs, such as the integrated submerged AMD and oscillating water chamber, have also been considered for better reduction in hydroelastic response of the VLFS under wave action. Fujikubo et al.[228] proposed a framing system with an antimotion heaving plate hinged to the VLFS. Their investigation showed that the framing breakwater system has sufficient wave dissipating capacity, with performance depending significantly on its submerged depth. Ikoma et al.[229] suggested using a combination of submerged vertical plates with a wave energy-absorbing air chamber to reduce the elastic deformation of the VLFS. They found that the wave drifting force could be reduced if wave energy is effectively absorbed by the OWC system.

A submerged horizontal AMD attached to the floating solar PV farms is considered to mitigate the hydroelastic response of the floating structure. The effect of rigidity of the hinged modules in mitigating the response of the OFPV is to be investigated by varying the thickness of the AMD.

8.3.2 Problem Definition

Figures 8.20(a) and 8.20(b) show the schematic diagram of the OFPV in elevated and plan views, respectively. Articulated plates with length L_{amd} are connected to the four sides of the OFPV. The articulated plates are connected to the OFPV with two types of connectors, i.e., the hinge and the rigid connectors where the latter is analogous to a welded AMD to the OFPV. As waves hit on the OFPV and AMDs, the OFPV deflects with amplitude w, whereas the hinged-connected AMD rotates at an angle Θ_y about the x-axis and Θ_x about the y-axis.

The considered OFPV in this study occupies an area of approximately 8,100 m^2 and form a 57×57 (number of rows × number of columns) grid layout. The total length of the OFPV is thus $L_{OFPV} = 90$ m. The structural properties of the OFPV made of high-density polyethylene (HDPE) modular units

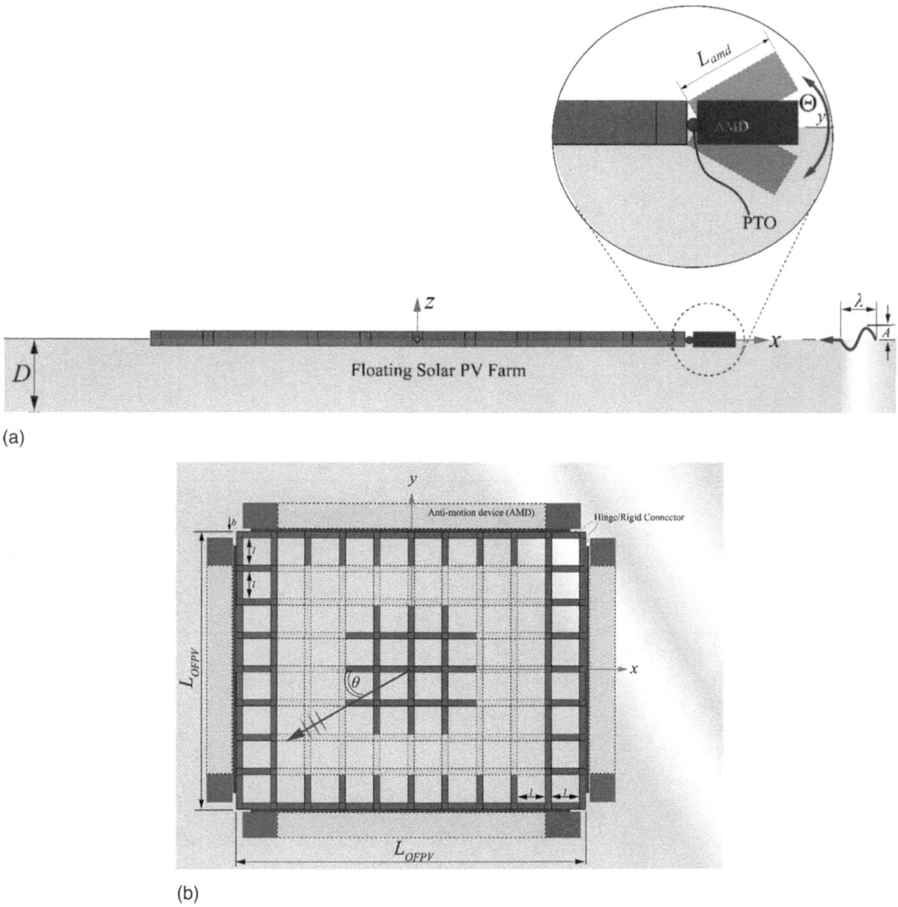

FIGURE 8.20
Schematic diagram showing (a) elevated view (b) plan view of OFPV connected to AMD.

are considered in Table 8.5. Three varying AMD length, i.e., $L_{amd} = 1, 2,$ and 5 m, and three different AMD thicknesses, i.e., $d_{amd} = 0.2, 0.4,$ and 0.6 m will be considered to investigate the effect the AMD length and AMD thickness has on the hydroelastic response of the OFPV.

8.3.3 Model Validation

Table 8.6 summarises the vibration frequency of the OFPV with AMDs, obtained from the PM and ABQ. For validation purposes, two OFPVs with different dimensions are considered, i.e., the 5×5 and 57×57 OFPVs. Each OFPV is rigidly and hinged-connected with AMD. The comparison between results predicted by PM with ABQ shows that the eigenvalue obtained

TABLE 8.5

Principal Dimensions, Properties, and Parameters of Case Studies

		Structural Parameters			Water Properties	
	Principal Dimensions	$\mathcal{E}$	ν	ρ_p	T(s)	θ(°)
OFPV	$L_{OFPV} \times L_{OFPV} = 90\,\text{m} \times 90\,\text{m}$ $d_{OFPV} = 0.2$ m	534 MPa	0	960 kg/m^3	3, 4, 5	0, –30, –45
AMD	$L_{amd} = 1, 2, 5$ m $d_{amd} = 0.2, 0.4, 0.6$ m	11.9 GPa	0.2	256.25 kg/m^3		

TABLE 8.6

Comparison of Vibration Frequency (in rad/s) of OFPV with Hinged-Connected AMD between PM with ABAQUS (ABQ)

	5 × 5 OFPV				**57 × 57 OFPV**			
	d_{amd} **= 0.2m**				d_{amd} **= 0.4m**			
	Rigid-connected AMD		**Hinged-connected AMD**		**Rigid-connected AMD**		**Hinged-connected AMD**	
Mode	**PM**	**ABQ**	**PM**	**ABQ**	**PM**	**ABQ**	**PM**	**ABQ**
RBM	0*		0**		0*		0**	
1	10.6160	10.3810	8.7886	8.3898	0.1062	0.0835	0.1055	0.0814
2	12.0603	12.7510	21.4276	20.9950	0.1311	0.1285	0.1358	0.1320
3	16.6509	16.8230	22.8637	22.0810	0.1341	0.1321	0.1401	0.1354
4	27.5259	27.8580	28.0795	27.3240	0.2515	0.2136	0.2505	0.2091
5	27.5320	27.8600	28.0797	27.2280	0.2515	0.2136	0.2505	0.2091
6	41.3579	41.8090	47.7833	46.2320	0.3659	0.3595	0.3805	0.3686
7	41.3579	41.9010	48.3629	49.6140	0.3659	0.3595	0.3805	0.3686
8	49.5542	49.8690	61.3563	58.5740	0.4637	0.3854	0.4615	0.37′55
9	50.4402	51.3960	63.1836	58.7450	0.4846	0.4384	0.4838	0.4321
10	71.5497	70.4330	68.7374	66.5370	0.4894	0.4444	0.4911	0.4387

* Number of rigid body mode (RBM) = 3
** Number of rigid body mode (RBM) = 7.

from present numerical scheme is in good agreement with its counterparts obtained from ABQ. The vibration frequency for the hinged-connected AMD is observed to be smaller when compared to the rigid-connected AMD. Also, when the AMD is connected to OFPV with larger dimensions, i.e., the 57×57 OFPV, the vibration frequencies are found to be significantly reduced.

8.3.4 Results and Discussion

The effect of rigid and hinge connectors on mitigating the hydroelastic response of the OFPV is studied, where the influence of varying L_{amd} and d_{amd} is considered.

8.3.4.1 Rigid-Connected AMD

The performance of the AMD rigidly connected to the OFPV is first considered. The study considers rigidly connected AMD with varying L_{amd} and d_{amd}. The summary of the χ_{OFPV} for head sea is given in Figure 8.21. The effect of AMD with varying L_{amd} is studied. The hydroelastic responses of the OFPV for different L_{amd}, i.e., L_{amd} = 1, 2, and 5 m and subjected to wave period $T = 3, 4, \text{and } 5\,\text{s}$ under head sea are presented in Figure 8.22. In general, the hydroelastic response of the OFPV shows similar characteristics where the side exposed to the wavefront has a higher response followed by a reduction in the response in the mid-body and an increase in the response at the rear end. The χ_{OFPV} value associates with each image in Figure 8.22 shows that the longer AMD is more effective in mitigating the hydroelastic response of the OFPV where $L_{amd} = 5\,\text{m}$ has a smaller χ_{OFPV} compared to its counterpart for $L_{amd} = 1\,\text{and}\,2\,\text{m}$. When comparing the effect of different wave periods, i.e., wavelength λ, the χ_{OFPV} reveals that the effectiveness of the AMD in mitigating the hydroelastic response of the OFPV diminishes with the increase of wavelength, i.e., higher wave period. For better clarity, the comparison of the deflection of the OFPV along $x / L = 0$ for different L_{amd} is presented in Figure 8.23. A reduction in hydroelastic response of the OFPV could be clearly observed when χ_{OFPV} given in Figure 8.21 is small.

8.3.4.2 Hinged-Connected AMD

The effect of the hinged-connected AMD on the hydroelastic response of the OFPV is studied. The wave periods considered is increased to $T = 8$ s in studying the effect of the three different AMD lengths, i.e., $L_{amd} = 1, 2, \text{and } 5\,\text{m}$. The χ_{OFPV} of the OFPV under the influence of hinged-connected AMD for head

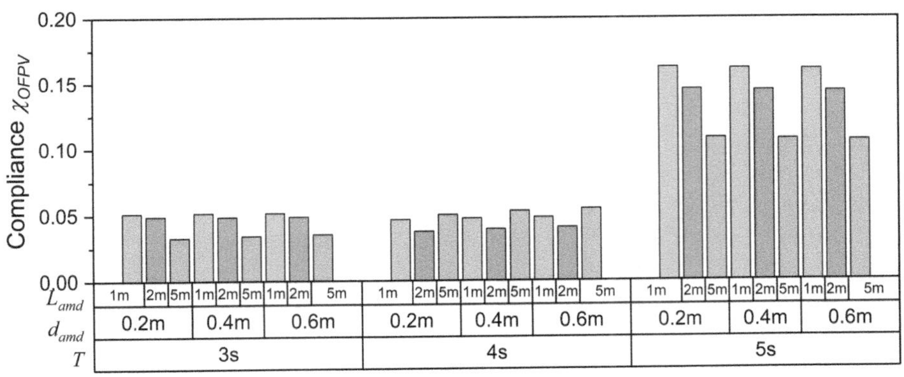

FIGURE 8.21
Comparison of compliance of OFPV rigidly connected with AMD for different AMD lengths. Head sea.

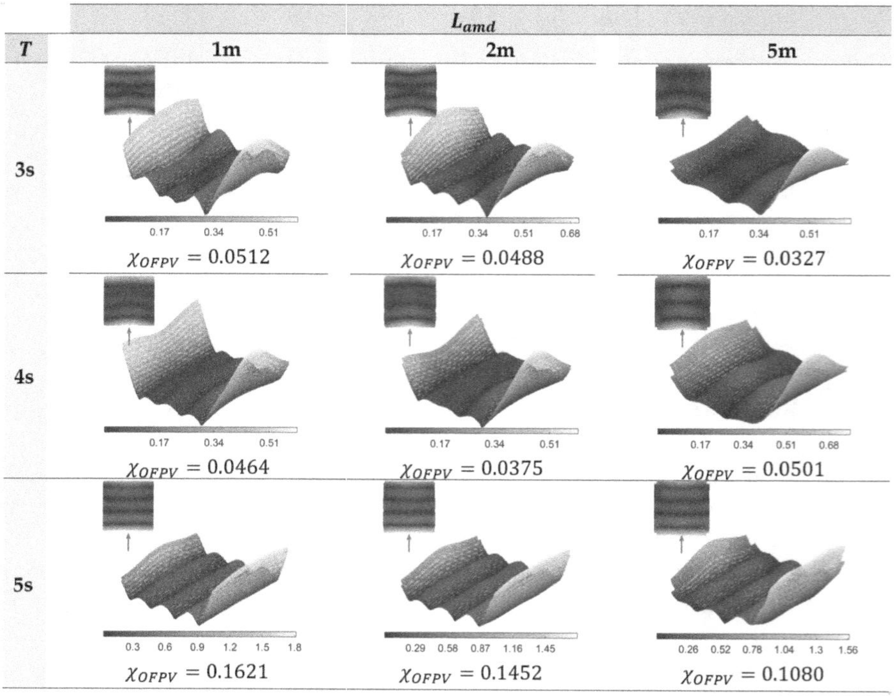

FIGURE 8.22
Hydroelastic response of OFPV rigidly connected with AMD for different AMD length L_{amd}. $d_{amd} = 0.2$ m. Head sea.

sea and oblique sea conditions are presented in Figure 8.24 for the three varying L_{amd}, i.e., $L_{amd} = 1, 2,$ and 5 m. In general, it can be seen that the effectiveness of the AMD in mitigating the hydroelastic response reduces with the increase of the wave period, evidenced from the increasing χ_{OFPV} as the wave periods get larger. This is because the AMD is more effective in attenuating the wave force when the ratio of the wavelength to AMD length is small as more radiated and diffracted waves will be generated from the AMD as compared to its counterpart of the larger ratio. It can be seen that the shorter AMD is more effective in reducing the hydroelastic response.

The deflection of the OFPV along $x / L = 0$ under head sea is presented in Figure 8.25 for wave period $T = 3, 4, 5, 6, 7,$ and 8 s. The results show that a significant reduction in the hydroelastic response, when waves approach from the head sea, could be achieved by using a shorter AMD, where the deflection of the OFPV with $L_{amd} = 1$m is relatively smaller as compared to its counterparts for L_{amd} = 2 and 5 m.

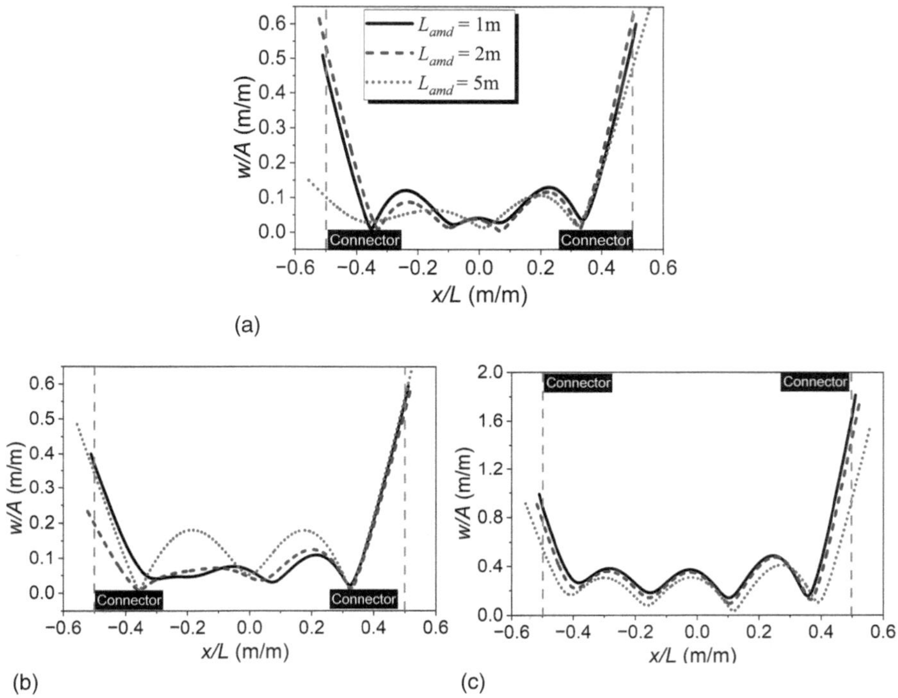

FIGURE 8.23
Comparison of hydroelastic response for OFPV with rigid-connected AMD for different AMD length L_{amd} along $/L = 0$. $d_{amd} = 0.2$m. Head sea. Waves travel from right to left.

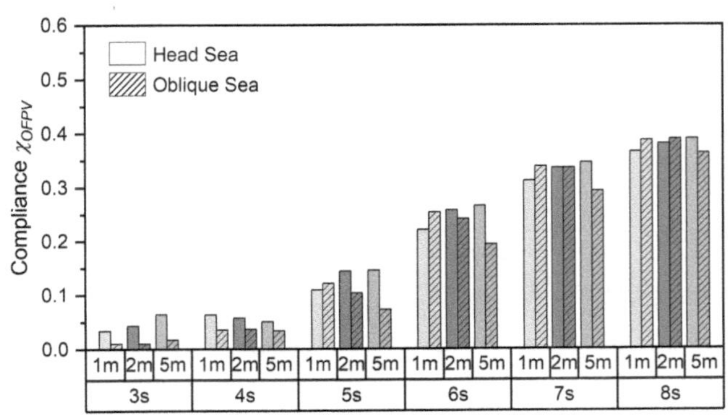

FIGURE 8.24
Comparison of compliance of OFPV hinge connected with AMD for different AMD lengths.

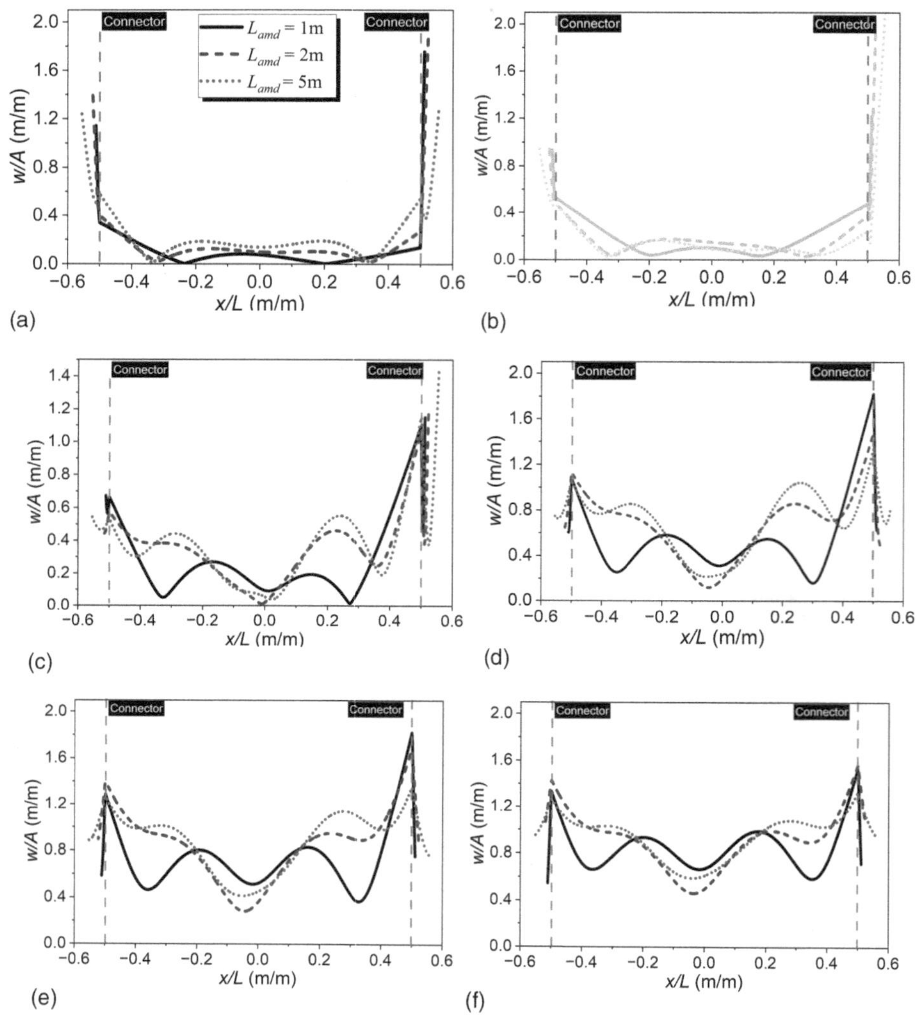

FIGURE 8.25
Comparison of hydroelastic response for OFPV with hinged-connected AMD for different AMD length L_{amd} along $/L = 0$. $d_{amd} = 0.2$ m. Head sea. Waves travel from right to left.

8.4 Integrated OFPV with WECs

8.4.1 State-of-the-Art Review

To be cost-effective, the offshore floating solar PV farms have to be deployed in large scale in the order of hectares. One of the means for efficient sea space utilisation is via the co-location of OFPV with WECs. This integrated approach allows for the simultaneous generation of electricity from solar radiation and wave motion. For example, the German company Sinn Power has developed a floating platform that combines wave, wind, and solar energy where the platform offers a flexible and expandable design, allowing for various combinations of energy conversion methods.[175] The integration of solar and wave energy offers vast benefits such as space-sharing and cost-sharing in which the floating solar platforms can create artificial space on the sea for ease of installation and maintenance of the WEC and additional renewable energy can be generated via the WECs.

A detail review of hybrid solar plant has been proposed[204, 230, 231] where concepts such as integration of floating PV farm with wind farm, hydropower, hydro energy storage, micro-hydrokinetic turbines, ocean energy system have been proposed. This paper is motivated by the co-location of WEC with OFPV following the various examples, as given by Nguyen et al.,[69] Tay,[165,166] and Nguyen,[169] where they proposed multi-modular articulated plates that are hinged attached to the OFPV. These articulated plates function both as an AMD and WEC, where the former mitigates the hydroelastic response arises in the large platform due to wave action, whereas the latter converts the kinetic energy from the plate rotational motion into electricity via PTO system. Appropriate PTO damping value has to be designed properly as having a large PTO damping does not effectively mitigate the hydroelastic response. This chapter looks into the performance of raft-type WECs attached to floating solar PV farm.

8.4.2 Problem Definition

Figure 8.26 shows the articulated-type WECs attached to the OFPV. The details of the OFPV are similar to that presented in Section 8.3. Each WEC has a length L_{wec}, width W_{wec} and thickness d_{wec}. The WECs are hinged-connected to the OFPV and PTO system is used to convert the mechanical motion into electricity. The effect of L_{wec} on the power generation of the WECs is studied where two L_{wec} are considered, i.e., $L_{wec} = 0.4$ m and $L_{wec} = 0.8$ m.

8.4.3 Model Validation

Table 8.7 summarises the vibration frequency of the OFPV with WECs, obtained from the PM and ABQ. Similar to validation of the OFPV integrated with AMD in Table 8.6, two OFPVs with different dimensions are considered, i.e., the 5×5 and 57×57 OFPVs. Each OFPV is rigidly and hinged connected with WEC. The comparison between the results predicted by PM and ABQ indicates that the eigenvalues obtained from the present numerical scheme closely match those obtained from ABQ.

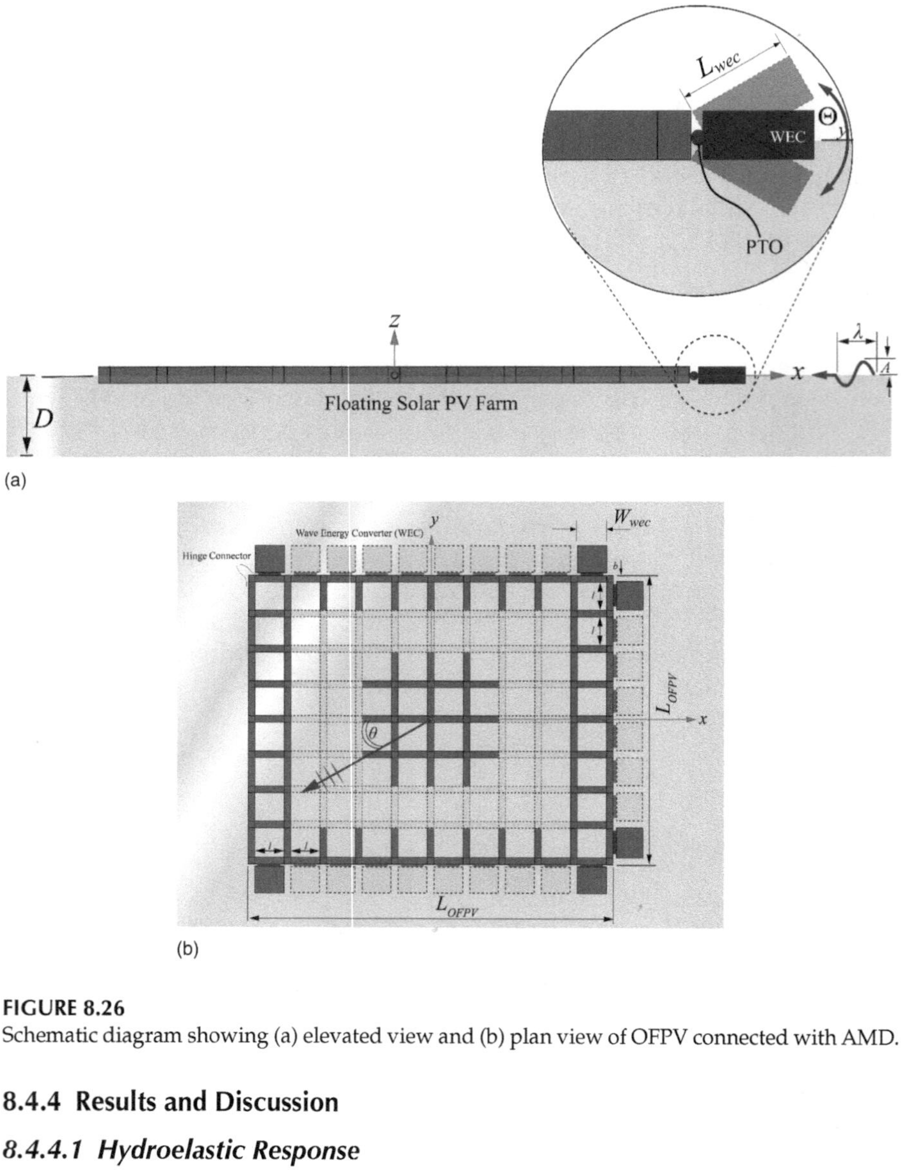

FIGURE 8.26
Schematic diagram showing (a) elevated view and (b) plan view of OFPV connected with AMD.

8.4.4 Results and Discussion

8.4.4.1 Hydroelastic Response

Figure 8.27 shows the hydroelastic response plotted along the centreline of the OFPV when connected with the WECs ($L_{wec} = 0.8$ m) under six wave periods, i.e., $T = 3, 4, 5, 6, 7,$ and 8s. It can be seen that the hydroelastic response of the OFPV increases with the wave period. However, the motion at the fore-end of the OFPV for the OFPV at small wave period, i.e., $T = 3$ and 4 s increases. This shows that energy is absorbed by the fore-end of the OFPV, similar to the findings of the OFPV integrated with AMD presented in Section 8.3.

TABLE 8.7

Comparison of Vibration Frequency (in rad/s) of OFPV with Hinged-Connected WECs between PM with ABAQUS (ABQ)

	5 × 5 OFPV, L_{wec} = 1 m		57 × 57 OFPV, L_{wec} = 0.4m	
	d_{wec} = 0.2m		d_{wec} = 0.2m	
	Hinged-connected AMD		Hinged-connected AMD	
Mode	**PM**	**ABQ**	**PM**	**ABQ**
RBM	0*		0**	
1	10.4899	9.8841	0.1094	0.0989
2	14.1794	13.6020	0.1350	0.1306
3	15.1729	14.8230	0.1396	0.1345
4	25.8778	24.4760	0.2587	0.2383
5	25.9673	24.7490	0.2587	0.2390
6	40.7873	39.1370	0.3785	0.3644
7	41.1305	40.0080	0.3797	0.3665
8	47.2400	44.4930	0.4767	0.4375
9	50.7801	48.0260	0.4967	0.4674
10	54.2572	52.3470	0.5037	0.4738

* Number of rigid body mode (RBM) = 3

** Number of rigid body mode (RBM) = 3 + number of hinged-connected AMD.

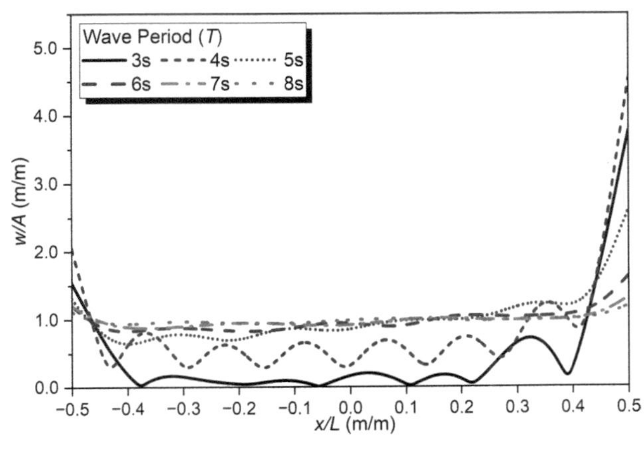

FIGURE 8.27
Hydroelastic response along $y/L = 0$ for OFPV with hinged-connected WEC under different wave periods. $d_{wec} = 0.2$ m and $L_{wec} = 0.8$ m. Head sea.

The effect of two different WEC lengths is investigated in Figure 8.28, where $L_{wec} = 0.4$ and 0.8 m are considered. The results show that the longer WEC has the capability to generate more energy when $T = 3$s but the performance reduces with the increase of wave periods, which may be due to the increase of mass inertia, therefore resulting in the WECs pitching in a smaller

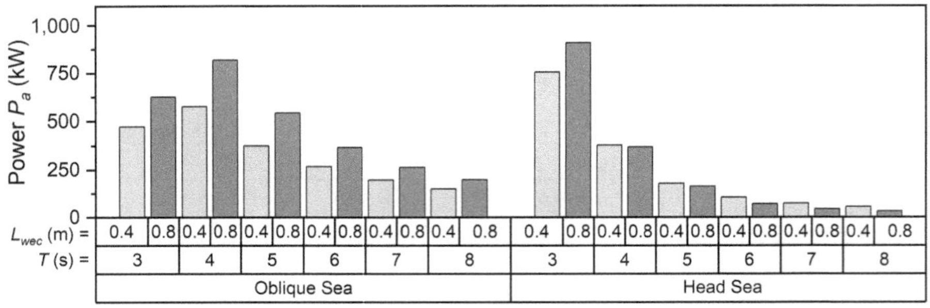

FIGURE 8.28
Comparison of absorbed power of OFPV with hinged-connected WEC for different L_{wec}. $d_{wec} = 0.2\,m$.

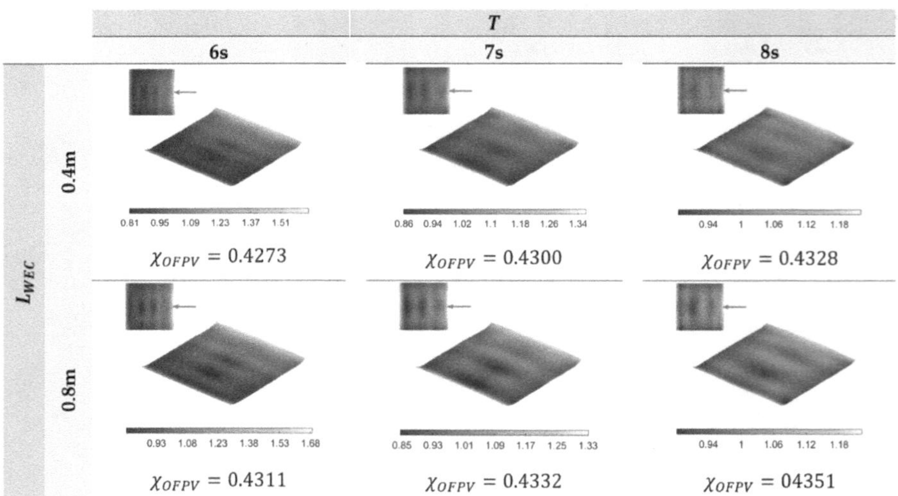

FIGURE 8.29
Hydroelastic response for OFPV with hinge-connected WEC for different L_{wec}. $d_{wec} = 0.2$ m and $B_{pto} = 150$ kN.m / s. Head sea

amplitude. The comparison of the hydroelastic response for the OFPV when connected with WEC of $L_{wec} = 0.4$ and 0.8 m is presented in Figure 8.29. The corresponding hydroelastic response along the centreline of the OFPV is also presented in Figure 8.30.

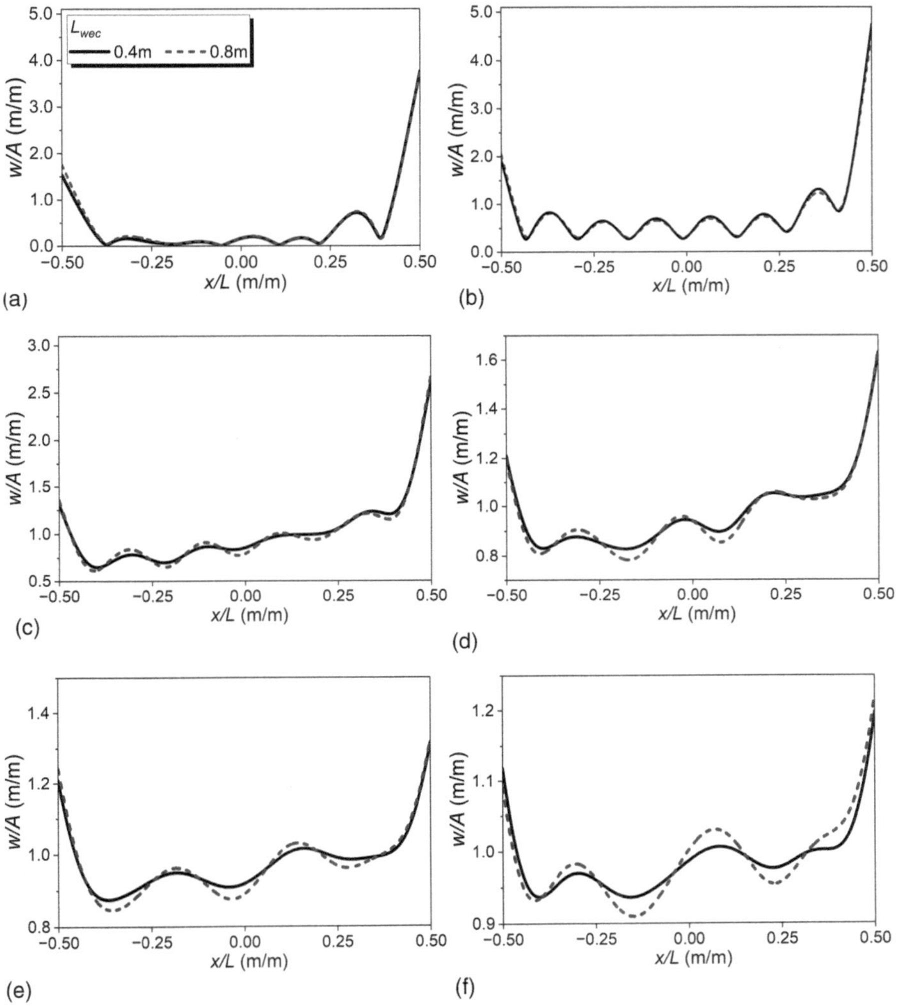

FIGURE 8.30
Comparison of hydroelastic response for OFPV with hinge-connected WEC for different WEC length L_{wec} along $/L = 0$. $d_{wec} = 0.2$ m. Head sea. Waves travel from right to left.

8.4.4.2 Absorbed Power

The energy extraction of the WECs for different B_{pto} is presented in Figure 8.31. In general, the P_a reduces with increase wave periods due to the WECs pitching at smaller amplitude when the wavelength λ is long. The energy generation of the WECs also increases when waves arrive from

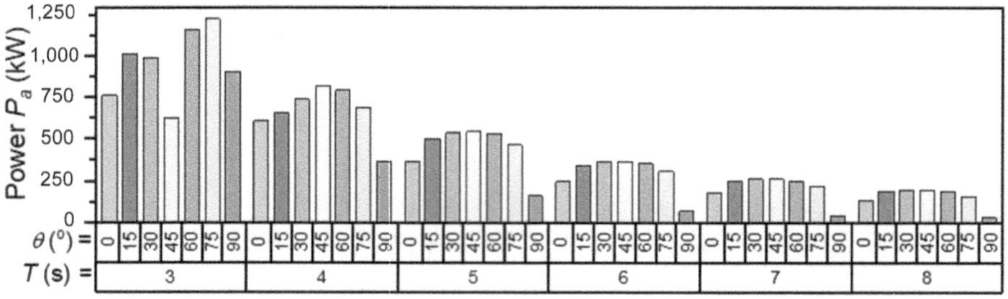

FIGURE 8.31
Comparison of absorbed power of OFPV with hinged-connected WEC for different θ. $d_{wec} = 0.2$ m and $L_{wec} = 0.8$ m.

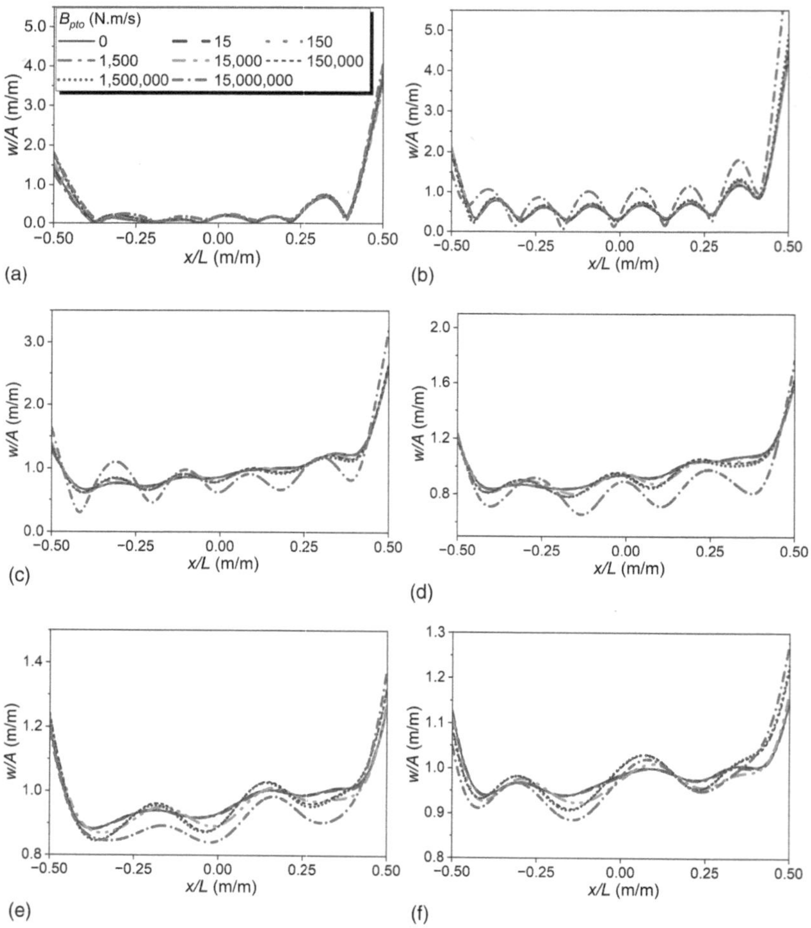

FIGURE 8.32
Comparison of hydroelastic response for OFPV with hinge-connected WEC for different B_{pto} along $/L = 0$. $d_{wec} = 0.2$ m and $L_{wec} = 0.8$ m. Head sea.

the oblique wave, achieving maximum value when θ is greater than $\theta = 45°$. The hydroelastic response along the centreline of the OFPV is presented in Figure 8.32 where it shows that greater B_{pto} results in smaller deflection of the OFPV.

Note

1 PM represents the present method in the chapter instead of the Pierson Moskowitz (PM) spectrum.

Appendix A: Classical Thin Plate Theory

The derivative of the governing equation for the classical thin plate theory (CPT), also known as the Kirchhoff-Love plate theory, is shown here.

A.1 Displacement Components

The deformation of the plate according to the CPT is shown in Figure A.1.

By assuming the mid-surface plane to represent a three-dimensional plate in two-dimensional form, the following kinematic assumptions are made in this theory:

- Straight lines normal to the mid-surface remain straight after deformation.
- Straight lines normal to the mid-surface remain normal to the mid-surface after deformation.
- The thickness of the plate does not change during a deformation.

The plane stress assumption is made by considering that the transverse stress σ_{zz} and out-of-plane shear stress τ_{xz} and τ_{yz} are negligibly small compared to the in-plane stress. By assuming harmonic motion, the displacement components are thus given by

$$U_x(x,y,z,t) = -z\theta^x(x,y)e^{\mathrm{i}\omega t}, \tag{A.1a}$$

$$U_y(x,y,z,t) = -z\theta^y(x,y)e^{\mathrm{i}\omega t}, \tag{A.1b}$$

$$U_z(x,y,z,t) = \hat{w}_0(x,y)e^{\mathrm{i}\omega t}, \tag{A.1c}$$

where t is the time variable, U_x, U_y the in-plane displacements and U_z the transverse displacement. $\hat{w}_0$ is the transverse deflection of a point on a mid-plane (i.e., $z = 0$), and θ^x, θ^y are the bending rotations of a transverse normal about the y- and x-axes, respectively, given as

$$\theta^x(x,y) = \frac{\partial \hat{w}_0}{\partial x}, \tag{A.2a}$$

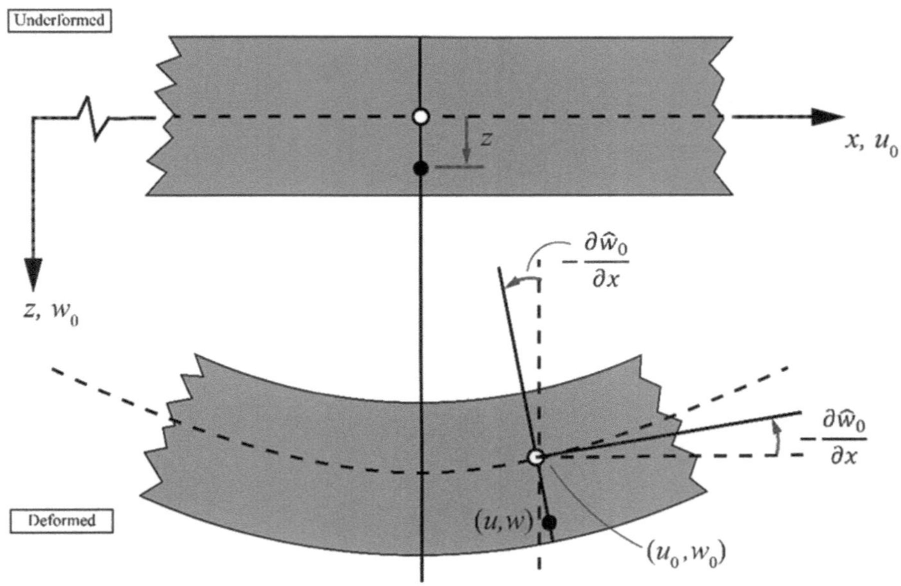

FIGURE A.1
Deformation of thin plate.

$$\theta^y(x,y) = \frac{\partial \hat{w}_0}{\partial y}, \tag{A.2b}$$

Equation (A.1) shows that the rotations (A.2) result in a negative in-plane displacement, as shown in Figure A.1.

A.2 Strain-Displacement (Compatibility) Relations

In view of Equations (A.1a)–(A.1c) and plane stress assumption, the linear components of the engineering strains can be expressed as

$$\epsilon_{xx} = -z\frac{\partial \theta^x}{\partial x} = -z\frac{\partial^2 w}{\partial x^2}, \tag{A.3a}$$

$$\epsilon_{yy} = -z\frac{\partial \theta^y}{\partial y} = z\frac{\partial^2 w}{\partial y^2}, \tag{A.3b}$$

$$\gamma_{xy} = z\left(-\frac{\partial \theta^x}{\partial y} - \frac{\partial \theta^y}{\partial x}\right) = -2z\left(\frac{\partial^2 w}{\partial x \partial y}\right), \tag{A.3c}$$

where ϵ_{xx}, ϵ_{yy} are the normal strains and γ_{xy} the shearing strain.

A.3 Hooke's Law

The plate is assumed to be an isotropic material that obeys the Hooke's law. Under plane stress conditions, Hooke's law takes the form

$$\sigma_{xx} = \frac{\mathcal{E}}{1-\nu^2}\left(\epsilon_{xx} + \nu\epsilon_{yy}\right), \tag{A.4a}$$

$$\sigma_{yy} = \frac{\mathcal{E}}{1-\nu^2}\left(\nu\epsilon_{xx} + \epsilon_{yy}\right), \tag{A.4b}$$

$$\tau_{xy} = \mathcal{G}\gamma_{xy}, \tag{A.4c}$$

where σ_{xx}, σ_{yy} are the normal stresses and τ_{xy} the shear stresses. $\mathcal{G} = \frac{\mathcal{E}}{\left[2(1+\nu)\right]}$ is the shear modulus, $\mathcal{E}$ the Young's modulus, and ν the Poisson ratio.

A.4 Stress Resultant-Displacement Relations

By using the Hooke's law (A.4) and strain-displacement relations (A.3), the stress resultant-displacement relations are given by

$$\begin{aligned} M_{xx} &= \int_{-\frac{d_p}{2}}^{\frac{d_p}{2}} \sigma_{xx} z \cdot dz = \int_{-\frac{d_p}{2}}^{\frac{d_p}{2}} \frac{\mathcal{E}}{1-\nu^2}\left(\epsilon_{xx} + \nu\epsilon_{yy}\right) \cdot dz \\ &= \mathcal{D}\left(\frac{\partial \theta^x}{\partial x} + \nu\frac{\partial \theta^y}{\partial y}\right) = \mathcal{D}\left(\frac{\partial^2 \hat{w}}{\partial x^2} + \nu\frac{\partial^2 \hat{w}}{\partial y^2}\right), \end{aligned} \tag{A.5a}$$

$$\begin{aligned} M_{yy} &= \int_{-\frac{D_p}{2}}^{\frac{d_p}{2}} \sigma_{yy} z \cdot dz = \int_{-\frac{d_p}{2}}^{\frac{d_p}{2}} \frac{\mathcal{E}}{1-\nu^2}\left(\epsilon_{yy} + \nu\epsilon_{xx}\right) \cdot dz \\ &= \mathcal{D}\left(\frac{\partial \theta^y}{\partial y} + \nu\frac{\partial \theta^x}{\partial x}\right) = \mathcal{D}\left(\frac{\partial^2 \hat{w}}{\partial y^2} + \nu\frac{\partial^2 \hat{w}}{\partial x^2}\right), \end{aligned} \tag{A.5b}$$

$$M_{xy} = \int_{-\frac{d_p}{2}}^{\frac{d_p}{2}} \tau_{xy} z \cdot dz = \int_{-\frac{d_p}{2}}^{\frac{d_p}{2}} \mathcal{G}\gamma_{xy} z \cdot dz = \frac{\mathcal{D}(1-\nu)}{2}\left(\frac{\partial \theta^x}{\partial y} + \frac{\partial \theta^y}{\partial x}\right)$$
$$= \mathcal{D}(1-\nu)\left(\frac{\partial^2 \hat{w}}{\partial x \partial y}\right), \tag{A.5c}$$

where M_{xx}, M_{yy} are the bending moments per unit length of plate, M_{xy} the twisting moment per unit length of plate, $\mathcal{E}$ the modulus of elasticity, ν the Poisson ratio, $\mathcal{D} = \mathcal{E}d_p^3)/\left[12(1-\nu \wedge 2)\right]$ the flexural rigidity, and d_p the thickness of the plate.

In addition, a pair of shear forces exists for the moment to be equilibrium. The shear stresses are derived from the moment equilibrium, given as

$$Q_x = \frac{\partial M_{xx}}{\partial x} + \frac{\partial M_{xy}}{\partial y} = \mathcal{D}\frac{\partial}{\partial x}\left(\frac{\partial^2 \hat{w}}{\partial x^2} + \frac{\partial^2 \hat{w}}{\partial y^2}\right), \tag{A.6a}$$

$$Q_y = \frac{\partial M_{xy}}{\partial y} + \frac{\partial M_{yy}}{\partial y} = \mathcal{D}\frac{\partial}{\partial y}\left(\frac{\partial^2 \hat{w}}{\partial x^2} + \frac{\partial^2 \hat{w}}{\partial y^2}\right), \tag{A.6b}$$

where Q_x and Q_y are the transverse shear forces per unit length of plate in the x- and y-axes.

A.5 Governing Equation

The governing equation of the CPT could be derived by using the variational approach or equilibrium approach.

A.5.1 Variational Approach

A.5.1.1 Energy Functionals

There are three types of energy functional involve in the derivative of the governing equation of CPT, i.e., the strain energy functional U_E, kinetic energy functional T_E, and the work done W_{nc}.

We first define the total strain energy functional U_0 during deformation of the plate as

$$U_0 = \frac{1}{2}\left(\sigma_{xx}\epsilon_{xx} + \sigma_{yy}\epsilon_{yy} + \tau_{xy}\gamma_{xy}\right) = \frac{1}{2}\cdot \mathbb{D}^f \cdot \{\epsilon\}^T, \tag{A.7}$$

where $\{\epsilon\}^T = \{\epsilon_{xx}\ \epsilon_{yy}\ \gamma_{xy}\}$, and $\mathbb{D}^f$ is the material property matrix for an isotropic plate given as

$$\mathbb{D}^f = \frac{\mathcal{E}}{1-\nu^2}\begin{bmatrix} 1 & \nu & 0 \\ \nu & 1 & 0 \\ 0 & 0 & \dfrac{1-\nu}{2} \end{bmatrix}. \tag{A.8}$$

The strain energy functional U_E (or the strain energy density) due to bending of the thin plate is then given by the integration of U_0 over the plate volume ∇ [124,125]:

$$U_E = \int_\nabla U_0 \cdot d\nabla = \frac{1}{2}\int_\nabla \{\epsilon\}^T \cdot \mathbb{D}^f \cdot \{\epsilon\} \cdot d\nabla. \tag{A.9}$$

The substitution of Equations (A.7) and (A.8) into Equation (A.9) and after the integration over the plate thickness, yields,

$$U_E = \frac{1}{2}\mathcal{D}\int_\Delta \begin{bmatrix} \left(\dfrac{\partial^2 \hat{w}}{\partial x^2}\right)^2 + 2\nu \dfrac{\partial^2 \hat{w}}{\partial x^2}\dfrac{\partial^2 \hat{w}}{\partial y^2} \\ + \left(\dfrac{\partial^2 \hat{w}}{\partial y^2}\right)^2 + 2(1-\nu)\left(\dfrac{\partial^2 \hat{w}}{\partial x \partial y}\right)^2 \end{bmatrix} \cdot d\Delta, \tag{A.10}$$

where Δ is the plate domain. The kinetic energy functional T_E of the vibrating Mindlin plate is given by

$$T_E = \frac{1}{2}\rho_p \int_\nabla \left(\frac{\partial U_z}{\partial t}\right)^2 \cdot d\nabla, \tag{A.11}$$

where ρ_p is the mass density (per unit volume) of the plate. Similarly, by substituting Equation (A.1) into Equation (A.11) and integrating through the thickness dimension, the kinetic energy T_E may be expressed as

$$T_E = -\frac{\omega^2}{2}\rho_p d_p \int_\Delta \hat{w}^2 \cdot d\Delta. \tag{A.12}$$

Lastly, the work done, W_{nc}, by the non-conservative forces is given by

$$W_{nc} = \int_\nabla \mathbf{P} \cdot U_z \cdot d\nabla = \rho_f \int_\nabla \left(\frac{\partial \Phi}{\partial t} + gU_z\right) U_z \cdot d\nabla = \int_\Delta \left(\mathrm{i}\omega\varphi + g\hat{w}\right) \cdot \hat{w} \cdot d\Delta, \tag{A.13}$$

where $\mathbf{P}$ is the pressure distribution on the bottom of the plate made up of hydrodynamic pressure $\rho_f\,\partial\Phi/\partial t$ and hydrostatic pressure $\rho_f g U_z$. The Lagrangian Π (known also as action functional) is thus given by

$$\Pi = T_E - U_E - W_{nc}. \tag{A.14}$$

A.5.1.2 Principle of Virtual Work (PVW)

The principle of virtual work states that the work done by external forces (external energy) on a mechanical system in equilibrium is equal to the work done by internal forces for any virtual displacement (internal energy) of the system that satisfies its constraints. The formulation for PVW could be done by using Hamilton's principle introduced by William Rowan Hamilton[238, 239] in the 19th century.

HamiltonRs principle is a powerful tool in classical mechanics that describes the motion of a system in terms of a single scalar function called action functional (Π). It states that the actual path taken by a system has a stationary action (i.e., $\delta\,\Pi = 0$) under small changes in the configuration of the system. As the virtual work quantities are scalar values, the quantities thus can be added algebraically. Algebraic equations allow complex mathematical problems to be solved numerically using computational methods that would be difficult or impossible to solve using purely analytical method.

Hamilton's principle requires that the variation of the action functional equals to zero, i.e.,

$$\delta\int_{t_i}^{t_f}\Pi\cdot dt, \tag{A.15}$$

where t_i denotes initial time, t_f the final time, and δ the variational operator. The principle implies that the system's trajectory is the one that makes the action functional stationary. In practical terms, this means the path is such that any small deviation would not lead to a decrease in the action.

By substituting Equations (A.10), (A.12), and (A.13) into Equation (A.15) yields

$$\int_{t_i}^{t_f}\left\{\begin{matrix}\rho_p d_p\int_\Delta\left(\dfrac{\partial\hat{w}}{\partial t}\dfrac{\partial\delta\hat{w}}{\partial t}\right)\cdot d\Delta-D\int_\Delta\left[\begin{matrix}\dfrac{\partial^2\hat{w}}{\partial x^2}\dfrac{\partial^2\delta\hat{w}}{\partial x^2}+2\nu\dfrac{\partial^2\hat{w}}{\partial x^2}\dfrac{\partial^2\delta\hat{w}}{\partial x^2}\dfrac{\partial^2\hat{w}}{\partial y^2}\dfrac{\partial^2\delta\hat{w}}{\partial y^2}\\+\dfrac{\partial^2\hat{w}}{\partial y^2}\dfrac{\partial^2\delta\hat{w}}{\partial y^2}+2(1-\nu)\left(\dfrac{\partial^2\hat{w}}{\partial x\partial y}\dfrac{\partial^2\delta\hat{w}}{\partial x\partial y}\right)\end{matrix}\right]\\\cdot d\Delta+\rho_f\int_\Delta\left(\dfrac{\partial\Phi}{\partial t}\delta\hat{w}+g\delta\hat{w}\right)\cdot d\Delta\end{matrix}\right\}\cdot dt. \tag{A.16}$$

By performing integration by parts on Equation (A.16) and grouping with respect to the variation terms, this results in

$$\delta w \int_{t_i}^{t_f} \left\{ \begin{array}{l} \rho_p d_p \int_{\Delta} \left(\frac{\partial^2 \hat{w}}{\partial t} \right) \cdot dA - \mathcal{D} \int_{\Delta} \left[\frac{\partial^4 \hat{w}}{\partial x^4} + \frac{\partial^4 \hat{w}}{\partial y^4} + 2(1-\nu) \left(\frac{\partial^4 \hat{w}}{\partial x^2 \partial y^2} \right) \right] \cdot \\ \qquad d\Delta + \rho_f \int_{\Delta} \left(\frac{\partial \Phi}{\partial t} + g\hat{w} \right) \cdot d\Delta \end{array} \right\} dt -$$
$$\delta w \int_{t_i}^{t_f} \oint_{\Gamma} \left(\frac{\partial \hat{w}}{\partial t} + \frac{\partial^2 \hat{w}}{\partial x^2} + \frac{\partial^2 \hat{w}}{\partial y^2} + \frac{\partial^2 \hat{w}}{\partial x \partial y} \right) \cdot d\Gamma = 0, \tag{A.17}$$

where Γ is the boundary path.

A.5.1.3 Governing Equation for Kirchhoff Plate

The governing equation for the CPT is thus given as

$$\mathcal{D} \left(\nabla^4 \hat{w} - 2\nu \frac{\partial^4 \hat{w}}{\partial x^2 \partial y^2} \right) = -\rho_p d_p \omega^2 \hat{w} + \mathrm{i} \rho_f \omega \varphi + \rho_f g \hat{w}, \tag{A.18}$$

where $\nabla^4(\cdot)$ is the biharmonic operator given as $\nabla^2 \cdot \nabla^2(\cdot) = \left(\frac{\partial^4}{\partial x^4} + 2\frac{\partial^4}{\partial x^2 \partial y^2} + \frac{\partial^4}{\partial y^4} \right)$.

A.5.2 Direct Equilibrium Method

The derivative of governing equation using the direct equilibrium method works by summation of moments and shear forces acting on the plate (shown in Figure A.2) given as

$$\frac{\partial Q_x}{\partial x} + \frac{\partial Q_y}{\partial y} + \bar{q} = 0, \tag{A.19a}$$

$$\frac{\partial M_{xx}}{\partial x} + \frac{\partial M_{xy}}{\partial y} = Q_x, \tag{A.19b}$$

$$\frac{\partial M_{yx}}{\partial x} + \frac{\partial M_{yy}}{\partial y} = Q_y. \tag{A.19c}$$

where $\bar{q}$ is the distributed load on the plate. By substituting Equations (A.5) and (A.6) into Equation (A.19), the equilibrium equation becomes

$$\frac{\partial}{\partial x} \left[\frac{\partial}{\partial x} \left(\frac{\partial^2 \hat{w}}{\partial x^2} + \frac{\partial^2 \hat{w}}{\partial y^2} \right) \right] + \frac{\partial}{\partial y} \left[\mathcal{D} \frac{\partial}{\partial y} \left(\frac{\partial^2 \hat{w}}{\partial x^2} + \frac{\partial^2 \hat{w}}{\partial y^2} \right) \right] + \bar{q} = 0, \tag{A.20a}$$

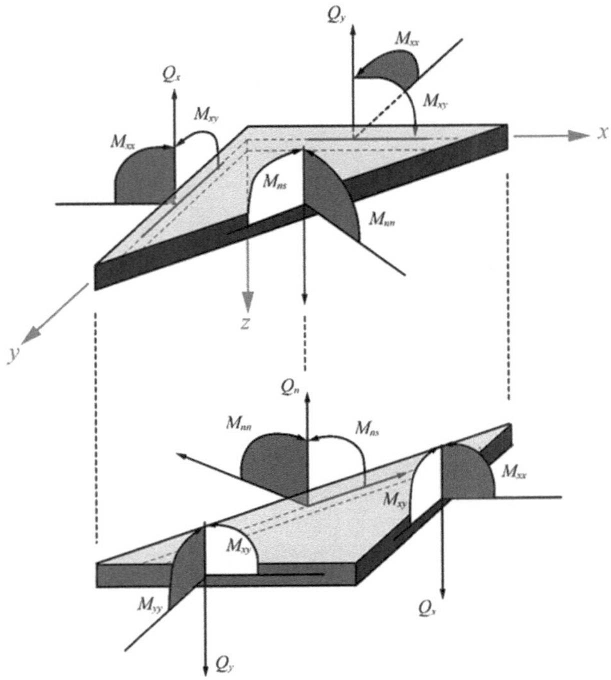

FIGURE A.2
Moment and shear force on plate.

$$\frac{\partial}{\partial x}\left[\mathcal{D}\left(\frac{\partial^2\hat{w}}{\partial x^2}+\nu\frac{\partial^2\hat{w}}{\partial y^2}\right)\right]+\frac{\partial}{\partial y}\mathcal{D}(1-\nu)\left(\frac{\partial^2\hat{w}}{\partial x\partial y}\right)=\mathcal{D}\frac{\partial}{\partial x}\left(\frac{\partial^2\hat{w}}{\partial x^2}+\frac{\partial^2\hat{w}}{\partial y^2}\right), \quad \text{(A.20b)}$$

$$\frac{\partial}{\partial x}\mathcal{D}(1-\nu)\left(\frac{\partial^2\hat{w}}{\partial x\partial y}\right)+\frac{\partial}{\partial y}\mathcal{D}\left(\frac{\partial^2\hat{w}}{\partial y^2}+\nu\frac{\partial^2\hat{w}}{\partial x^2}\right)=\mathcal{D}\frac{\partial}{\partial y}\left(\frac{\partial^2\hat{w}}{\partial x^2}+\frac{\partial^2\hat{w}}{\partial y^2}\right). \quad \text{(A.20c)}$$

Solving the simultaneous Equation (A.20) yields the governing equation:

$$\mathcal{D}\left(\frac{\partial^4\hat{w}}{\partial x^4}+2\frac{\partial^4\hat{w}}{\partial x^2\partial y^2}+\frac{\partial^4\hat{w}}{\partial y^4}\right)+\bar{q}=0, \quad \text{(A.21)}$$

which is the same as Equation (A.18), where $\bar{q}=-\omega^2\rho_p d_p\hat{w}+\rho_f\left(\mathrm{i}\omega\varphi+g\hat{w}\right)$.

Appendix B: First-Order Shear Deformation Plate Theory

The governing equation for the first-order shear deformation plate theory (FSDT; or the Mindlin plate theory) is derived here.

B.1 Displacement Components

The displacement of a Mindlin plate is shown in Figure B.1.

In FSDT, the displacement component can be expressed in a similar way as for the CPT, whereby assuming that the normal to the undeformed midplane remains straight and unstretched in length but not necessarily normal to the deformed midplane, the displacement components are given by Equation (B.1.):

$$U_x(x,y,z,t) = -z\psi^x(x,y)e^{i\omega t}, \tag{B.1a}$$

$$U_y(x,y,z,t) = -z\psi^y(x,y)e^{i\omega t}, \tag{B.1b}$$

$$U_z(x,y,z,t) = \hat{w}(x,y)e^{i\omega t}. \tag{B.1c}$$

The variables are the same as in Appendix A for CPT unless otherwise stated. Here, ψ^x, ψ^y are the bending rotations of a transverse normal about the y - and x -axes, respectively, given as

$$\psi^x(x,y) = \frac{\partial \hat{w}}{\partial x} + \phi^x, \tag{B.2a}$$

$$\psi^x(x,y) = \frac{\partial \hat{w}}{\partial y} + \phi^y. \tag{B.2b}$$

As compared to the CPT, the bending rotations contains the additional ϕ^x, ϕ^y, which are the additional rotations due to transverse shear deformation about the y - and x -axes. ϕ^x and ϕ^y exist due to the lack of orthogonality of the normal with the middle plane after deformation.

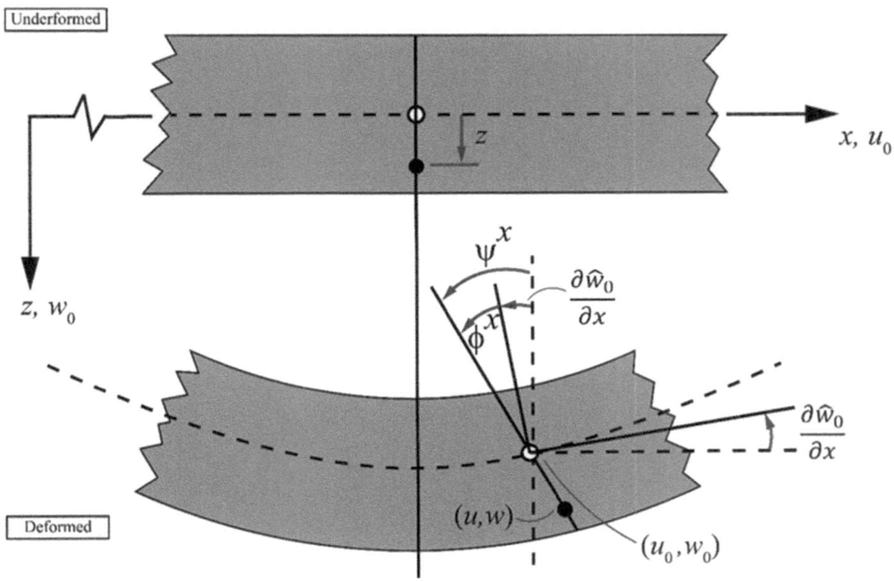

FIGURE B.1
Deformation of Mindlin plate.

B.2 Strain-Displacement Relations

The strain-displacement relations are the same as in CPT but with additional shearing strains γ_{xz} and γ_{yz}:

$$\epsilon_{xx} = -z\frac{\partial \psi^x}{\partial x}, \tag{B.3a}$$

$$\epsilon_{yy} = -z\frac{\partial \psi^y}{\partial y}, \tag{B.3b}$$

$$\gamma_{xy} = -z\left(\frac{\partial \psi^x}{\partial y} + \frac{\partial \psi^y}{\partial x}\right), \tag{B.3c}$$

$$\gamma_{xz} = -\psi^x + \frac{\partial \hat{w}}{\partial x}, \tag{B.3d}$$

$$\gamma_{yz} = -\psi^y + \frac{\partial \hat{w}}{\partial y}, \tag{B.3e}$$

where ϵ_{xx}, ϵ_{yy} are the normal strains and γ_{xy}, γ_{xz}, γ_{yz} the shearing strains.

B.3 Hooke's Law

The plate is assumed to be an isotropic material that obeys the Hooke's law. Under plane stress condition, Hooke's law takes the form

$$\sigma_{xx} = \frac{\mathcal{E}}{1-\nu^2}\left(\epsilon_{xx} + \nu\epsilon_{yy}\right), \tag{B.4a}$$

$$\sigma_{yy} = \frac{\mathcal{E}}{1-\nu^2}\left(\nu\epsilon_{xx} + \epsilon_{yy}\right), \tag{B.4b}$$

$$\tau_{xy} = \mathcal{G}\gamma_{xy}, \tag{B.4c}$$

$$\tau_{xz} = \mathcal{G}\gamma_{xz}, \tag{B.4d}$$

$$\tau_{yz} = \mathcal{G}\gamma_{yz}. \tag{B.4e}$$

B.4 Stress Resultant-Displacement Relations

By using Hooke's law (B.4) and strain-displacement relations (B.3), the stress resultant-displacement relations are given by

$$M_{xx} = \int_{-\frac{d_p}{2}}^{\frac{d_p}{2}} \sigma_{xx} z \cdot dz = \int_{-\frac{d_p}{2}}^{\frac{d_p}{2}} \frac{\mathcal{E}}{1-\nu^2}\left(\epsilon_{xx} + \nu\epsilon_{yy}\right) \cdot dz = \mathcal{D}\left(\frac{\partial \psi^x}{\partial x} + \nu\frac{\partial \psi^y}{\partial y}\right), \tag{B.5a}$$

$$M_{yy} = \int_{-\frac{d_p}{2}}^{\frac{d_p}{2}} \sigma_{yy} z \cdot dz = \int_{-\frac{d_p}{2}}^{\frac{d_p}{2}} \frac{\mathcal{E}}{1-\nu^2}\left(\epsilon_{yy} + \nu\epsilon_{xx}\right) \cdot dz = \mathcal{D}\left(\frac{\partial \psi^y}{\partial y} + \nu\frac{\partial \psi^x}{\partial x}\right), \tag{B.5b}$$

$$M_{xy} = \int_{-\frac{d_p}{2}}^{\frac{d_p}{2}} \tau_{xy} z \cdot dz = \int_{-\frac{d_p}{2}}^{\frac{d_p}{2}} \mathcal{G}\gamma_{xy} z \cdot sdz = \frac{\mathcal{D}(1-\nu)}{2}\left(\frac{\partial \psi^x}{\partial y} + \frac{\partial \psi^y}{\partial x}\right), \tag{B.5c}$$

$$Q_x = \kappa^2 \int_{-\frac{d_p}{2}}^{\frac{d_p}{2}} \tau_{xz} \cdot dz = \kappa^2 \mathcal{G} d_p\left(\psi^x + \frac{\partial \hat{w}}{\partial x}\right), \tag{B.5d}$$

$$Q_y = \kappa^2 \int_{-\frac{d_p}{2}}^{\frac{d_p}{2}} \tau_{yz} \cdot dz = \kappa^2 \mathcal{G} d_p\left(\psi^y + \frac{\partial \hat{w}}{\partial y}\right). \tag{B.5e}$$

The transverse shear forces Q_x, Q_y per unit length are derived from the shear stresses τ_{xz}, τ_{yz}, instead of from the shear moment M_{xy} as in the CPT. The stress resultant according to Equation (B.5a–e) only involves up to the first-order derivative of the plate deflection, thus contributing to the term 'first-order shear deformation plate theory'. This has great advantage when solving the plate equation as the deflection only needs to be first-order differentiable and thus improve the accuracy in the stress-resultants. A shear correction factor κ^2, typically around 5/6, is introduced to compensate for the error due to the assumption of a constant shear strain (and hence constant shear stress) throughout the plate thickness.

B.5 Governing Equation

B.5.1 Variational Approach

B.5.1.1 Energy Functionals

The total strain energy U_0 during deformation of the plate is given as

$$U_0 = \frac{1}{2}\left[\sigma_{xx}\epsilon_{xx} + \sigma_{yy}\epsilon_{yy} + \tau_{xy}\gamma_{xy} + \tau_{xz}\gamma_{xz} + \tau_{yz}\gamma_{yz}\right] = \frac{1}{2}\cdot \mathbb{D}^{FSDT}\cdot\{\epsilon\}^T, \tag{B.6}$$

where $\{\epsilon\}^T = \left[\epsilon_{xx}\ \epsilon_{yy}\ \gamma_{xy}\ \gamma_{xz}\ \gamma_{yz}\right]$ and $\mathbb{D}^{FSDT}$ is the material property matrix for an isotropic plate given as

$$\mathbb{D}^{FSDT} = \begin{bmatrix} \frac{\mathcal{E}}{1-\nu^2} & \frac{\nu\mathcal{E}}{1-\nu^2} & 0 & 0 & 0 \\ \frac{\nu\mathcal{E}}{1-\nu^2} & \frac{\mathcal{E}}{1-\nu^2} & 0 & 0 & 0 \\ 0 & 0 & \mathcal{G} & 0 & 0 \\ 0 & 0 & 0 & \kappa^2\mathcal{G} & 0 \\ 0 & 0 & 0 & 0 & \kappa^2\mathcal{G} \end{bmatrix}. \tag{B.7}$$

The strain energy functional U_E (or the strain energy density) due to bending of the Mindlin plate is then given by the integration of U_0 over the plate volume ∇ [124,125]:

$$U_E = \int_\nabla U_0 \cdot d\nabla = \frac{1}{2}\int_\nabla \{\epsilon\}^T \cdot \mathbb{B}\cdot\{\epsilon\}\cdot d\nabla. \tag{B.8}$$

The substitution of Equations (B.6) and (B.7) into Equation (B.8) and after the integration over the plate thickness yields

$$U_E = \frac{1}{2}\int_{\Delta} \left\{ \begin{array}{l} \mathcal{D}\left[\left(\frac{\partial \psi^x}{\partial x} + \nu\frac{\partial \psi^y}{\partial y}\right)^2 + \left(\frac{\partial \psi^y}{\partial y} + \nu\frac{\partial \psi^x}{\partial x}\right)^2\right] \\ + \frac{\mathcal{D}(1-\nu)}{2}\left(\frac{\partial \psi^x}{\partial y} + \frac{\partial \psi^y}{\partial x}\right)^2 \\ +\kappa^2 \mathcal{G} d_p\left[\left(\psi^x + \frac{\partial \hat{w}}{\partial x}\right)^2 + \left(\psi^y + \frac{\partial \hat{w}}{\partial y}\right)^2\right] \end{array} \right\} \cdot d\Delta. \tag{B.9}$$

The kinetic energy functional T_E of the vibrating Mindlin plate is given by

$$T_E = \frac{1}{2}\int_V \rho_p\left[\left(\frac{\partial U_x}{\partial t}\right)^2 + \left(\frac{\partial U_y}{\partial t}\right)^2 + \left(\frac{\partial U_z}{\partial t}\right)^2\right] \cdot dV, \tag{B.10}$$

where ρ_p is the mass density (per unit volume) of the plate. Similarly, by substituting Equation (B.1) into Equation (B.10) and integrating through the thickness dimension, the kinetic energy T_E may be expressed as

$$T_E = -\frac{\omega^2 \rho_p}{2}\int_{\Delta}\left\{d_p(\hat{w})^2 + \frac{d_p^3}{12}\left[\left(\psi^x\right)^2 + \left(\psi^y\right)^2\right]\right\} \cdot d\Delta. \tag{B.11}$$

The work done, W_{nc}, by the non-conservative forces is the same as for CPT, i.e., Equation (A.13).

B.5.1.2 Principle of Virtual Work (PVW)

Substituting Equations (B.9), (B.11), (A.13), and (A.14) into Equation (A.15) and writing the partial derivative in subscript notation yields

$$\delta\int_{t_i}^{t_f}\int_{\Delta} \left\{ \begin{array}{l} -\mathcal{D}\left(\psi_x^x \cdot \delta\psi_x^x + \psi_y^y \cdot \delta\psi_y^y + \nu\psi_x^x \cdot \delta\psi_y^y + \nu\psi_y^y \cdot \delta\psi_y^x\right) \\ -\frac{\mathcal{D}(1-\nu)}{2}\left(\psi_y^x \cdot \delta\psi_y^x + \psi_x^y \cdot \delta\psi_y^x + \psi_x^y \cdot \delta\psi_x^y + \psi_y^x \cdot \delta\psi_x^y\right) \\ -\kappa^2 \mathcal{G} d_p\left[\left(\psi^x + w_x\right)\left(\delta\psi^x + \delta w_x\right) + \left(\psi_y + w_y\right)\left(\delta\psi_y + \delta w_y\right)\right] \\ -\rho_p d_p \dot{\hat{w}} \cdot \delta\dot{\hat{w}} + \frac{\rho_p d_p^3}{12}\left(\dot{\psi}^x \cdot \delta\dot{\psi}^x + \dot{\psi}^y \cdot \delta\dot{\psi}^y\right) \\ -\left(\rho_p \ddot{\hat{w}} \cdot \delta\hat{w} + \rho_p g \hat{w} \cdot \delta\hat{w}\right) \end{array} \right\} \cdot d\Delta \cdot dt, \tag{B.12}$$

where the subscript denotes partial derivative, e.g., $\psi_x^x = \dfrac{\partial \psi^x}{\partial x}, \psi_{xx}^x = \dfrac{\partial^2 \psi^x}{\partial x^2}, \psi_{xy}^x = \dfrac{\partial^2 \psi^x}{\partial x \partial y}$ and $\dot{\hat{w}} = \dfrac{\partial \hat{w}}{\partial t}$, so on and so forth. By performing integration by parts on Equation (B.12) and grouping with respect to the variation terms, this results in

$$
\begin{aligned}
&\delta\psi^x \int_{t_i}^{t_f} \int_\Delta \left[\mathcal{D}\left(\psi_{xx}^x + \upsilon\psi_{xy}^y\right) + \frac{\mathcal{D}(1-\upsilon)}{2}\left(\psi_{yy}^x + \psi_{xy}^y\right) - k^2\mathcal{G}d_p\left(\psi^x + \hat{w}_x\right) - \frac{\rho_p d_p^3}{12}\ddot{\psi}^x \right] \cdot d\Delta \cdot dt + \\
&\delta\psi^y \int_{t_i}^{t_f} \int_\Delta \left[\mathcal{D}\left(\psi_{yy}^y + \upsilon\psi_{xy}^y\right) + \frac{\mathcal{D}(1-\upsilon)}{2}\left(\psi_{xx}^y + \psi_{yx}^x\right) - k^2\mathcal{G}d_p\left(\psi^y + \hat{w}_y\right) - \frac{\rho_p d_p^3}{12}\ddot{\psi}^y \right] \cdot d\Delta \cdot dt + \\
&\delta w \int_{t_i}^{t_f} \int_\Delta \left[k^2\mathcal{G}d_p\left(\psi_x^x + w_{xx} + \psi_y^y + w_{yy}\right) - \rho_p d_p \ddot{w} - \rho_p\left(\dot{\varphi} + g\hat{w}\right) \right] \cdot d\Delta \cdot dt + \\
&\delta\psi^x \left\{ -\int_{t_i}^{t_f} \oint_\Gamma \left[\mathcal{D}\left(\psi_x^x + \upsilon\psi_y^y\right) \right] \cdot dy \cdot dt + \int_{t_i}^{t_f} \oint_\Gamma \left[\frac{\mathcal{D}(1-\upsilon)}{2}\left(\psi_y^x + \psi_x^y\right) \right] \cdot dx \cdot dt \right\} + \\
&\delta\psi^y \left\{ -\int_{t_i}^{t_f} \oint_\Gamma \left[\mathcal{D}\left(\psi_y^y + \upsilon\psi_x^x\right) \right] \cdot dx \cdot dt + \int_{t_i}^{t_f} \oint_\Gamma \left[\frac{\mathcal{D}(1-\upsilon)}{2}\left(\psi_y^x + \psi_x^y\right) \right] \cdot dy \cdot dt \right\} + \\
&\delta w \left\{ \int_{t_i}^{t_f} \oint_\Gamma \left[k^2\mathcal{G}d_p\left(\psi^x + \hat{w}_x\right) \right] \cdot dy \cdot dt - \int_{t_i}^{t_f} \oint_\Gamma \left[k^2\mathcal{G}d_p\left(\psi^y + \hat{w}_y\right) \right] \cdot dx \cdot dt \right\} = 0,
\end{aligned}
\tag{B.13}
$$

where Γ is the boundary path.

B.5.1.3 Governing Equation for Mindlin Plate

The following three governing equations of motion are obtained after omitting the factor $e^{i\omega t}$.

$$
\kappa^2\mathcal{G}d_p\left[\left(\frac{\partial^2\hat{w}}{\partial x^2} + \frac{\partial^2\hat{w}}{\partial y^2}\right) + \left(\frac{\partial\psi^x}{\partial x} + \frac{\partial\psi^y}{\partial y}\right)\right] = -\omega^2\rho_p d_p\hat{w} + \rho_f\left(\mathrm{i}\omega\varphi + g\hat{w}\right), \tag{B.14a}
$$

$$
\frac{\mathcal{D}(1-\nu)}{2}\left(\frac{\partial^2\psi^x}{\partial x^2} + \frac{\partial^2\psi^x}{\partial y^2}\right) + \frac{\mathcal{D}(1+\nu)}{2}\left(\frac{\partial^2\psi^x}{\partial x^2} + \frac{\partial^2\psi^y}{\partial x\partial y}\right) - \kappa^2\mathcal{G}d_p\left(\psi^x + \frac{\partial\hat{w}}{\partial x}\right) = -\frac{\rho_p d_p^3}{12}\omega^2\psi^x, \tag{B.14b}
$$

$$
\frac{\mathcal{D}(1-\nu)}{2}\left(\frac{\partial^2\psi^y}{\partial x^2} + \frac{\partial^2\psi^y}{\partial y^2}\right) + \frac{\mathcal{D}(1+\nu)}{2}\left(\frac{\partial^2\psi^x}{\partial x\partial y} + \frac{\partial^2\psi^y}{\partial y^2}\right) - \kappa^2\mathcal{G}d_p\left(\psi^y + \frac{\partial\hat{w}}{\partial y}\right) = -\frac{\rho_p d_p^3}{12}\omega^2\psi^y. \tag{B.14c}
$$

B.5.2 Direct Equilibrium Approach

The derivative of the governing equation using the direct equilibrium method can be expressed as the summation of moments and shear forces acting on the plate (shown in Figure A.2) given as

$$\frac{\partial Q_x}{\partial x}+\frac{\partial Q_y}{\partial y}=-\rho_p d_p \omega^2 \hat{w}+\rho_f\left(\mathrm{i}\omega\varphi+g\hat{w}\right), \tag{B.15a}$$

$$\frac{\partial M_{xx}}{\partial x}+\frac{\partial M_{xy}}{\partial y}-Q_x=-\frac{\rho_p d_p^3}{12}\omega^2\psi^x, \tag{B.15b}$$

$$\frac{\partial M_{yx}}{\partial x}+\frac{\partial M_{yy}}{\partial y}-Q_y=-\frac{\rho_p d_p^3}{12}\omega^2\psi^y. \tag{B.15c}$$

By substituting Equation (B.5a–e) into Equation (B.15a–c), we obtain the same governing equation as Equation (B.14a–c).

Appendix C: Approximate Methods

The approximation method such as the FEM, finite difference method (FDM), or BEM could be used to solve a complex function that has no analytical solution or is difficult to solve analytically. The details presented here should be taken as a reference to facilitate the derivatives of equations in this book. Readers interested in FEM, FDM, or BEM should refer to reference books targeting these topics specifically.

Consider a function

$$f(x_0) = \mathscr{b} \rightarrow R_\Omega = f(x_0) - \mathscr{b} = 0. \tag{C.1}$$

The residual R_Ω can be written as

$$R_\Omega = f(x_0) - \mathscr{b} = 0, \tag{C.2}$$

where the residual $R_\Omega = 0$ if the exact solution x_0 is evaluated for the function $f(x)$.

The function $f(x)$ also satisfies the boundary conditions

$$S(x_0) = s \text{ on } \Gamma_1, \tag{C.3}$$

$$G(x_0) = g \text{ on } \Gamma_2, \tag{C.4}$$

where S and G are the essential and natural boundary conditions on surface Γ_1 and Γ_2, respectively, with $\Gamma \ni (\Gamma_1 \cup \Gamma_2)$.

The residuals R_{Γ_1} and R_{Γ_2} on boundary Γ_1 and Γ_2 are given as

$$R_{\Gamma_1} = S(x_0) - \mathscr{s} = 0 \text{ on } \Gamma_1, \tag{C.5a}$$

$$R_{\Gamma_2} = G(x_0) - \mathscr{g} = 0 \text{ on } \Gamma_2. \tag{C.5b}$$

As the exact solution x_0 usually does not exist or is difficult to obtain analytically, we can use an approximate solution x, defined as a polynomial function:

$$x = \phi = \phi_0 + \sum_{k=1}^{\tilde{n}} \alpha_k \phi_k, \tag{C.6}$$

where ϕ_0 is a known function that satisfies the essential boundary condition on Γ, ϕ_k the $\tilde{p}$ sets of a linearly independent set of known analytical functions that satisfies the natural boundary condition, and α_k the unknown constants to be determined.

If we substitute x in Equation (C.6) into Equation (C.1), we may realise that $R_\Omega \neq 0$, but a more accurate result could be obtained by selecting an approximate solution x that produces a residual close to zero – i.e., $R_\Omega \to 0$. Often, the condition that $R_\Omega = 0$ is hard to achieve. Therefore, the weighted residual method is introduced in the next appendix. The residuals at the boundary, i.e., $R_{\Gamma_1} = R_{\Gamma_2} = 0$ as the selected function satisfies the boundary conditions on the surface Γ.

C.1 Method of Weighted Residual

The weighted residual method, also known as the Rayleigh-Ritz method, could be applied to Equation (C.1) by multiplying a weighted function $\bar{w}$ to the residual R and integrating over the domain Ω:

$$\int_\Omega \bar{w} \cdot R_\Omega \cdot d\Omega + \int_\Gamma \bar{w} \cdot R_\Gamma \cdot d\Omega = 0. \tag{C.7}$$

If we use the same approximate function x in Equation (C.6) that satisfies the boundary conditions, i.e., $R_\Gamma = 0$, we will have Equation (C.8). Therefore, we have Equation (C.8) upon substituting Equation (C.2) into Equation (C.7):

$$\int_\Omega \bar{w} \cdot R_\Omega \cdot d\Omega = \int_\Omega w \cdot \left[f(x) - b \right] \cdot d\Omega = 0. \tag{C.8}$$

To impose the condition that $R_\Omega = 0$ would over-enforce the condition resulting in an indeterminate system, i.e., a system of simultaneous equations which has more than one solution. Instead, we could impose the condition $R_\Omega \approx 0$ such that Equation (C.8) is satisfied. Hence, the weighted function $\bar{w}$ is applied to the residual R_Ω so that the average residual is zero.

This could be done by substituting Equation (C.6) into Equation (C.8):

$$\int_\Omega \bar{w}_j \cdot \left[f\left(\phi_0 + \sum_{k=1}^{\tilde{p}} \alpha_k \phi_k \right) - b \right] \cdot d\Omega = \int_\Omega \bar{w}_j \cdot \left[f(\phi_0) + f\left(\sum_{k=1}^{\tilde{p}} \alpha_k \phi_k \right) - b \right] \cdot d\Omega = 0, \tag{C.9}$$

which upon arrangement can also be written as

$$\sum_{k=1}^{\tilde{n}} \alpha_k \cdot \int_{\Omega} \bar{w}_j \cdot f(\phi_k) \cdot d\Omega = \int_{\Omega} \bar{w}_j \cdot b \cdot d\Omega - \int_{\Omega} w_j \cdot f(\phi_0) \cdot d\Omega, \tag{C.10}$$

where the term $\int_{\Omega} \bar{w}_j \cdot f(\phi_k) \cdot d\Omega = K_{ij}$, which is an important function for formulating matrices in FEM:

$$K_{jk} = \int_{\Omega} \bar{w}_j \cdot f(\phi_k) \cdot d\Omega. \tag{C.11}$$

It should be noted that the number of weighted functions $\bar{w}_j$ is the same as the number of $\tilde{n}$ in Equation (C.6). This is to say, by taking different $\bar{w}_j$ up to $\bar{w}_{\tilde{n}}$, there will be $\tilde{n}$ sets of simultaneous equations to be solved for the $\tilde{n}$ numbers of unknowns α_k. The choice of $\bar{w}_j = x^{j-1}$.

C.2 Collocation Method

The collocation method is also a form of weighted residual method, but instead of taking $\bar{w}_j = x^{j-1}$, the collocation method imposes the condition such that the residual at a certain location is zero, and the solution at that particular point is exact.

In the collocation method, the weighted function $\bar{w}_j$ is defined as

$$\bar{w}_j = \delta(x - x_j), \tag{C.12}$$

where the Dirac delta function has a property such that

$$\delta(x - x_j) = \begin{cases} 0 & \text{for } x \neq x_j \\ \infty & \text{for } x = x_j \end{cases}. \tag{C.13}$$

The integration of the Dirac delta function over a domain will give us

$$\int_{\Omega} \delta(x - x_j) d\Omega = \begin{cases} 0 & \text{for } x \neq x_j \\ 1 & \text{for } x = x_j \end{cases}, \tag{C.14}$$

which is a useful function when evaluating the integral of the weighted residual function.

By applying Equation (C.12) into Equation (C.10) and applying the condition given by Equation (D.15), this gives us

$$\sum_{k=1}^{\tilde{p}} \alpha_k \cdot \int_\Omega \delta(x - x_j) \cdot f(\phi_k) \cdot d\Omega = \int_\Omega \delta(x - x_j) \cdot b \cdot d\Omega - \int_\Omega \delta(x - x_j) \cdot f(\phi_0) \cdot d\Omega, \tag{C.15}$$

$$\sum_{k=1}^{\tilde{p}} \alpha_k \cdot f\left[\phi_k(x_j)\right] = b - f(\phi_0), \tag{C.16}$$

where, similarly, $K_{jk} = f\left[\phi_k(x_j)\right]$ and $b_j = b - f(\phi_0)$. So we have

$$\sum_{k=1}^{\tilde{p}} \alpha_k K_{jk} = b_j. \tag{C.17}$$

By choosing a specific location x_j and imposing the condition such that $R_\Omega(x_j) = 0$, the solutions will be exact at x_j.

C.3 Bubnov-Garlekin Method

The Bubnov-Garlekin method, known simply as the Garlekin method, uses a combination of an orthogonal trial and weighted functions based on variational formulation (weak formulation by Bubnov) and the weighted integral (strong formulation by Garlerkin) where the weighted functions need not be orthogonal.

The weighted function is assumed to be

$$w_j = \frac{\partial}{\partial \alpha_j} \phi(x), \tag{C.18}$$

where $\phi(x)$ is taken to be the same as in Equation (C.6), i.e.,

$$w_j = \frac{\partial}{\partial \alpha_j} \left[\alpha_0 \phi_0 + \sum_{k=1}^{\tilde{p}} \alpha_k \phi_k \right] = \phi_j(x), j = 1, 2, \ldots \tilde{p}. \tag{C.19}$$

ϕ_j must satisfy the boundary conditions $G(x)$ and $S(x)$, whereas w_j must be linearly independent. On top of that, $w_j(x)$ and $\phi_j(x)$ must be chosen from the first P function of a complete set of functions.

If we substitute Equation (C.19) into Equation (C.10):

$$\sum_{k=1}^{\tilde{n}} \alpha_k \cdot \int_{\Omega} \phi_j \cdot f(\phi_k) \cdot d\Omega = \int_{\Omega} \phi_j \cdot b \cdot d\Omega - \int_{\Omega} \phi_j \cdot f(\phi_0) \cdot d\Omega. \tag{C.20}$$

and taking

$$K_{jk} = \int_{\Omega} \phi_j \cdot f(\phi_k) \cdot d\Omega, \tag{C.21}$$

$$b_j = \int_{\Omega} \phi_j \cdot b \cdot d\Omega - \int_{\Omega} \phi_j \cdot f(\phi_0) \cdot d\Omega, \tag{C.22}$$

where Equation (C.21) involves the integral of sets of orthogonal functions in which Equation (C20) can be written in the same form as Equation (C.17).

C.3.1 Example C.1: Exact Solution

Consider a simple second-order ODE

$$\frac{d^2u(x)}{dx^2} + u(x) = 1 \text{ for } 0 \le x \le 1, \tag{C.23}$$

with the boundary conditions $u(0) = 0$ and $u(1) = 0$.

The second-order ODE is a linear non-homogeneous ODE and has complementary and particular solutions.

C.3.1.1 Complementary Solution

The complementary solution is obtained by solving the homogeneous part of the equation given as $\frac{d^2u(x)}{dx^2} + u(x) = 0$.

By solving the characteristic equation

$$\imath^2 + 1 = 0. \tag{C.24}$$

We have the roots $\imath$ given as, $\imath = \pm\sqrt{-1}$. The complementary solution for imaginary roots $\pm\imath$ is given by

$$u_c = A\cos(x) + B\sin(x). \tag{C.25}$$

C.3.1.2 Particular Solution

As the RHS of Equation (C.23) is zero, the particular solution is assumed to be a constant C:

$$u_p(x) = C. \tag{C.26}$$

Substituting Equation (C.26) into Equation (C.23), $C = 1$. Thus, the particular solution is given as

$$u_p(x) = 1. \tag{C.27}$$

C.3.1.3 General Solution

The general solution is given as

$$u(x) = u_c(x) + u_p(x) = A\cos(x) + B\sin(x) + 1. \tag{C.28}$$

The unknown constants A and B can be solved by using the boundary conditions:

Boundary condition (i): $\mathrm{u}(0) = 0$,

$$\mathrm{A}\cdot 1 + \mathrm{B}\cdot 0 + 1 = 0 \rightarrow \mathrm{A} = -1, \tag{C.29}$$

Boundary condition (ii): $\mathrm{u}(1) = 0$,

$$-1\cdot\cos(1) + B\cdot\sin(1) + 1 = 0 \rightarrow B = \frac{\cos(1) - 1}{\sin(1)}. \tag{C.30}$$

Substituting Equations (C.29) and (C.30) into Equation (C.28), the general solution yields

$$u(x) = -\cos(x) + \left[\frac{\cos(1) - 1}{\sin(1)}\right]\cdot\sin(x) + 1. \tag{C.31}$$

C.3.2 Example C.2: Method of Weighted Residual (MWR)

To solve using the MWR, assume a trial function

$$u_{approx}(x) = c_1 x(1 - x). \tag{C.32}$$

Substituting the trial function $u_{approx}(x)$ into the ODE (C.23) yields the residual R:

$$R(x) = -2c_1 + c_1x(1-x) - 1. \tag{C.33}$$

In the weighted residual method, we choose a weight function. For simplicity, we take $\bar{w}(x) = x(1-x)$ and apply the MWR as in Equation (C.8):

$$\int_0^1 \bar{w} \cdot R(x) \cdot dx = \int_0^1 \left[x(1-x)\right] \cdot \left[-2c_1 + c_1x(1-x) - 1\right] \cdot dx. \tag{C.34}$$

Upon evaluation, this yields

$$c_1 = \frac{2}{3}. \tag{C.35}$$

The approximate solution using MWR is

$$u_{approx}(x) = \frac{2}{3}x(1-x). \tag{C.36}$$

C.3.3 Example C.3: Collocation Method

In the collocation method, we enforce that the residual Equation (C.33) is zero at selected collocation points. Typically, these collocation points are chosen within the domain. For demonstration, we select the collocation point at $x = 0.5$. $R(0.5)$, thus becoming

$$R(0.5) = -2c_1 + c_1(0.5)(1-0.5) - 1 = 0. \tag{C.37}$$

This yields

$$c_1 = -\frac{4}{7}. \tag{C.38}$$

The approximate solution is thus

$$u_{approx}(x) = -\frac{4}{7}(x)(1-x). \tag{C.39}$$

C.3.4 Example C.4: Bubnov-Galerkin Method

In the Galerkin method, we require that the residual be orthogonal to the weight function $\bar{w}(x)$, which is chosen as the same function as the basis function used in the trial function of Equation (C.32):

$$\bar{w}(x) = (x)(1-x). \tag{C.40}$$

The Galerkin condition is given by

$$\int_0^1 \bar{w}(x)\cdot R(x)\cdot dx = 0. \tag{C.41}$$

Substitute $R(x)$ into the integral gives us the same as Equation (C.34).

Equation (C.34) can be expanded as

$$\int_0^1 -2c_1\cdot\left[x(1-x)\right]\cdot dx + \int_0^1 c_1 x(1-x)\cdot\left[x(1-x)\right]\cdot dx - \int_0^1 x(1-x)\cdot dx. \tag{C.42}$$

Upon evaluation,

$$-\frac{c_1}{3} + c_1\cdot\frac{1}{30} - \frac{1}{6} = 0 \rightarrow c_1 = -\frac{5}{9}. \tag{C.43}$$

Substitute Equation (C.43) into Equation (C.32):

$$u_{approx}(x) = -\frac{5}{9}x(1-x). \tag{C.44}$$

C.3.4.1 Summary and Comparison of Results

Exact solutions:

$$u(x) = -\cos(x) + \left[\frac{\cos(1)-1}{\sin(1)}\right]\cdot\sin(x) + 1 \tag{C.45}$$

MWRs:

$$u_{approx}(x) = -\frac{2}{3}x(1-x) \tag{C.46}$$

Collocation method:

$$u_{approx}(x) = -\frac{4}{7}(x)(1-x) \tag{C.47}$$

Bubnov-Galerkin method:

$$u_{approx}(x) = -\frac{5}{9}x(1-x) \tag{C.48}$$

The comparison of $u_{approx}(x)$ with exact solution $u(x)$ are given in Table C.1 and Figure C.1. The percentage error in the table is computed from

$$\text{Absolute Error}(\%) = \frac{|u_{approx}(x) - u(x)|}{u(x)}\%. \quad (C.49)$$

TABLE C.1

Comparison of Approximate Solution with Exact Solution and Their Percentage Errors

		Method of Weighted Residual		Collocation Method		Galerkin Method	
x	Exact Solution $u(x)$	$u_{approx}(x)$	Absolute Error	$u_{approx}(x)$	Absolute Error	$u_{approx}(x)$	Absolute Error
0.00	0.0000	0.0000	0.00%	0.0000	0.00%	0.0000	0.00%
0.10	−0.0495	0.0600	10.95%	−0.0514	0.19%	−0.0500	0.05%
0.20	−0.0886	0.1067	19.53%	−0.0914	0.28%	−0.0889	0.03%
0.30	−0.1168	0.1400	25.68%	−0.1200	0.32%	−0.1167	0.01%
0.40	−0.1338	0.1600	29.38%	−0.1371	0.33%	−0.1333	0.05%
0.50	−0.1395	0.1667	30.62%	−0.1429	0.34%	−0.1389	0.06%
0.60	−0.1338	0.1600	29.38%	−0.1371	0.33%	−0.1333	0.05%
0.70	−0.1168	0.1400	25.68%	−0.1200	0.32%	−0.1167	0.01%
0.80	−0.0886	0.1067	19.53%	−0.0914	0.28%	−0.0889	0.03%
0.90	−0.0495	0.0600	10.95%	−0.0514	0.19%	−0.0500	0.05%
1.00	0.0000	0.0000	0.00%	0.0000	0.00%	0.0000	0.00%

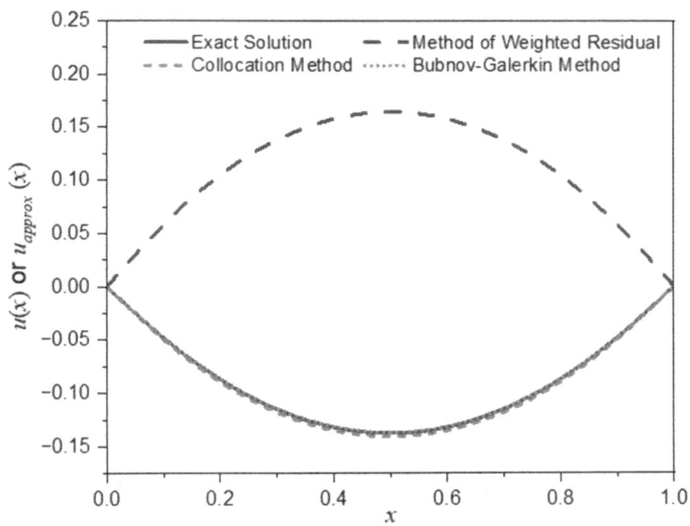

FIGURE C.1
Exact solution and approximate solution to ODE.

It can be seen that the $u_{approx}(x)$ predicted by the Galerkin method has the smaller percentage error, followed by the collocation method. The $u_{approx}(x)$ predicted by MWR is completely off when compared with the exact solution $u(x)$.

C.3.4.2 Residual

The residual $R(x)$ for the three approximate solutions summarised as

$$R(x) = -2c_1 + c_1 x(1-x) - 1, \text{where } c_1 = \begin{cases} \dfrac{2}{3} \text{ for weighted residual method} \\ -\dfrac{4}{7} \text{ for collocation method} \\ -\dfrac{5}{9} \text{ for Bubnov} - \text{Galerkin method} \end{cases} . \tag{C.50}$$

Table C.2 shows the residual $R(x)$ of the three approximation methods. A comparison of $R(x)$ for the collocation and Bubnov-Galerkin method is plotted in Figure C.1. It can be seen that, for the collocation method, the $R(x)$ is zero at $x = 0.5$ due to the imposed condition in (C.37). The figure also shows that the Galerkin method has a higher accuracy as compared to the collocation method.

TABLE C.2

Residual of Approximation Methods

x	MWR	Collocation Method	Bubnov-Galerkin Method
0.00	−2.3333	0.1429	0.1111
0.10	−2.2733	0.0914	0.0611
0.20	−2.2267	0.0514	0.0222
0.30	−2.1933	0.0229	−0.0056
0.40	−2.1733	0.0057	−0.0222
0.50	−2.1667	0.0000	−0.0278
0.60	−2.1733	0.0057	−0.0222
0.70	−2.1933	0.0229	−0.0056
0.80	−2.2267	0.0514	0.0222
0.90	−2.2733	0.0914	0.0611
1.00	−2.3333	0.1429	0.1111

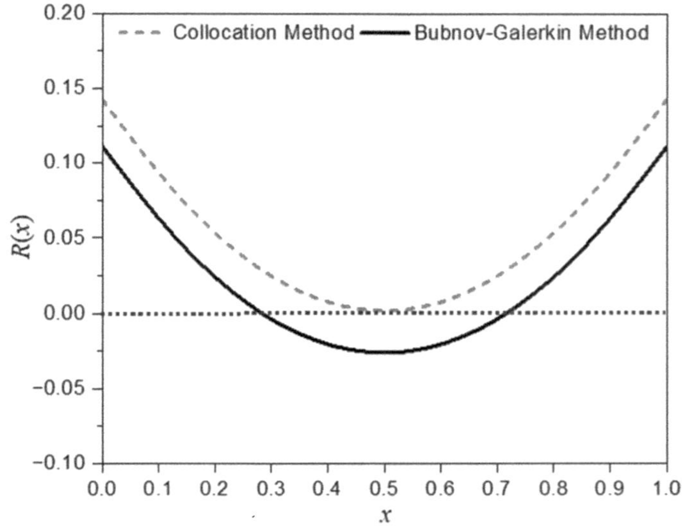

FIGURE C.2
Residual for collocation and Bubnov-Galerkin methods.

Appendix D: Shape Function

The numerical method requires the discretisation of domain Ω or surface Γ into N_e elements with N_d number of nodes in each element. The displacement at its nodes (known as nodal displacement) is denoted by $\mathbf{u_d}$, whereas at the element by $\mathbf{u_e}$. The geometry of the element could be of various shapes depending on the model being built and accuracy of the simulation required for the analysis. A three-dimensional element would require more computational time compared to a two-dimensional element. In this book, the model is represented in two dimensions, such as the VLFS modelled as plate structure in the FEM or the consideration of surface domain Γ in the BEM. Therefore, focus is given to element in two dimensions here.

D.1 Four-Node Quadrilateral Rectangular Element

We consider a four-node quadrilateral rectangular element. Here, we define the local coordinate system for an element in reference to the global coordinate system for the entire structure. By using polynomial interpolation, the element displacement is expressed as a summation of nodal displacement:

$$\mathbf{u_e} = \sum_{j=1}^{N_d} N_j(x,y,z)\cdot(u_d)_j = \mathrm{n}(x,y,z)\cdot\{\mathbf{u_d}\}, \tag{D.1}$$

where n is the shape function given as

$$\mathrm{n} = \begin{bmatrix} N_1 & 0 & 0 & 0 \\ 0 & N_2 & 0 & 0 \\ 0 & 0 & \ddots & 0 \\ 0 & 0 & 0 & N_{N_d} \end{bmatrix}, \tag{D.2}$$

and the nodal displacement is given as

$$\{\mathbf{u_d}\} = \left\{(u_d)_1 \ (u_d)_2 \ \cdots \ (u_d)_{N_d}\right\}^T. \tag{D.3}$$

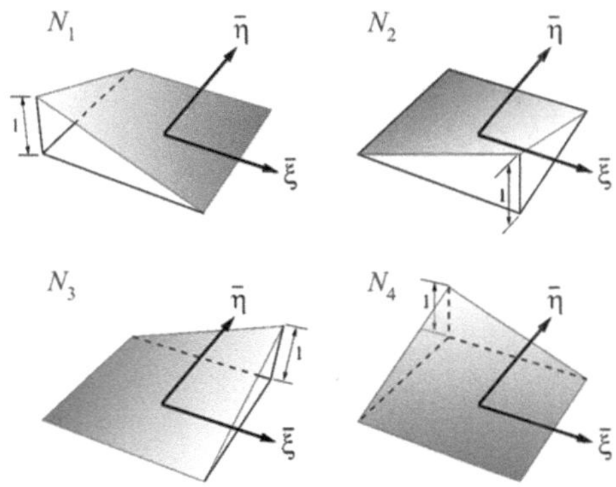

FIGURE D.1
Four-node quadrilateral element.

The construction of the shape function requires that N_i only has a value of 1 in node i and zero at other nodes, as shown in Figure D.1. The polynomial function could be used as interpolation functions. Two different well-known polynomials are presented here, i.e., the Lagrange polynomial and serendipity polynomial.

D.2 Lagrange Polynomial

The easiest way to construct the shape function from the Langrage polynomial is by using Pascal's triangle.

Assuming that the element displacement $\mathbf{u}_e$ can be written as an expansion of polynomial ϕ written in the same form as Equation (B.4),

$$\mathbf{u}_e = \phi = \sum_{j}^{\tilde{p}} \alpha_j \phi_j, \tag{D.4}$$

where ϕ can be represented by the binomial expansion of order $\tilde{p}$ and has to be complete, i.e., including all the terms required for binomial expansion up to order $\tilde{p}$. Pascal's triangle is shown in Figure D.2.

Say, for example, that a binomial quadratic expansion or order $\tilde{p} = 1$ is used for constructing the interpolation function. For completeness, the number of terms required according to Pascal's triangle is 4 or given as $n_d = \left(\tilde{p} + 1\right)^2$. Therefore, the polynomial ϕ is given as

$$\mathbf{u}_e = \phi = \alpha_1 + \alpha_2 \bar{\xi} + a_3 \bar{\eta} + \alpha_4 \bar{\xi}\bar{\eta}, \tag{D.5}$$

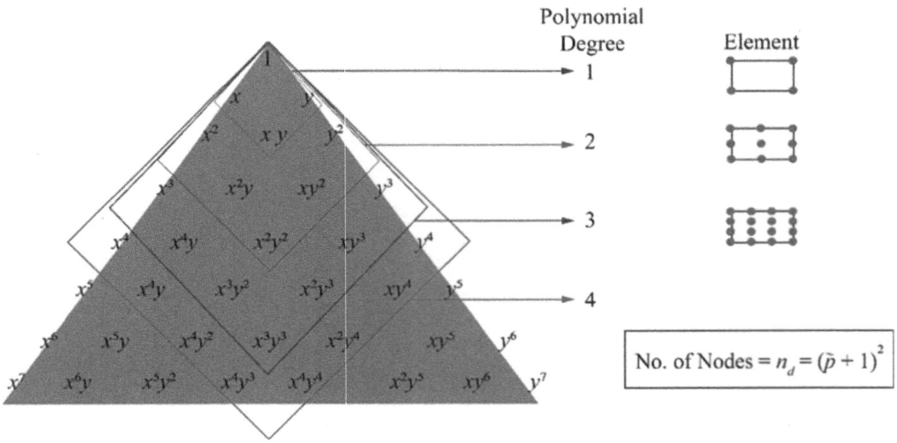

FIGURE D.2
Pascal's triangle for square and triangle element, which are complete.

or in matrix form,

$$\mathbf{u_e} = \begin{bmatrix} 1 & \bar{\xi} & \bar{\eta} & \bar{\xi}\bar{\eta} \end{bmatrix} \begin{bmatrix} \alpha_1 & \alpha_2 & \alpha_3 & \alpha_4 \end{bmatrix}^T. \tag{D.6}$$

If we assume a quadrilateral rectangular plate having $n_d = 4$, where the nodes are located at $(\bar{\xi}, \bar{\eta}) = (1,1), (-1,1), (-1,-1)$ and $(1,-1)$, we can form four equations as follows:

$$\begin{aligned} \phi(1,1) &= \alpha_1 + \alpha_2 + a_3 + \alpha_4, \\ \phi(-1,1) &= \alpha_1 - \alpha_2 + a_3 - \alpha_4, \\ \phi(-1,-1) &= \alpha_1 - \alpha_2 - a_3 + \alpha_4, \\ (u_d)_1 = \phi(1,-1) &= \alpha_1 + \alpha_2 - a_3 - \alpha_4. \end{aligned} \tag{D.7}$$

Writing in matrix form,

$$\begin{bmatrix} \phi_1 \\ \phi_2 \\ \phi_3 \\ \phi_4 \end{bmatrix} = \begin{bmatrix} 1 & 1 & 1 & 1 \\ 1 & -1 & 1 & -1 \\ 1 & -1 & -1 & 1 \\ 1 & 1 & -1 & -1 \end{bmatrix} \begin{bmatrix} \alpha_1 \\ \alpha_2 \\ \alpha_3 \\ \alpha_4 \end{bmatrix}. \tag{D.8}$$

Solving for α gives us

$$\begin{bmatrix} \alpha_1 \\ \alpha_2 \\ \alpha_3 \\ \alpha_4 \end{bmatrix} = \frac{1}{4} \begin{bmatrix} 1 & 1 & 1 & 1 \\ 1 & -1 & -1 & 1 \\ 1 & 1 & -1 & -1 \\ 1 & -1 & 1 & -1 \end{bmatrix} \begin{bmatrix} \phi_1 \\ \phi_2 \\ \phi_3 \\ \phi_4 \end{bmatrix}. \tag{D.9}$$

Substituting Equation (D.8) back into Equation (D.6) and taking $\phi = u_d$ gives us

$$\mathbf{u_e} = \frac{1}{4} \begin{bmatrix} 1 & \bar{\xi} & \bar{\eta} & \bar{\xi}\bar{\eta} \end{bmatrix} \begin{bmatrix} 1 & 1 & 1 & 1 \\ 1 & -1 & -1 & 1 \\ 1 & 1 & -1 & -1 \\ 1 & -1 & 1 & -1 \end{bmatrix} \begin{bmatrix} (u_d)_1 \\ (u_d)_2 \\ (u_d)_3 \\ (u_d)_4 \end{bmatrix}. \tag{D.10}$$

which upon expansion gives us

$$\mathbf{u_e} = \frac{\left(1+\bar{\xi}+\bar{\eta}+\bar{\xi}\bar{\eta}\right)}{4}(u_d)_1 + \frac{\left(1-\bar{\xi}+\bar{\eta}-\bar{\xi}\bar{\eta}\right)}{4}(u_d)_2 + \frac{\left(1-\bar{\xi}-\bar{\eta}+\bar{\xi}\bar{\eta}\right)}{4}(u_d)_3 - \frac{\left(1+\bar{\xi}-\bar{\eta}-\bar{\xi}\bar{\eta}\right)}{4}(u_d)_4 = \mathbb{N}\left(\bar{\xi},\bar{\eta}\right)\cdot\mathbf{u_d}. \tag{D.11}$$

Therefore, we have the interpolation function, which upon simplification gives us

$$N_1 = \frac{1}{4}\left(1+\bar{\xi}\right)\left(1+\bar{\eta}\right),$$

$$N_2 = \frac{1}{4}\left(1-\bar{\xi}\right)\left(1+\bar{\eta}\right),$$

$$N_3 = \frac{1}{4}\left(1-\bar{\xi}\right)\left(1-\bar{\eta}\right),$$

$$N_4 = \frac{1}{4}\left(1+\bar{\xi}\right)\left(1-\bar{\eta}\right). \tag{D.12}$$

D.3 Nine-Node Quadrilateral Rectangular Element

Similarly, the interpolation function for polynomial is a nine-node quadrilateral plate that can be obtained by the method mentioned earlier, presented here as

$$\mathbf{u}_e = \phi = \alpha_1 + \alpha_2\bar{\xi} + a_3\bar{\eta} + \alpha_4\bar{\xi}\bar{\eta} + \alpha_5\bar{\xi}^2 + \alpha_6\bar{\eta}^2 + \alpha_7\bar{\xi}^2\bar{\eta} + \alpha_8\bar{\xi}\bar{\eta}^2 + \alpha_9\bar{\xi}^2\bar{\eta}^2. \quad \text{(D.13)}$$

The quadrilateral rectangular plate with $n_d = 9$ in Equation (D.13), where the nodes are located at $(\bar{\xi}, \bar{\eta}) = (1,1), (-1,1), (-1,-1)$, $(1,-1), (0,1), (-1,0), (0,-1)$, (1,0), and $(0,0)$, forms nine equations as follows:

$$\begin{gathered}
\phi(1,1) = \alpha_1 + \alpha_2 + a_3 + \alpha_4 + \alpha_5 + \alpha_6 + \alpha_7 + \alpha_8 + \alpha_9, \\
\phi(-1,1) = \alpha_1 - \alpha_2 + a_3 - \alpha_4 + \alpha_5 + \alpha_6 + \alpha_7 - \alpha_8 + \alpha_9, \\
\phi(-1,-1) = \alpha_1 - \alpha_2 - a_3 + \alpha_4 + \alpha_5 + \alpha_6 - \alpha_7 - \alpha_8 + \alpha_9, \\
\phi(1,-1) = \alpha_1 + \alpha_2 - a_3 - \alpha_4 + \alpha_5 + \alpha_6 - \alpha_7 + \alpha_8 + \alpha_9, \\
\phi(0,1) = \alpha_1 + a_3 + \alpha_6, \phi(-1,0) = \alpha_1 - \alpha_2 + \alpha_5, \\
\phi(0,-1) = \alpha_1 - a_3 + \alpha_6, \phi(1,0) = \alpha_1 + \alpha_2 + \alpha_5, \phi(0,0) = \alpha_1.
\end{gathered} \quad \text{(D.14)}$$

Writing in matrix form,

$$\begin{bmatrix} \phi_1 \\ \phi_2 \\ \phi_3 \\ \phi_4 \\ \phi_5 \\ \phi_6 \\ \phi_7 \\ \phi_8 \\ \phi_9 \end{bmatrix} = \begin{bmatrix} 1 & 1 & 1 & 1 & 1 & 1 & 1 & 1 & 1 \\ 1 & -1 & 1 & -1 & 1 & 1 & 1 & -1 & 1 \\ 1 & -1 & -1 & 1 & 1 & 1 & -1 & -1 & 1 \\ 1 & 1 & -1 & -1 & 1 & 1 & -1 & 1 & 1 \\ 1 & 0 & 1 & 0 & 0 & 1 & 0 & 0 & 0 \\ 1 & -1 & 0 & 0 & 1 & 0 & 0 & 0 & 0 \\ 1 & 0 & -1 & 0 & 0 & 1 & 0 & 0 & 0 \\ 1 & 1 & 0 & 0 & 1 & 0 & 0 & 0 & 0 \\ 1 & 0 & 0 & 0 & 0 & 0 & 0 & 0 & 0 \end{bmatrix} \begin{bmatrix} \alpha_1 \\ \alpha_2 \\ \alpha_3 \\ \alpha_4 \\ \alpha_5 \\ \alpha_6 \\ \alpha_7 \\ \alpha_8 \\ \alpha_9 \end{bmatrix}. \quad \text{(D.15)}$$

Solving for α gives us

$$\begin{bmatrix} \alpha_1 \\ \alpha_2 \\ \alpha_3 \\ \alpha_4 \\ \alpha_5 \\ \alpha_6 \\ \alpha_7 \\ \alpha_8 \\ \alpha_9 \end{bmatrix} = \frac{1}{4} \begin{bmatrix} 0 & 0 & 0 & 0 & 0 & 0 & 0 & 0 & 1 \\ 0 & 0 & 0 & 0 & 0 & -2 & 0 & 2 & 0 \\ 0 & 0 & 0 & 0 & 2 & 0 & -2 & 0 & 0 \\ 1 & -1 & 1 & -1 & 0 & 0 & 0 & 0 & 0 \\ 0 & 0 & 0 & 0 & 0 & 2 & 0 & 2 & -1 \\ 0 & 0 & 0 & 0 & 2 & 0 & 2 & 0 & -1 \\ 1 & 1 & -1 & -1 & -2 & 0 & 2 & 0 & 0 \\ 1 & -1 & -1 & 1 & 0 & 2 & 0 & -2 & 0 \\ 1 & 0 & 0 & 0 & 0 & 0 & 0 & 0 & 0 \end{bmatrix} \begin{bmatrix} (u_d)_1 \\ (u_d)_2 \\ (u_d)_3 \\ (u_d)_4 \\ (u_d)_5 \\ (u_d)_6 \\ (u_d)_7 \\ (u_d)_8 \\ (u_d)_9 \end{bmatrix}. \tag{D.16}$$

Substituting Equation (D.8) back into Equation (D.6) gives us

$$\mathbf{u_e} = \frac{1}{4}\begin{bmatrix} 1 & \bar{\xi} & \bar{\eta} & \bar{\xi}\bar{\eta} & \bar{\xi}^2 & \bar{\eta}^2 & \bar{\xi}^2\bar{\eta} & \bar{\xi}\bar{\eta}^2 & \bar{\xi}^2\bar{\eta}^2 \end{bmatrix} \times \begin{bmatrix} 0 & 0 & 0 & 0 & 0 & 0 & 0 & 0 & 1 \\ 0 & 0 & 0 & 0 & 0 & -2 & 0 & 2 & 0 \\ 0 & 0 & 0 & 0 & 2 & 0 & -2 & 0 & 0 \\ 1 & -1 & 1 & -1 & 0 & 0 & 0 & 0 & 0 \\ 0 & 0 & 0 & 0 & 0 & 2 & 0 & 2 & -1 \\ 0 & 0 & 0 & 0 & 2 & 0 & 2 & 0 & -1 \\ 1 & 1 & -1 & -1 & -2 & 0 & 2 & 0 & 0 \\ 1 & -1 & -1 & 1 & 0 & 2 & 0 & -2 & 0 \\ 1 & 0 & 0 & 0 & 0 & 0 & 0 & 0 & 0 \end{bmatrix} \begin{bmatrix} (u_d)_1 \\ (u_d)_2 \\ (u_d)_3 \\ (u_d)_4 \\ (u_d)_5 \\ (u_d)_6 \\ (u_d)_7 \\ (u_d)_8 \\ (u_d)_9 \end{bmatrix}, \tag{D.17}$$

which upon expansion gives us

$$\begin{aligned} \mathbf{u_e} = {} & \frac{\left(\bar{\xi}^2\bar{\eta}^2 + \bar{\xi}\bar{\eta}^2 + \bar{\xi}^2\bar{\eta} + \bar{\xi}\bar{\eta}\right)}{4}(u_d)_1 + \frac{\left(-\bar{\xi}\bar{\eta} + \bar{\xi}^2\bar{\eta} - \bar{\xi}\bar{\eta}^2 + \bar{\xi}^2\bar{\eta}\right)}{4}(u_d)_2 \\ & + \frac{\left(\bar{\xi}^2\bar{\eta}^2 - \bar{\xi}\bar{\eta}^2 - \bar{\xi}^2\bar{\eta} + \bar{\xi}\bar{\eta}\right)}{4}(u_d)_3 + \frac{\left(-\bar{\xi}\bar{\eta} - \bar{\xi}^2\bar{\eta} + \bar{\xi}\bar{\eta}^2 + \bar{\xi}^2\bar{\eta}^2\right)}{4}(u_d)_4 \\ & + \frac{\left(\bar{\eta} - \bar{\xi}^2\bar{\eta} + \bar{\eta}^2 - \bar{\xi}^2\bar{\eta}^2\right)}{2}(u_d)_5 + \frac{\left(-\bar{\xi}^2\bar{\eta}^2 + \bar{\xi}\bar{\eta}^2 + \bar{\xi}^2 - \bar{\xi}\right)}{2}x_6 \\ & + \frac{\left(-\bar{\xi}^2\bar{\eta}^2 + \bar{\eta}^2 + \bar{\xi}^2\bar{\eta} - \bar{\eta}\right)}{2}(u_d)_7 + \frac{\left(\bar{\xi} - \bar{\xi}\bar{\eta}^2 + \bar{\xi}^2 - \bar{\xi}^2\bar{\eta}^2\right)}{2}(u_d)_8 \\ & + \left(\bar{\xi}^2\bar{\eta}^2 - \bar{\eta}^2 + \bar{\xi}^2 + 1\right)(u_d)_9 = \mathbb{N}(\xi,\eta)\cdot\{\mathbf{u_d}\}. \end{aligned} \tag{D.18}$$

Therefore, we have the interpolation function, which upon simplification gives us

$$N_1 = \frac{1}{4}\bar{\xi}\left(1+\bar{\xi}\right)\left(1+\bar{\eta}\right)\bar{\eta},$$

$$N_2 = -\frac{1}{4}\bar{\xi}\left(1-\bar{\xi}\right)\left(1+\bar{\eta}\right)\bar{\eta},$$

$$N_3 = \frac{1}{4}\bar{\xi}\left(1-\bar{\xi}\right)\left(1-\bar{\eta}\right)\bar{\eta},$$

$$N_4 = -\frac{1}{4}\bar{\xi}\left(1+\bar{\xi}\right)\left(1+\bar{\eta}\right)\bar{\eta}, \tag{D.19}$$

$$N_5 = \frac{1}{2}\left(1+\bar{\xi}\right)\left(1-\bar{\xi}\right)\bar{\eta},$$

$$N_6 = -\frac{1}{2}\bar{\xi}\left(1-\bar{\xi}\right)\left(1-\bar{\eta}\right)\bar{\eta},$$

$$N_7 = -\frac{1}{2}\left(1+\bar{\xi}\right)\left(1-\bar{\xi}\right)\left(1-\bar{\eta}\right)\bar{\eta},$$

$$N_8 = \frac{1}{2}\bar{\xi}\left(1+\bar{\xi}\right)\left(1-\bar{\eta}\right)\bar{\eta},$$

$$N_9 = \left(1-\bar{\xi}^2\right)\left(1-\bar{\eta}^2\right).$$

The nine-node quadrilateral plate element is shown in Figure D.3.

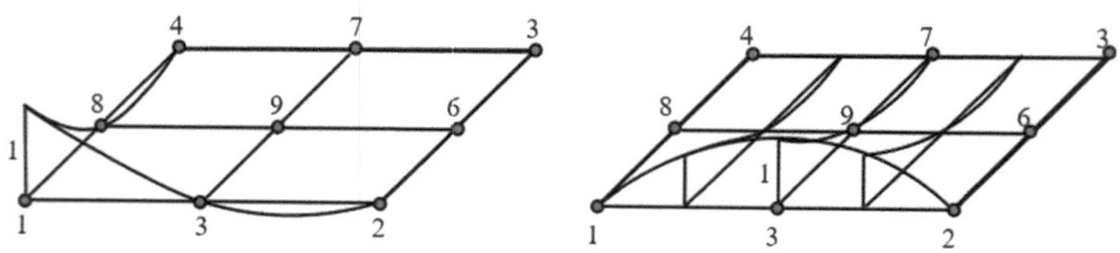

FIGURE D.3
Nine-node quadrilateral element.

D.4 Eight-Node Serendipity Rectangular Element

The serendipity element is similar to the nine-node quadrilateral element except for the internal element that is being removed from the element. This could reduce the computational time as the size of the matrix is reduced and does not compromise the accuracy of the results. Using the same method, the interpolation function is given as

$$
\begin{aligned}
N_1 &= \frac{1}{4}(1+\xi)(1+\eta) - \frac{1}{2}(N_5+N_8),\\
N_2 &= \frac{1}{4}(1-\xi)(1+\eta) - \frac{1}{2}(N_5+N_6),\\
N_3 &= \frac{1}{4}(1-\xi)(1-\eta) - \frac{1}{2}(N_6+N_7),\\
N_4 &= \frac{1}{4}(1+\xi)(1-\eta) - \frac{1}{2}(N_7+N_8),\\
N_5 &= \frac{1}{2}(1-\xi^2)(1+\eta),\\
N_6 &= \frac{1}{2}(1-\xi)(1-\eta^2),\\
N_7 &= \frac{1}{2}(1-\xi^2)(1-\eta),\\
N_8 &= \frac{1}{2}(1+\xi)(1-\eta^2).
\end{aligned} \tag{D.20}
$$

The eight-node serendipity rectangular element is given in Figure D.4.

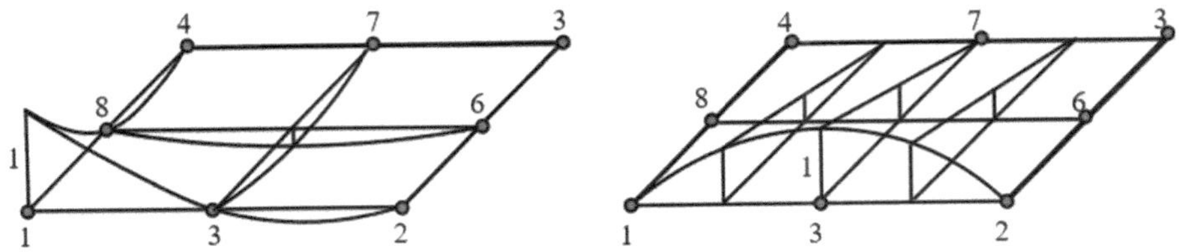

FIGURE D.4
Eight-node serendipity rectangular element.

Appendix E: Evaluation of Integral

The formulation of FEM or BEM requires the evaluation of an integral over a domain or surface as it is derived based on the weighted integral methods as outlined in Appendix C. Here, some common integrals that involve the interpolation function are described.

As shown in Figure E.1, an element in the x-y coordinate system is represented by element in the $\bar{\xi}$-$\bar{\eta}$ coordinates, therefore it requires the transformation from the x-axis to the $\bar{\xi}$-$\bar{\eta}$-axis. Taking a four-node quadrilateral element as an example,

$$x = \begin{bmatrix} N_1 & 0 & 0 & 0 \\ 0 & N_2 & 0 & 0 \\ 0 & 0 & \ddots & \vdots \\ 0 & 0 & \cdots & N_{N_d} \end{bmatrix} \begin{bmatrix} x_1 \\ x_2 \\ x_3 \\ x_4 \end{bmatrix}, y = \begin{bmatrix} N_1 & 0 & 0 & 0 \\ 0 & N_2 & 0 & 0 \\ 0 & 0 & \ddots & \vdots \\ 0 & 0 & \cdots & N_{N_d} \end{bmatrix} \begin{bmatrix} y_1 \\ y_2 \\ y_3 \\ y_4 \end{bmatrix}. \quad \text{(E.1)}$$

E.1 Jacobian Matrix

In order to transform from the actual coordinate system to natural coordinate system, we need to use the Jacobian matrix, defined as

$$\det J = \begin{vmatrix} \dfrac{\partial x}{\partial \bar{\xi}} & \dfrac{\partial x}{\partial \bar{\eta}} \\ \dfrac{\partial y}{\partial \bar{\xi}} & \dfrac{\partial y}{\partial \bar{\eta}} \end{vmatrix}. \quad \text{(E.2)}$$

The determinant J denotes how much the element in the actual coordinates is stretched when transform to the natural coordinates. Now we can use the determinant J to evaluate the integral. The integration of function $f(x)$ over a surface Γ can be written as

$$\int_{\Gamma} f(\bar{\xi}, \bar{\eta}) \cdot d\Gamma = \int_{-1}^{1} \int_{-1}^{1} f(\bar{\xi}, \bar{\eta}) \cdot |J| \cdot d\bar{\xi} \cdot d\bar{\eta}. \quad \text{(E.3)}$$

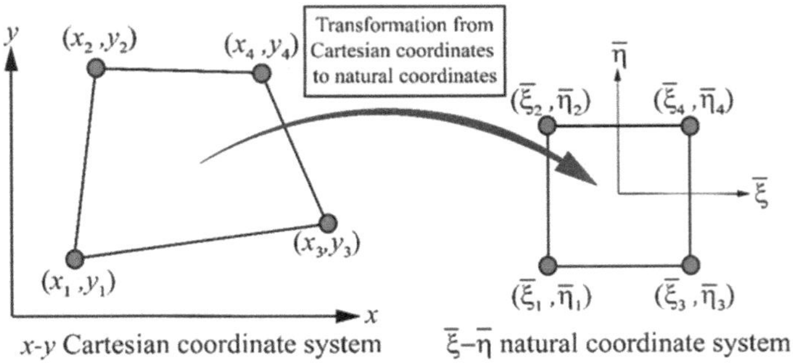

FIGURE E.1
Transformation of element from x - y Cartesian coordinate system to $\bar{\xi}$ - $\bar{\eta}$ natural coordinate system.

E.2 Gauss Quadrature

Now, we introduce the Gauss quadrature to evaluate the integral, i.e., area under curve or volume under surface. For a generic integral given as follows

$$\int_{-1}^{1}\int_{-1}^{1} g\left(\bar{\xi},\bar{\eta}\right)\cdot d\bar{\xi}\cdot d\bar{\eta} = \sum_{k=1}^{N_g} g\left(\bar{\xi}_k,\bar{\eta}_k\right)\cdot \bar{w}_k, \tag{E.4}$$

where $\bar{w}_k$ is the weight of the Gauss quadrature and $(\bar{\xi}_k, \bar{\eta}_k)$ the interpolation points for the Gauss quadrature. N_g is the number of interpolation points for the Gauss quadrature. The list of $\bar{w}_k$ and $(\bar{\xi}_k, \bar{\eta}_k)$ for different elements are given in Table E.1. The evaluation of the integral has been approximated using the Gauss quadrature as shown in the RHS of Equation (E.4).

Similarly, by using the Gaussian quadrature, Equation (E.3) can be evaluated as follows:

$$\int_{-1}^{1}\int_{-1}^{1} f\left(\bar{\xi},\bar{\eta}\right)\cdot |J|\cdot d\bar{\xi}\cdot d\bar{\eta} = \sum_{k=1}^{N_g} f\left(\bar{\xi}_k,\bar{\eta}_k\right)\cdot \left|J\left(\bar{\xi}_k,\bar{\eta}_k\right)\right|\cdot \bar{w}_k. \tag{E.5}$$

To evaluate only the area for an element Δ_e,

$$\Delta_e = \sum_{k=1}^{N_g} \left|J\left(\bar{\xi}_k,\bar{\eta}_k\right)\right|\cdot \bar{w}_k. \tag{E.6}$$

TABLE E.1

Gauss Quadrature Weights and Integration Points

Number of Points	Points, $\bar{\xi}_k$ or $\bar{\eta}_k$	Weights, $\bar{w}_k$
1	0	2
2	$\pm\sqrt{\frac{1}{3}} = \pm 0.5774$	1
3	0	$\frac{8}{9} = 0.8889$
	$\pm\sqrt{\frac{3}{5}} = \pm 0.7746$	$\frac{5}{9} = 0.5556$
4	$\pm\sqrt{\frac{3}{5} - \frac{2}{7}\sqrt{\frac{6}{5}}} = \pm 0.3400$	$\frac{18+\sqrt{30}}{36} = 0.6522$
	$\pm\sqrt{\frac{3}{7} + \frac{2}{7}\sqrt{\frac{6}{5}}} = \pm 0.8611$	$\frac{18-\sqrt{30}}{36} = 0.3479$
	0	$\frac{128}{225} = 0.5689$
5	$\pm\frac{1}{3}\sqrt{5 - 2\sqrt{\frac{10}{7}}} = \pm 1.3767$	$\frac{322+13\sqrt{70}}{900} = 0.4786$
	$\pm\frac{1}{3}\sqrt{5 + 2\sqrt{\frac{10}{7}}} = \pm 1.9479$	$\frac{322-13\sqrt{70}}{900} = 0.2369$

E.2.1 Example E.1

A trapezoid shown in Figure E.1 is to be transformed into its natural coordinates in a four-node quadrilateral element. The coordinate for the trapezoid is given by

$$(x_1, y_1) = (3,3), (x_2, y_2) = (-1,3), (x_3, y_3) = (-6,-3), (x_4, y_4) = (1,-3). \quad \text{(E.7)}$$

The Gauss quadrature interpolation points are $(\bar{\xi}_k, \bar{\eta}_k) = \left(\pm\frac{1}{\sqrt{3}}, \pm\frac{1}{\sqrt{3}}\right)$ and the weights $\bar{w}_k = 1$.

We know that x and y can be written as

$$x = N_1 x_1 + N_2 x_2 + N_3 x_3 + N_4 x_4, \quad \text{(E.8a)}$$

$$y = N_1 y_1 + N_2 y_2 + N_3 y_3 + N_4 y_4, \quad \text{(E.8b)}$$

where the shape function N_1, N_2, N_3, and N_4 are given in Equation (D.12). Substituting Equation (E.7) into Equation (E.8),

$$x = -\frac{3}{2}(1-\bar{\eta})(1-\bar{\xi}) - \frac{1}{4}(1+\bar{\eta})(1-\bar{\xi}) + \frac{1}{4}(1-\bar{\eta})(1+\bar{\xi}) + \frac{3}{4}(1+\bar{\eta})(1+\bar{\xi}), \tag{E.9a}$$

$$y = -\frac{3}{4}(1-\bar{\eta})(1-\bar{\xi}) + \frac{3}{4}(1+\bar{\eta})(1-\bar{\xi}) - \frac{3}{4}(1-\bar{\eta})(1+\bar{\xi}) + \frac{3}{4}(1+\bar{\eta})(1+\bar{\xi}). \tag{E.9b}$$

The area from (E.6) can be expanded as

$$\begin{aligned} Area &= \left|J(\bar{\xi}_1, \bar{\eta}_1)\right| \bar{w}_1 + \left|J(\bar{\xi}_2, \bar{\eta}_2)\right| \bar{w}_2 + \left|J(\bar{\xi}_3, \bar{\eta}_3)\right| \bar{w}_3 + \left|J(\bar{\xi}_4, \bar{\eta}_4)\right| \bar{w}_4 \\ &= \left|J\left(-\frac{1}{\sqrt{3}}, -\frac{1}{\sqrt{3}}\right) + J\left(-\frac{1}{\sqrt{3}}, +\frac{1}{\sqrt{3}}\right) + J\left(+\frac{1}{\sqrt{3}}, -\frac{1}{\sqrt{3}}\right) + J\left(+\frac{1}{\sqrt{3}}, +\frac{1}{\sqrt{3}}\right)\right| = 33. \end{aligned} \tag{E.10}$$

E.3 Global Matrices

The area for the rest of the elements could be computed the same way and assembled into a global matrix:

$$\Delta = \begin{bmatrix} \Delta_{e1} & 0 & 0 & 0 \\ 0 & \Delta_{e2} & 0 & 0 \\ 0 & 0 & \ddots & \vdots \\ 0 & 0 & \cdots & \Delta_{eN_e} \end{bmatrix}. \tag{E.11}$$

The mass matrix given in Equation (E.11) has a lump-sum mass for each element and is useful for evaluating Green's function over the surface using the constant panel method. For the finite element method, the mass matrices have to be distributed to each node.

References

1. National Geographic Society. *Tropics. National Geographic*; 2022 [cited 2 February 2023]; Available from: https://education.nationalgeographic.org/resource/tropics
2. World Population Review. *Tropical Countries* 2023. https://worldpopulationreview.com/country-rankings/tropical-countries; 2023 [cited 2 February 2023]; Available from: https://worldpopulationreview.com/country-rankings/tropical-countries
3. Beck HE, McVicar TR, Vergopolan N, Berg A, Lutsko NJ, Dufour A, et al. High-resolution (1 km) Köppen-Geiger maps for 1901–2099 based on constrained CMIP6 projections. *Scientific Data*. 2023;10(1):724.
4. Susan L, Foy A. *Tropical Savannas*. Radford University; 2022 [cited 2 February 2023]; Available from: https://php.radford.edu/~swoodwar/biomes/?page_id=105
5. Smith JMB. *Tropical Rainforest*. 2022 [cited 2 February 2023]; Available from: https://www.britannica.com/science/tropical-rainforest
6. Jaganmoham M. *Monthly Rainfall in the United Kingdom 2014-2022*. Statista; 2022 [cited 2 February 2023]; Available from: https://www.statista.com/statistics/584914/monthly-rainfall-in-uk/
7. Macreadie P, Baird M, Trevathan-Tackett S, Larkum A, Ralph P. Quantifying and modelling the carbon sequestration capacity of seagrass meadows–a critical assessment. *Marine Pollution Bulletin*. 2014;83(2):430–9.
8. NOAA. Inter-Tropical Convergence Zone. *National Oceanic and Atmospheric Administration (NOAA), US Department of Commerce*; 2022 [cited 2 February 2023]; Available from: https://www.noaa.gov/jetstream/tropical/convergence-zone#:~:text=Therefore%2C%20the%20ITCZ%20is%20responsible,makes%20for%20two%20dry%20seasons
9. Met Office. *Intertropical Convergence Zone (ITCZ)*. Met Office UK; 2022 [cited 2 February 2023]; Available from: https://www.metoffice.gov.uk/weather/learn-about/weather/atmosphere/intertropical-convergence-zone#:~:text=The%20Intertropical%20Convergence%20Zone%2C%20or,frequent%20thunderstorms%20and%20heavy%20rain
10. United Nations. The World Cities in 2018. *United Nations*; 2018 [cited 31 January 2023]; Available from: https://www.un.org/en/development/desa/population/publications/pdf/urbanization/the_worlds_cities_in_2018_data_booklet.pdf
11. UN@Habitat. *UN-Habitat and Partners Unveil OCEANIX Busan, the World's First Prototype Floating City. UN@Habitat*; 2022 [cited 31 January 2023]; Available from: https://unhabitat.org/news/27-apr-2022/un-habitat-and-partners-unveil-oceanix-busan-the-worlds-first-prototype-floating
12. United Nations Development Programme. Human Development Report 2021-22: Uncertain Times, Unsettled Lives: Shaping Our Future in a Transforming World: UNDP; 2022.

13. Waterstudio. *Maldives Floating City, A Benchmark for Vibrant Communities*. Waterstudio; 2022 [cited 31 January 2023]; Available from: https://www.waterstudio.nl/projects/maldives-floating-city-a-benchmark-for-vibrant-communities-beyond-the-waterfront/
14. Ritchie H, Roser M. *Urbanization. Our World in Data*; 2018 [cited 31 January 2023]; Available from: https://ourworldindata.org/urbanization
15. James Cook University. *The Future of Tropical Cities*. 2016 [cited 10 Oct 2024]; Available from: https://habitat3.org/wp-content/uploads/Side-Event-Report-Submitted-by-Lead-Organization-The-Future-of-Tropical-Cities.pdf
16. Thornton A. 10 Cities are Predicted to Gain Megacity Status by 2030. *World Economic Forum*; 2019 [cited 1 February 2023].
17. Kolesnikov-Jessop S. Singapore Looks to Store Oil in Caverns Under the Sea. *The New York Times*; 2011 [cited 1 February 2023]; Available from: https://www.nytimes.com/2011/02/21/business/energy-environment/21iht-renoilstore21.html
18. Othman L. World's Biggest Underground District Cooling Network Now at Marina Bay. *Today*; 2016 [cited 1 February 2023]; Available from: https://www.todayonline.com/singapore/plant-underground-district-cooling-network-marina-bay-commissioned#:~:text=The%20underground%20centralised%20system%20has,cooling%20towers%E2%80%9D%2C%20SP%20said
19. Seneviratne K. Farming in the Sky in Singapore. *Our World*; 2012 [cited 1 February 2023]; Available from: https://ourworld.unu.edu/en/farming-in-the-sky-in-singapore
20. Liu V. Nine Multi-storey Carpark Rooftops in Singapore to be Converted to Urban Farms. *The Straits Times*; 2020 [cited 1 February 2022]; Available from: https://www.straitstimes.com/singapore/nine-multi-storey-carpark-rooftops-in-singapore-to-be-converted-to-urban-farms
21. Powell MA. Singapore's Lost Coast: Land Reclamation, National Development and the Erasure of Human and Ecological Communities, 1822-Present. *Environment and History*. 2021;27(4):635–63.
22. Heller P. Walking *On Water: The Significance of Land Reclamation in Hong Kong and Singapore*. Forbes; 2021 [cited 1 February 2023]; Available from: https://www.forbes.com/sites/philipheller/2021/05/26/walking-on-water-the-significance-of-land-reclamation-in-hong-kong-and-singapore/?sh=321cbbdd4ef2
23. United Nations. *Global Population Growth and Sustainable Development*. United Nations. 2021.
24. IPCC. *Climate Change 2013: The Physical Science Basis. Contribution of Working Group I to the Fifth Assessment Report of the Intergovernmental Panel on Climate Change*. United Kingdom and New Work, NY, USA: Cambridge University Press; 2013.
25. Center for Liveable Cities, Urban Land Institute. *Building Climate Resilience in Cities Worldwide*. Singapore: Center for Liveable Cities (CLC) and Urban Land Institute (ULI); 2020.
26. Nicholls R, Hanson S, Herweijer C, Patmore N, Hallegatte S, Corfee-Morlot J, et al. *Ranking of the World's Cities Most Exposed to Coastal Flooding Today and in the Future*. Paris: Organisation for Economic Co-operation and Development (OECD). 2007.

27. World Energy Council. *World Energy Resources: Marine Energy*. London, UK: World Energy Council; 2016.
28. Salter S. *Progress on Edinburgh Ducks. Hydrodynamic of Ocean Wave-Energy Utilisation: IUTAM Symposium Lisbon/Portugal 1985*. Berlin Heidelberg: Springer; 1986. 35–50.
29. Tay ZY, Venugopal V. Optimization of spacing for oscillating wave surge converter arrays using genetic algorithm. *Journal of Waterway, Port, Coastal, and Ocean Engineering*. 2017;143(2):04016019.
30. Tay ZY, Venugopal V. Hydrodynamic interactions of oscillating wave surge converters in an array under random sea state. *Ocean Engineering*. 2017;145:382–94.
31. Tay ZY, Venugopal V. The impact of energy extraction of wave energy converter arrays on wave climate under multi-directional seas. *Journal of Ocean Engineering and Marine Energy*. 2019;5:51–72.
32. Falnes J. A review of wave-energy extraction. *Marine Structures*. 2007;20(4):185–201.
33. Aderinto T, Li H. Ocean wave energy converters: Status and challenges. *Energies*. 2018;11(5):1250.
34. Drew B, Plummer AR, Sahinkaya MN. *A Review of Wave Energy Converter Technology*. London, England: Sage Publications Sage UK; 2009.
35. WaveRoller. AW-Energy Winning Project Secures Funding: AW-Energy's Technical Expertise is at the Forefront in an Industry Initiative to Unlock Future Multi-GW Wavefarms. AW-Energy; 2-2-. n.d. [cited 1 February 2023]; Available from: https://aw-energy.com/winning-project-secures-funding-aw-energys-technical-expertise-is-at-the-forefront-in-an-industry-initiative-to-unlock-future-multi-gw-wavefarms/
36. CorPower. *CorPower Extends Equity Funding to 20.3 MEUR for Commercial Scale Demonstration. CorPower Ocean*; 2021 [cited 1 February 2023]; Available from: https://corpowerocean.com/corpower-extends-equity-funding-to-20-3-meur-for-commercial-scale-demonstration/#:~:text=After%20securing%20a%20private%20placement,of%20its%20wave%20energy%20technology.&text=The%20investment%20round%20was%20led,family%20offices%20and%20private%20investors
37. Scotsman T. How can Scotland become a world leader in wave energy?: *The Scotsman*; 2020 [cited 1 February]; Available from: https://www.scotsman.com/news/environment/how-can-scotland-become-world-leader-wave-energy-2912603
38. Babarit A, Duclos G, Clément AH. Comparison of latching control strategies for a heaving wave energy device in random sea. *Applied Ocean Research*. 2004;26(5):227–38.
39. Cargenie Clean Energy. *CETO Technology*. 2019 [cited 31 January 2021]; Available from: https://www.carnegiece.com/technology/
40. Whittaker T, Collier D, Folley M, Osterried M, Henry A, Crowley M, editors. The development of Oyster—a shallow water surging wave energy converter. *Proceedings of the 7th European Wave and Tidal Energy Conference*; 2007.
41. Haren P, Mei CC. Wave power extraction by a train of rafts: Hydrodynamic theory and optimum design. *Applied Ocean Research*. 1979;1(3):147–57.
42. Yemm R, Pizer D, Retzler C, Henderson R. Pelamis: Experience from concept to connection. *Philosophical Transactions of the Royal Society A: Mathematical, Physical and Engineering Sciences*. 2012;370(1959):365–80.

43. Rendell D. *Waves and Water Dynamics*. Pearson Publication Inc.; 2011 [cited 2 February 2023]; Available from: https://slideplayer.com/user/5750528/
44. Felix A.V. Hernández-Fontes J, Lithgow D, Mendoza E, Posada G, Ring M, et al. Wave energy in tropical regions: Deployment challenges, environmental and social perspectives. *Journal of Marine Science and Engineering*. 2019;7(7):219.
45. Sánchez AS, DeJong P, Torres EA. Wind, solar and waves' energy in the Ne region of Brazil: analysis of meteorological parameters related to the load curve of electric demand (In Portuguese). CONEM 2012-7th Brazilian National Congress of Mechanical Engineering; São Luis, Maranhão, Brazil.2012.
46. Pillai IR, Banerjee R. Renewable energy in India: Status and potential. *Energy*. 2009;34(8):970–80.
47. Muzathik AM. *Wave Energy Potential of Penisular Malaysia*. Terengganu, Malaysia: Universiti Malaysia Terengganu, 2010.
48. Babarit A. A database of capture width ratio of wave energy converters. *Renewable Energy*. 2015;80:610–28.
49. Patel RP, Nagababu G, Kumar SVA, Seemanth M, Kachhwaha SS. Wave resource assessment and wave energy exploitation along the Indian coast. *Ocean Engineering*. 2020;217:107834.
50. Kumar VS, Anoop T. Wave energy resource assessment for the Indian shelf seas. *Renewable Energy*. 2015;76:212–9.
51. IGB Staff. Ocean Wave Energy: A Significant Renewable Energy and Manufacturing Opportunity for India. India Global Business (IGB); 2020 [cited 1 February 2023]; Available from: https://www.indiaglobalbusiness.com/guestcolumns/ocean-wave-energy-a-significant-renewable-energy-and-manufacturing-opportunity-for-india
52. Jiang B, Wu G, Ding J, Ma C, Fang Y, Wang X, editors. Assessment of the wave energy resource in the South China Sea. *Proceedings of the Institution of Civil Engineers-Maritime Engineering*; 2019: Thomas Telford Ltd.
53. Zheng C-w, Pan J, Li J-x. Assessing the China Sea wind energy and wave energy resources from 1988 to 2009. *Ocean Engineering*. 2013;65:39–48.
54. Yaakob O, Hashim FE, Omar KM, Din AHM, Koh KK. Satellite-based wave data and wave energy resource assessment for South China Sea. *Renewable Energy*. 2016;88:359–371.
55. IEA. *Global Energy Demand Rose by 2.3% in 2018, its Fastest Pace in the Last Decade*. International Energy Agency; 2019 [cited 1 February 2023]; Available from: https://www.iea.org/news/global-energy-demand-rose-by-23-in-2018-its-fastest-pace-in-the-last-decade
56. IISD. *Global Energy Demand in 2018 Grew at Fastest Pace in a Decade*. The International Institute for Sustainable Development (IISD); 2019 [cited 1 February 2023]; Available from: https://sdg.iisd.org/news/global-energy-demand-in-2018-grew-at-fastest-pace-in-a-decade/#:~:text=26%20March%20 2019%3A%20A%20report,in%20energy%2Drelated%20CO2%20emissions
57. Cozzi L, Gould T, Bouckart S, Crow D, Kim T, Mcglade C, et al. World energy outlook 2020. *International Energy Agency: Paris, France*. 2020:1–461.
58. Dudley B. BP statistical review of world energy 2016. *British Petroleum Statistical Review of World Energy*, Bplc editor, UK: Pureprint Group Limited. 2019.
59. Olivier JG, Schure K, Peters J. Trends in global CO_2 and total greenhouse gas emissions. *PBL Netherlands Environmental Assessment Agency*. 2017;5:1–11.

60. EPA. Inventory of U.S. *Greenhouse Gas Emissions and Sinks 1999–2020*. Washington DC, US: United States Environmental Protection Agency (EPA)2022.
61. IEA. *Global Energy Review: CO_2 Emissions in 2021*. Paris, France: International Energy Agency 2022.
62. United Nations. *Transforming our world: the 2030 Agenda for Sustainable Development*. United Nations: New York, NY, USA. 2015.
63. IMDC. *Se@Ports: Wave Energy Converters in Port Infrastructure*. IMDC; 2018 [cited 1 February 2023]; Available from: https://imdc.be/en/reference/se-ports-wave-energy-converters-in-port-infrastructure
64. Eco Wave Power. *Eco Wave Power Commences Real Conditions Test Runs of its Newest EWP-EDF One Wave Energy Project at Jaffa Port, Israel*. Eco Wave Power; 2022 [cited 1 February 2023]; Available from: https://www.ecowavepower.com/eco-wave-power-commences-real-conditions-test-runs-of-its-newest-ewp-edf-one-wave-energy-project-at-jaffa-port-israel/
65. Cheng Y, Dai S, Dai S, Ji C, Collu M, Yuan Z, et al. Energy conversion and hydrodynamic analysis of multi-degree-of-freedom wave energy converters integrated into a semi-submersible platform. *Energy Conversion and Management*. 2022;252:115075.
66. Tay ZY, Htoo NL, Konovessis D. A Comparison of the Capture Width and Interaction Factors of WEC Arrays That Are Co-Located with Semi-Submersible-, Spar- and Barge-Supported Floating Offshore Wind Turbines. *Journal of Marine Science and Engineering*. 2024; 12(11):2019. https://doi.org/10.3390/jmse12112019
67. Karimirad M, Koushan K, editors. WindWEC: Combining wind and wave energy inspired by hywind and wavestar. *2016 IEEE International Conference on Renewable Energy Research and Applications (ICRERA)*; 2016: IEEE.
68. Blue21. Space@Sea. Blue21; 2019 [cited 3 February 2023]; Available from: https://www.blue21.nl/portfolio/spaceatsea/
69. Nguyen HP, Wang CM, Tay ZY, Luong VH. Wave energy converter and large floating platform integration: A review. *Ocean Engineering*, 213: 107768.
70. Wang CM, Tay ZY, Takagi K, Utsunomiya T. Literature review of methods for mitigating hydroelastic response of VLFS under wave action. *Applied Mechanics Reviews*. 2010;63(3).
71. Zheng K, Sun J, Guan C, Shao W. Analysis of the global swell and wind sea energy distribution using WAVEWATCH III. *Advances in Meteorology*. 2016;2016:1–9.
72. Caribbean Tsunami Information Center. *What Causes Tsunamis. Caribbean Tsunami Information Center*; [cited 21 March 2023].
73. Babanin A. *Breaking and Dissipation of Ocean Surface waves*: Cambridge University Press; 2011.
74. Denny M. *Air and Water: The Biology and Physics of Life's Media*: Princeton, New Jersey: Princeton University Press; 1993.
75. Dean RG, Dalrymple RA. *Water Wave Mechanics for Engineers and Scientists: World Scientific Publishing Company*; Singapore: World Scientific Publishing Company, 1991.
76. Rabinovich AB. Seiches and harbor oscillations. *Handbook of Coastal and Ocean Engineering*: Singapore: World Scientific; 2010, 193–236.
77. Earthquake Hazards Group. Seismic Seiches: Definition of Seismic Seiches and Examples Dating Back to Their First Public Mention in 1755.: USGS;

[cited 22 March 2023]; Available from: https://www.usgs.gov/programs/earthquake-hazards/seismic-seiches
78. Merian J *On the Motion of Drippable Liquids in Containers: Thesis (in German).* Basel: Schweighauser; 1828.
79. Le Méhauté B. *An Introduction to Hydrodynamics and Water Waves*: Berlin, Germany: Springer Science & Business Media; 2013.
80. Zhao K, Wang Y, Liu PL-F. A guide for selecting periodic water wave theories-Le Méhauté (1976)'s graph revisited. *Coastal Engineering*. 2024;188:104432.
81. Faltinsen O. *Sea Loads on Ships and Offshore Structures*: Cambridge, United Kingdom: Cambridge University Press; 1993.
82. Goda Y. *Random Seas and Design of Maritime Structures.* Singapore: World Scientific Publishing Company; 2010.
83. Pierson Jr WJ, Moskowitz L. A proposed spectral form for fully developed wind seas based on the similarity theory of SA Kitaigorodskii. *Journal of Geophysical Research*. 1964;69(24):5181–90.
84. Hasselmann K, Barnett TP, Bouws E, Carlson H, Cartwright DE, Enke K, et al. Measurements of wind-wave growth and swell decay during the Joint North Sea Wave Project (JONSWAP). *Ergaenzungsheft zur Deutschen Hydrographischen Zeitschrift, Reihe A*. 1973.
85. Ochi MK, Hubble EN. Six-parameter wave spectra. *Coastal Engineering*. 1976;1(15):301–28.
86. Duarte T, Gueydon S, Jonkman J, Sarmento A. Computation of wave loads under multidirectional sea states for floating offshore wind turbines. *ASME 2014 33rd International Conference on Ocean, Offshore and Arctic Engineering: American Society of Mechanical Engineers*; 2014. pp. V09BTA023–V09BT09A.
87. John F. On the motion of floating bodies. *I. Communications on Pure and Applied Mathematics*. 1949;2(1):13–57.
88. John F. On the motion of floating bodies II. Simple harmonic motions. *Communications on Pure and Applied Mathematics*. 1950;3(1):45–101.
89. Wehausen JV, Laitone EV. *Surface Waves*: Berlin, Germany: Springer; 1960.
90. Newman JN. *Marine Hydrodynamics*: Cambridge, MA: The MIT Press; 2018.
91. Newman JN. Algorithms for the free-surface Green function. *Journal of Engineering Mathematics*. 1985;19(1):57–67.
92. Newman JN. Wave effects on deformable bodies. *Applied Ocean Research*. 1994;16(1):47–59.
93. Noblesse F. Potential theory of steady motion of ships: Massachusetts Institute of Technology, Department of Ocean Engineering; 1978.
94. Chakrabarti SK. *Hydrodynamics of Offshore Structures*: Southampton, United Kingdom: WIT press; 1987.
95. Mei CC. *The Applied Dynamics of Ocean Surface Waves*: Singapore: World Scientific; 1989.
96. Bishop RED, Bishop RE, Price W. *Hydroelasticity of Ships*: Cambridge, United Kingdom: Cambridge University Press; 1979.
97. Ertekin RC, Riggs HR, Che XL, Du SX. Efficient methods for hydroelastic analysis of very large floating structures. *Journal of Ship Research*. 1993;37(01):58–76.
98. Suzuki H, Yoshida K, editors. Design flow and strategy for safety of very large floating structure. *Proceedings of Int Workshop on Very Large Floating Structures*, VLFS; 1996.

99. Suzuki H. Overview of Megafloat: Concept, design criteria, analysis, and design. *Marine Structures*. 2005;18(2):111–32.
100. Yago K, Endo H. On the hydoroelastic response of box-shaped floating structure with shallow draft tank test with large scale model. *Journal of the Society of Naval Architects of Japan*. 1996;1996(180):341–52.
101. Kashiwagi M. A B-spline Galerkin scheme for calculating the hydroelastic response of a very large floating structure in waves. *Journal of Marine Science and Technology*. 1998;3:37–49.
102. Kashiwagi M. Research on hydroelastic responses of VLFS: Recent progress and future work. *International Journal of Offshore and Polar Engineering*. 2000;10(02):81–90.
103. Utsunomiya T, Watanabe E, Taylor RE, editors. Wave response analysis of a box-like VLFS close to a breakwater. *Proceedings of the 1998 17th International Conference on Offshore Mechanics and Arctic Engineering*, OMAE; 1998: ASME.
104. Ohmatsu S, editor. Numerical calculation of hydroelastic behavior of a pontoon-type VLFS in waves. *Proceedings of the 17th International Conference on Offshore Mechanics and Arctic Engineering*; 1998.
105. Ohmatsu S. Numerical calculation method for the hydroelastic response of a pontoon-type very large floating structure close to a breakwater. *Journal of Marine Science and Technology*. 2000;5:147–60.
106. Meylan MH, Squire VA. Response of a circular ice floe to ocean waves. *Journal of Geophysical Research: Oceans*. 1996;101(C4):8869–84.
107. Meylan MH. The forced vibration of a thin plate floating on an infinite liquid. *Journal of Sound and Vibration*. 1997;205(5):581–91.
108. Meylan MH. A variational equation for the wave forcing of floating thin plates. *Applied Ocean Research*. 2001;23(4):195–206.
109. Inman DJ. *Engineering Vibration*. Upper Saddle. NJ: Pearson Education Inc; 2008.
110. Newman JN. The theory of ship motions. *Advances in Applied Mechanics*. 1979;18:221–83.
111. Cruz J, Sykes R, Siddorn P, Taylor RE. Estimating the loads and energy yield of arrays of wave energy converters under realistic seas. *IET Renewable Power Generation*. 2010;4(6):488–97.
112. Venugopal V, Tay ZY, Nemalidinne TR. Numerical Modelling Techniques for Wave Energy Converters in Arrays. *Ocean Wave Energy Systems: Hydrodynamics, Power Takeoff and Control Systems*: Cham, Switzerland: Springer; 2021. pp. 281–322.
113. Budal K. Theory for absorption of wave power by a system of interacting bodies. *Journal of Ship Research*. 1977;21(4):248–253.
114. Green G. An essay on the application of mathematical analysis to the theories of electricity and magnetism. *für die reine und angewandte Mathematik (Crelle's Journal)* 1889;39:73–89.
115. Cannell DM, Lord N. George Green, mathematician and physicist 1793–1841. *The Mathematical Gazette*. 1993;77(478):26–51.
116. Lamb H. *Hydrodynamics*: Cambridge, United Kingdom: Cambridge University Press; 1924.
117. Linton CM. Rapidly convergent representations for Green'functions for Laplace'equation. *Proceedings of the Royal Society of London Series A: Mathematical, Physical and Engineering Sciences*. 1999;455(1985):1767–97.

118. Haskind L. The exciting forces and moments on floating bodies. *Transactions of the Royal Institution of Naval Architects*. 1957;99:314–7.
119. Wang CM and Tay ZY, 2010. Hydroelastic analysis and response of pontoon-type very large floating structures. *Fluid Structure Interaction II: Modelling, Simulation, Optimization*: Berlin, Heidelberg: Springer Berlin Heidelberg; 2010, pp. 103–30.
120. Kim JW, Webster WC. The drag on an airplane taking off from a floating runway. *Journal of Marine Science and Technology*. 1998;3:76–81.
121. Watanabe E, Utsunomiya T, Wang C. Hydroelastic analysis of pontoon-type VLFS: A literature survey. *Engineering Structures*. 2004;26(2):245–56.
122. Kashiwagi M. Transient responses of a VLFS during landing and take-off of an airplane. *Journal of Marine Science and Technology*. 2004;9:14–23.
123. Suzuki H, Yasuzawa Y, Fujikubo M, Okada S, Endo H, Hattori Y, et al. Structural response and design of large scale floating structure. *Proceedings of the 1997 16th International Conference on Offshore Mechanics and Arctic Engineering Part 1-B (of 6)*; 13–17 April 1997; Yokohama, Japan: ASME; 1997. pp. 131–7.
124. Mindlin R. Influence of rotatory inertia and shear on flexural motions of isotropic, elastic plates. 1951.
125. Liew K-M, Xiang Y, Kitipornchai S, Wang C. *Vibration of Mindlin Plates: Programming the p-version Ritz Method*: Oxford, United Kingdom: Elsevier; 1998.
126. Reddy JN. *Theory and Analysis of Elastic Plates and Shells*: Florida, United States: CRC Press; 2006.
127. Sarpkaya T, Isaacson M, Wehausen J. *Mechanics of Wave Forces on Offshore Structures*. New York City, New York: Van Nostrand Reinhold, 1981.
128. Taylor RE, Waite J. The dynamics of offshore structures evaluated by boundary integral techniques. *International Journal for Numerical Methods in Engineering*. 1978;13(1):73–92.
129. Mamidipudi P, Webster W. The motions performance of a mat-like floating airport. *Hydro-elasticity in Marine Technology*: Trondheim, Norway: Routledge; 2022. pp. 363–75.
130. Wu C, Watanabe E, Utsunomiya T. An eigenfunction expansion-matching method for analyzing the wave-induced responses of an elastic floating plate. *Applied Ocean Research*. 1995;17(5):301–10.
131. Nagata S, Yoshida H, Fujita T, Isshiki H, editors. The analysis of wave-induced response of an elastic floating plate in a sea with a breakwater. *ISOPE International Ocean and Polar Engineering Conference*; 1998: Montreal, Canada: ISOPE.
132. Petyt M. *Introduction to Finite Element Vibration Analysis*: Cambridge, United Kingdom: Cambridge University Press; 2010.
133. Cook RD. *Concepts and Applications of Finite Element Analysis*: John wiley & sons; 2007.
134. Brebbia CA, Telles JCF, Wrobel LC. *Boundary Element Techniques: Theory and Applications in Engineering*: Springer Science & Business Media; 2012.
135. Brebbia CA, Walker S. *Boundary Element Techniques in Engineering*: Amsterdam, Netherlands: Elsevier; 2016.
136. Linton C. Rapidly convergent representations for Green'functions for Laplace'equation. *Proceedings of the Royal Society of London Series A: Mathematical, Physical and Engineering Sciences*. 1999;455(1985):1767–97.

137. Tay ZY, Wang CD and Wang CM. Hydroelastic response of a box-like floating fuel storage module modeled using non-conforming quadratic-serendipity Mindlin plate element. *Engineering Structures*. 2007;29(12):3503–14.
138. Alam M-R. Nonlinear analysis of an actuated seafloor-mounted carpet for a high-performance wave energy extraction. *Proceedings of the Royal Society A: Mathematical, Physical and Engineering Sciences*. 2012;468(2146):3153–71.
139. Chaplin JR, Farley FJM, Prentice ME, Rainey RCT, Rimmer SJ, Roach AT. Development of the ANACONDA all-rubber WEC. *Proceedings of the 7th European Wave and Tidal Energy Conference (EWTEC 2007)*, Page: 1–8, Porto, Portugal, Conference: European Wave and Tidal Energy Conference, 2007.
140. Jean P, Wattez A, Ardoise G, Melis C, Van Kessel R, Fourmon A, et al., editors. Standing wave tube electro active polymer wave energy converter. *Electroactive Polymer Actuators and Devices (EAPAD) 2012*; 2012: Washington, United States: SPIE.
141. Algie C, Ryan S, Fleming A. Predicted power performance of a submerged membrane pressure-differential wave energy converter. *International Journal of Marine Energy*. 2017;20:125–34.
142. Michailides C. Power performance and dynamic response of the WLC wave energy converter based on hydroelastic analysis. *International Journal of Marine Energy*. 2017;19:83–94.
143. Zhang H, Xu D, Zhao H, Xia S, Wu Y. Energy extraction of wave energy converters embedded in a very large modularized floating platform. *Energy*. 2018;158:317–29.
144. Kahfi K. Indonesia has lost land equal to size of Jakarta in last 15 years. *TheJakartaPost*; 2019 [cited 25 March 2020]; Available from: https://www.thejakartapost.com/news/2019/07/07/indonesia-has-lost-land-equal-to-size-of-jakarta-in-last-15-years.html
145. Headland JR, Alfageme S, Smith E, Kotulak P. Coastal Structure Design for Shore Protection and Sand Retention: Practical Aspects. *Coastal Sediments'* 07. 2007:2432–45.
146. Schoonees T, Mancheño AG, Scheres B, Bouma TJ, Silva R, Schlurmann T, et al. Hard structures for coastal protection, towards greener designs. *Estuaries and Coasts*. 2019;42(7):1709–29.
147. Pilarczyk KW, Press C. *Coastal Protection*. The Netherlands: AA Balkema Rotterdam; 1990.
148. Shin E, Oh Y. Coastal erosion prevention by geotextile tube technology. *Geotextiles and Geomembranes*. 2007;25(4–5):264–77.
149. Hornsey W, Carley J, Coghlan I, Cox R. Geotextile sand container shoreline protection systems: Design and application. *Geotextiles and Geomembranes*. 2011;29(4):425–39.
150. Van Rijn LC. Coastal erosion and control. *Ocean & Coastal Management*. 2011;54(12):867–87.
151. Williams A, Geiger P, McDougal W. Flexible floating breakwater. *Journal of Waterway, Port, Coastal, and Ocean Engineering*. 1991;117(5):429–50.
152. Drimer N, Agnon Y, Stiassnie M. A simplified analytical model for a floating breakwater in water of finite depth. *Applied Ocean Research*. 1992;14(1):33–41.
153. Hales LZ. *Floating Breakwaters: State-of-the-Art Literature Review*. Springfield, Virginia: Coastal Engineering Research Center, US Army Corps of Engineers, 1981.

154. Dai J, Wang CM, Utsunomiya T, Duan W. Review of recent research and developments on floating breakwaters. *Ocean Engineering*. 2018;158:132–51.
155. McCartney B. Floating breakwater design. *Journal of Waterway, Port, Coastal and Ocean Engineering*. 1985;111(2):304–18.
156. Sun B, Zhang H, Li C, Li Z. Research progress and prospect of wave attenuation performance and integration of wave-energy converter of floating breakwater. *Journal of Environmental Engineering*. 2023;149(4):03123001.
157. Howe D, Nader J-R, Macfarlane G. Experimental investigation of multiple oscillating water column wave energy converters integrated in a floating breakwater: energy extraction performance. *Applied Ocean Research*. 2020;97:102086.
158. He F, Huang Z, Law AW-K. An experimental study of a floating breakwater with asymmetric pneumatic chambers for wave energy extraction. *Applied Energy*. 2013;106:222–31.
159. Zhao X, Ning D, Zou Q, Qiao D, Cai S. Hybrid floating breakwater-WEC system: A review. *Ocean Engineering*. 2019;186:106126.
160. Ning D, Zhao X, Göteman M, Kang H. Hydrodynamic performance of a pile-restrained WEC-type floating breakwater: An experimental study. *Renewable Energy*. 2016;95:531–41.
161. Martinelli L, Ruol P, Favaretto C, editors. Hybrid structure combining a wave energy converter and a floating breakwater. *The 26th International Ocean and Polar Engineering Conference*; 2016 June 2016; Rhodes, Greece: OnePetro.
162. Cabral T, Clemente D, Rosa-Santos P, Taveira-Pinto F, Morais T, Belga F, et al. Performance assessment of a hybrid wave energy converter integrated into a harbor breakwater. *Energies*. 2020;13(1):236.
163. Chen M, Xiao P, Zhang Z, Sun L, Li F. Effects of the end-stop mechanism on the nonlinear dynamics and power generation of a point absorber in regular waves. *Ocean Engineering*. 2021;242:110123.
164. Favaretto C, Martinelli L, Ruol P, Cortellazzo G, editors. Investigation on possible layouts of a catamaran floating breakwater behind a wave energy converter. *The 27th International Ocean and Polar Engineering Conference*; 2017 June 2017; San Francisco, California, USA: OnePetro.
165. Mustapa MA, Yaakob OB, Ahmed YM, Rheem C-K, Koh KK, Adnan FA. Wave energy device and breakwater integration: A review. *Renewable and Sustainable Energy Reviews*. 2017;77:43–58.
166. Michailides C, Angelides DC. Modeling of energy extraction and behavior of a Flexible Floating Breakwater. *Applied Ocean Research*. 2012;35:77–94.
167. Tay ZY. Energy generation from anti-motion device of very large floating structure. *Proceedings of the Twelfth European Wave and Tidal Energy Conference*; 2017. Cork, Ireland.
168. Tay ZY. Energy extraction from an articulated plate anti-motion device of a very large floating structure under irregular waves. *Renewable Energy*. 2019;130:206–22.
169. Wilkinson L, Whittaker T, Thies P, Day S, Ingram D. The power-capture of a nearshore, modular, flap-type wave energy converter in regular waves. *Ocean Engineering*. 2017;137:394–403.
170. Noad IF, Porter R. Modelling an articulated raft wave energy converter. *Renewable Energy*. 2017;114:1146–59.

171. Nguyen HP, Wang CM, Flocard F, Pedroso DM. Extracting energy while reducing hydroelastic responses of VLFS using a modular raft wec-type attachment. *Applied Ocean Research*. 2019;84:302–16.
172. Liu C, Yang Q, Bao G. Performance investigation of a two-raft-type wave energy converter with hydraulic power take-off unit. *Applied Ocean Research*. 2017;62:139–55.
173. Tay ZY. Performance and wave impact of an integrated multi-raft wave energy converter with floating breakwater for tropical climate. *Ocean Engineering*. 2020;218:108136.
174. Tay ZY, Lee L. Performance of integrated vertical raft-type WEC and floating breakwater. *Ocean Systems Engineering*. 2022;12(1):39–61.
175. Cascajo R, García E, Quiles E, Correcher A, Morant F. Integration of marine wave energy converters into seaports: A case study in the port of Valencia. *Energies*. 2019;12(5):787.
176. Contestabile P, Ferrante V, Di Lauro E, Vicinanza D. Full-scale prototype of an overtopping breakwater for wave energy conversion. *Coastal Engineering Proceedings* 2017;1(35):12.
177. SINN Power. *Wave Energy Converter Installation in Heraklion: First Preparation Phase Successfully Completed*. SINN Power; 2021 [cited 4 May 2023]; Available from: https://www.sinnpower.com/post/2018/06/11/wave-energy-converter-installation-in-heraklion-first-preparation-phase-successfully-comp
178. Tethys. Pico Power Plant. Tethys; [cited]; Available from: https://tethys.pnnl.gov/project-sites/pico-power-plant
179. Ghisu G, Carabotta P. *Ocean Energy exploitation in Italy: Ongoing R&D activities*. National Agency for New Technologies, Energy and Sustainable Economic Development (ENEA), Rome, Italy.
180. Tay ZY, Wang CM, Utsunomiya T. Hydroelastic responses and interactions of floating fuel storage modules placed side-by-side with floating breakwaters. *Marine Structures*. 2009;22(3):633–58.
181. MacCamy R, Fuchs RA. *Wave Forces on Piles: A Diffraction Theory*: Washington, United States: US Beach Erosion Board; 1954.
182. Dong GH, Zheng YN, Li YC, Teng B, Guan CT, Lin DF. Experiments on wave transmission coefficients of floating breakwaters. *Ocean Engineering*. 2008;35(8–9):931–8.
183. Elchahal G, Younes R, Lafon P. The effects of reflection coefficient of the harbour sidewall on the performance of floating breakwaters. *Ocean Engineering*. 2008;35(11–12):1102–12.
184. Chen Z-j, Wang Y-x, Dong H-y, Zheng B-x. Time-domain hydrodynamic analysis of pontoon-plate floating breakwater. *Water Science and Engineering*. 2012;5(3):291–303.
185. Baldwin S, Bindewald G, Brown A, Chen C, Cheung K, Clark C, et al. *Quadrennial Technology Review: An Assessment of Energy Technologies and Research Opportunities*. Washington, DC, USA: US Department of Energy 2015.
186. Nguyen HP, Wang C, Tay ZY, Luong VH. Wave energy converter and large floating platform integration: A review. *Ocean Engineering*. 2020;213:107768.

187. Zhang H, Zhou B, Vogel C, Willden R, Zang J, Zhang L. Hydrodynamic performance of a floating breakwater as an oscillating-buoy type wave energy converter. *Applied Energy*. 2020;257:113996.
188. Zhang Y, Li M, Zhao X, Chen L. The effect of the coastal reflection on the performance of a floating breakwater-WEC system. *Applied Ocean Research*. 2020;100:102117.
189. Liu H, Yan F, Jing F, Ao J, Han Z, Kong F. Numerical and experimental investigation on a moonpool-buoy wave energy converter. *Energies*. 2020;13(9):2364.
190. Tay ZY. Effect of resonance and wave reflection in semi-enclosed moonpool on performance enhancement of point absorber arrays. *Ocean Engineering*. 2022;243:110182.
191. Lee C-H, Newman JN. *WAMIT User Manual*. Massachusetts, United States: WAMIT, Inc. 2006.
192. Molin B, Zhang X, Huang H, Remy F. On natural modes in moonpools and gaps in finite depth. *Journal of Fluid Mechanics*. 2018;840:530–54.
193. Wang CY, Wang CM. *Structural Vibration: Exact Solutions for Strings, Membranes, Beams, and Plates*: Florida, United States: CRC Press; 2013.
194. Molin B, Remy F, Camhi A, Ledoux A, editors. Experimental and numerical study of the gap resonances in-between two rectangular barges. *Proceedings of the 13th Congress of the International Maritime Association of the Mediterranean (IMAM 2009)*; 2009. Istanbul, Turkey.
195. Molin B, Remy F, Kimmoun O, Stassen Y. Experimental study of the wave propagation and decay in a channel through a rigid ice-sheet. *Applied Ocean Research*. 2002;24(5):247–60.
196. Garrett CJR. Wave forces on a circular dock. *Journal of Fluid Mechanics*. 1971;46(1):129–39.
197. Holland JH. *Adaptation in Natural and Artificial Systems: An Introductory Analysis with Applications to Biology, Control, and Artificial Intelligence*: Massachusetts, United States: MIT press; 1992.
198. Goldberg DE, Deb K. *A Comparative Analysis of Selection Schemes Used in Genetic Algorithms: Foundations of Genetic Algorithms*: Elsevier; 1991. pp. 69–93.
199. Tay ZY, Wei Y. Power enhancement of pontoon-type wave energy convertor via hydroelastic response and variable power take-off system. *Journal of Ocean Engineering and Science*. 2019;5(1):1–18.
200. Farret FA, Simões MG. *Integration of Alternative Sources of Energy*: New Jersey, United States: Wiley Online Library; 2006.
201. Hertwich E, de Larderel JA, Arvesen A, Bayer P, Bergesen J, Bouman E, et al. *Green Energy Choices: The benefits, risks, and trade-offs of low-carbon technologies for electricity production*. Paris, France: United Nations Environment Programme (UNEP), 2016.
202. Kabeyi MJB, Olanrewaju OA. Sustainable energy transition for renewable and low carbon grid electricity generation and supply. *Frontiers in Energy Research*. 2022;9:743114.
203. Tay ZY, Konovessis D. Sustainable energy propulsion system for sea transport to achieve United Nations sustainable development goals: A review. *Discover Sustainability*. 2023;4(1):20.

204. Gorjian S, Sharon H, Ebadi H, Kant K, Scavo FB, Tina GM. Recent technical advancements, economics and environmental impacts of floating photovoltaic solar energy conversion systems. *Journal of Cleaner Production*. 2021;278:124285.
205. Kumar M, Niyaz HM, Gupta R. Challenges and opportunities towards the development of floating photovoltaic systems. *Solar Energy Materials and Solar Cells*. 2021;233:111408.
206. Solomin E, Sirotkin E, Cuce E, Selvanathan SP, Kumarasamy S. Hybrid floating solar plant designs: A review. *Energies*. 2021;14(10):2751.
207. Friel D, Karimirad M, Whittaker T, Doran W, Howlin E, editors. A review of floating photovoltaic design concepts and installed variations. *4th International Conference on Offshore Renewable Energy CORE2019 proceedings,* Glasgow: ASRANet Ltd, UK, 30 Aug 2019; 2019: ASRANet Ltd.
208. Vidović V, Krajačić G, Matak N, Stunjek G, Mimica M. Review of the potentials for implementation of floating solar panels on lakes and water reservoirs. *Renewable and Sustainable Energy Reviews*. 2023;178:113237.
209. Sahu A, Yadav N, Sudhakar K. Floating photovoltaic power plant: A review. *Renewable and Sustainable Energy Reviews*. 2016;66:815–24.
210. Farrar LW, Bahaj AS, James P, Anwar A, Amdar N. Floating solar PV to reduce water evaporation in water stressed regions and powering water pumping: Case study Jordan. *Energy Conversion and Management*. 2022;260:115598.
211. Oliveira-Pinto S, Stokkermans J, editors. Marine floating solar plants: An overview of potential, challenges and feasibility. *Proceedings of the Institution of Civil Engineers-Maritime Engineering*; 2020: Thomas Telford Ltd.
212. Xiao X, Yang Y. Could the oceans host floating solar power plants? *Nature*. 2022;607(7918):239.
213. Golroodbari SZ, van Sark W. Simulation of performance differences between offshore and land-based photovoltaic systems. *Progress in Photovoltaics: Research and Applications*. 2020;28(9):873–86.
214. Shi Y, Wei Y, Tay ZY, Chen Z. Hydroelastic analysis of offshore floating photovoltaic based on frequency-domain model. *Ocean Engineering*. 2023;289:116213.
215. Tay ZY. Three-dimensional hydroelasticity of multi-connected modular offshore floating solar photovoltaic farm. *Journal of Marine Science and Engineering*. 2023;11(10):1968.
216. Suzuki H, Riggs H, Fujikubo M, Shugar T, Seto H, Yasuzawa Y, et al., editors. Very large floating structures. *International Conference on Offshore Mechanics and Arctic Engineering*; 2007. San Diego, California, United States.
217. Kyozuka Y, Kato S, Nakagawa H. A numerical study on environmental impact assessment of mega-float of Japan. *Marine Structures*. 2001;14(1–2):147–61.
218. Fujikubo M, Yamamura K, Xiao TY. Structural safety evaluation of pontoon-type mega-float considering failure of breakwater. *Journal of the Society of Naval Architects of Japan*. 2001;2001(190):337–45.
219. Aristaghes P, Blanchet V, Compte D, Weulersse A. *Offshore Urban Extension in Monaco: the Anse du Portier Project*. Sevilla: PIANC, 2018.
220. Wang CM, Tay ZY. Very large floating structures: Applications, research and development. *Procedia Engineering*. 2011;14:62–72.

221. Zhou B, Wang Y, Zheng Z, Jin P, Ning D. Power generation and wave attenuation of a hybrid system involving a heaving cylindrical wave energy converter in front of a parabolic breakwater. *Energy*. 2023;282:128364.
222. Ning DZ, Zhao XL, Chen LF, Zhao M. Hydrodynamic performance of an array of wave energy converters integrated with a pontoon-type breakwater. *Energies*. 2018;11(3):685.
223. Cheng Y, Du W, Dai S, Ji C, Collu M, Cocard M, et al. Hydrodynamic characteristics of a hybrid oscillating water column-oscillating buoy wave energy converter integrated into a π-type floating breakwater. *Renewable and Sustainable Energy Reviews*. 2022;161:112299.
224. Khabakhpasheva T, Korobkin A. Hydroelastic behaviour of compound floating plate in waves. *Journal of Engineering Mathematics*. 2002;44:21–40.
225. Kim BW, Kyoung JH, Hong SY, Cho SK, editors. Investigation of the effect of stiffness distribution and structure shape on hydroelastic responses of very large floating structures. *ISOPE International Ocean and Polar Engineering Conference*; 2005: Seoul, Korea: ISOPE.
226. Wang CM, Riyansyah M, Choo YS, editors. Reducing hydroelastic response of interconnected floating beams using semi-rigid connections. *International Conference on Offshore Mechanics and Arctic Engineering*; 2009. Honolulu, Hawaii, United States.
227. Takagi K, Shimada K, Ikebuchi T. An anti-motion device for a very large floating structure. *Marine Structures*. 2000;13(4–5):421–36.
228. Ohta H, Torii T, Hayashi N, Watanabe E, Utsunomiya T, Sekita K, et al., editors. Effect of attachment of a horizontal/vertical plate on the wave response of a VLFS. *Proceedings of the Third International Workshop on Very Large Floating Structures*; 1999. Honolulu, Hawaii, United States.
229. Masanobu S, Kato S, Maeda K, Namba Y, editors. Response of the mega-float equipped with novel wave energy absorber. *International Conference on Offshore Mechanics and Arctic Engineering*; 2003. Cancun, Mexico.
230. Fujikubo M, Takaki M, Xiao T-Y, Yanagihara D, Kanda M, Imai Y, editors. Structural Response And Strength of A New Type VLFS Using Submerged Plates. *The Twelfth International Offshore and Polar Engineering Conference (ISOPE)*; 2002; Kitakyushu, Japan: ISOPE.
231. Ikoma T, Maeda, HJ, Masuda, K, Rheem, C-K, Arita, M, editor. Effects of submerged vertical plates and air chamber units in hydroelastic response reductions. *The Twelfth International Offshore and Polar Engineering Conference*; 2002; Kitakyushu, Japan: ISOPE.
232. Lee N, Grunwald U, Rosenlieb E, Mirletz H, Aznar A, Spencer R, et al. Hybrid floating solar photovoltaics-hydropower systems: Benefits and global assessment of technical potential. *Renewable Energy*. 2020;162:1415–27.
233. Domakonda VK, Farooq S, Chinthamreddy S, Puviarasi R, Sudhakar M, Boopathi S. *Sustainable Developments of Hybrid Floating Solar Power Plants: Photovoltaic System. Human Agro-Energy Optimization for Business and Industry*: Pennsylvania, United States: IGI Global; 2023. pp. 148–67.
234. Gao RP, Tay ZY, Wang CM and Koh CG. Hydroelastic response of very large floating structure with a flexible line connection. *Ocean Engineering*. 2011;38(17–18):1957–66.

235. Tay ZY. Energy generation enhancement of arrays of point absorber wave energy converters via Moonpool's resonance effect. *Renewable Energy*. 2022;188:.830–48.
236. Tay ZY. Artificial neural network framework for prediction of hydroelastic response of very large floating structure. *Applied Ocean Research* 2023;139:103701.
237. Tay ZY. Performance of integrated FB and WEC for offshore floating solar photovoltaic farm considering the effect of hydroelasticity. *Ocean Engineering*. 2024;312:119165.
238. Hamilton WR. Theory of systems of rays. *The Transactions of the Royal Irish Academy*. 1828:69–174.
239. Hamilton WR. *On a general method in dynamics*: Richard Taylor United Kingdom; 1834.

Index

Pages in *italics* refer to figures and pages in **bold** refer to tables.

Kolibri 360 GmbH
Ferdinand-Jühlke-Straße 7
99095 Erfurt, Deutschland
produktsicherheit@kolibri360.de